Handbook of Intelligent and Sustainable Manufacturing

Intelligent and sustainable manufacturing is a broad category of manufacturing that employs computer-integrated manufacturing, high levels of adaptability and rapid design changes, digital information technology, and more flexible technical workforce training. Other goals sometimes include fast changes in production levels based on demand, optimization of the production system, efficient production, and recyclability. This handbook provides compiled knowledge of intelligent and sustainable manufacturing within the context of Industry 4.0. along with tools, principles, and strategies.

Handbook of Intelligent and Sustainable Manufacturing: Tools, Principles, and Strategies offers recent developments, future outlooks, and advanced and analytical modeling techniques of intelligent and sustainable manufacturing with examples backed up by experimental and numerical data. It bridges the gap between R&D in intelligent and sustainable manufacturing–related fields and presents case studies and solutions alongside social and green environmental impact. The handbook includes a wide range of advanced tools and applications with modeling results and explains how different internet technologies integrate the manufacturing approach with people, products, and complex systems. By encompassing advanced technologies such as digital twins, big data informatics, artificial intelligence, nature-inspired algorithms, IoT, Industry 4.0, simulation approaches, analytical strategies, quality tools, roots and pillars, diagnostic tools, and methodical strategies, this handbook provides the most up-to-date and advanced information source available.

This handbook will help industries and organizations to implement intelligent manufacturing and move towards the sustainability of manufacturing practices. It will also serve as a reference for senior graduate-level courses in mechanical, production, industrial, and aerospace engineering and a value-added asset to libraries of all technical institutions.

Advancements in Intelligent and Sustainable Technologies and Systems

Series Editor: Ajay Kumar

This book series aims to provide a platform for academicians, researchers, professionals, and individuals to participate and to provide novel systematic theoretical, experimental, computational work in the form of edited text or reference books, monographs in the area of intelligent and sustainable technologies and systems from engineering, management, applied science, healthcare, etc. domains. This book series will educate and inform the readers with a comprehensive overview of advancements in intelligent and sustainable techniques and systems with novel intelligent tools and algorithms to move industries from different domains from a data-centric community to a sustainable world. The book series covers ideas and innovations to help the research community and professionals to understand fundamentals, opportunities, challenges, future outlook, layout, lifecycle, and framework of intelligent and sustainable technologies and systems for different sectors. It serves as a guide for computer science, mechanical, manufacturing, electrical, electronics, civil, automobile, industrial engineering, biomedical, healthcare, and management professionals.

If you are interested in writing or editing a book for the series or would like more information, please contact Cindy Carelli, cindy.carelli@taylorandfrancis.com

5G-Based Smart Hospitals and Healthcare Systems
Evaluation, Integration, and Deployment
Edited by Arun Kumar, Sumit Chakravarty, Aravinda K., and Mohit Kumar Sharma

Handbook of Intelligent and Sustainable Smart Dentistry
Nature and Bio-Inspired Approaches, Processes, Materials, and Manufacturing
Edited by Ajay Kumar, Namrata Dogra, Sarita, Surbhi Bhatia, M S Sidhu

Handbook of Intelligent and Sustainable Manufacturing
Tools, Principles, and Strategies
Edited by Ajay Kumar, Parveen Kumar, Yang Liu, and Rakesh Kumar

Handbook of Intelligent and Sustainable Manufacturing
Tools, Principles, and Strategies

Edited by
Ajay Kumar, Parveen Kumar,
Yang Liu, and Rakesh Kumar

CRC Press
Taylor & Francis Group
Boca Raton London New York

CRC Press is an imprint of the
Taylor & Francis Group, an **informa** business

First edition published 2025
by CRC Press
2385 NW Executive Center Drive, Suite 320, Boca Raton FL 33431

and by CRC Press
4 Park Square, Milton Park, Abingdon, Oxon, OX14 4RN

CRC Press is an imprint of Taylor & Francis Group, LLC

ISBN: 978-1-032-51983-8 (hbk)
ISBN: 978-1-032-52276-0 (pbk)
ISBN: 978-1-003-40587-0 (ebk)

DOI: 10.1201/9781003405870

Typeset in Times
by Apex CoVantage, LLC

Contents

Chapter 4   Digital Twins and IoT for Sustainable Manufacturing:
                A Survey of Current Practices and Future Directions 62

*Raja Lavanya, G. Bhavani, C. Santhiya, G. Vinoth
Chakkaravarthy, and Parveen Kumar*

Chapter 5   Overview of Cyber Security in Intelligent and Sustainable
                Manufacturing 77

*Aditi Paul, Somnath Sinha, Parveen Kumar, Shirasthi
Choudhary, Krishna Samdani, and Saumya Mishra*

Chapter 7 Identifying and Prioritizing Quality 4.0 Practices in
Sustainable Manufacturing Using Rough Number-Based
AHP-MABAC 117

Rana Basu, Manik Chandra Das, Ajay Kumar, and Bijan Sarkar

Chapter 8 Intelligent Manufacturing in the Context of Industry
4.0 in Belarus: Overview and Perspectives 148

Yuliya Pranuza

Chapter 16 Intelligent and Sustainable Manufacturing Applications in the
Automotive Industry.. 287

Saurabh Tege and Parveen Kumar

Preface

Intelligent and sustainable manufacturing is a broad category of manufacturing that employs computer-integrated manufacturing, high levels of adaptability and rapid design changes, digital information technology, and more flexible technical workforce training. Other goals sometimes include fast changes in production levels based on demand, optimization of the production system, efficient production and recyclability. This book is expected to provide compiled knowledge on intelligent and sustainable manufacturing in the context of Industry 4.0 to involved stakeholders from users in the industry to academics and researchers.

This book discusses advancements in technology and management tools for intelligent and sustainable manufacturing from an Industry 4.0 perspective from various mechanical, automobile, industrial, and aerospace engineering domains and how this technology interacts with green manufacturing. It highlights the advancements involved in proper utilization of intelligent manufacturing techniques, including various optimization and simulation approaches. Written in a didactic style, it offers a guide and insights into intelligent and sustainable manufacturing process. Each chapter provides in-depth technical information on intelligent and sustainable manufacturing theory and its advancement. The book shows how intelligent and sustainable manufacturing may help us solve human and environmental issues in the future and suggests where current research may lead.

This book presents a collection of studies on state-of-art techniques developed specifically for intelligent and sustainable manufacturing, with an emphasis on using various analytical strategies, computational and simulation approaches, and artificial intelligence approaches to develop innovative intelligent and sustainable manufacturing techniques. The book offers an ideal reference guide for academic researchers and industrial engineers in the fields of digital manufacturing. It can also be used as a comprehensive reference source for university students in automobile, production, industrial, and mechanical engineering and in the medical sector.

The book consists of 16 chapters that describe perspectives of intelligent and sustainable manufacturing. Chapter 1, "Human–robot/Machine Interaction for Sustainable Manufacturing: Industry 5.0 Perspectives", discusses the role of human–robot/machine interaction (HRMI) on the transformation of Industry 4.0 to Industry 5.0 towards improved sustainable manufacturing with a personalized and flexible approach. Chapter 2, "Nature-Inspired Optimization Techniques of Industry 4.0 for Sustainable Manufacturing", discusses recent nature-inspired optimization algorithms like the Adam optimizer, stochastic gradient descent (SGD), water wave optimization, the seed-based plant propagation algorithm, and bumblebee mating optimization (BBMO), along with their advantages and disadvantages. Chapter 3, "Framework for Implementation of Disruptive Emerging ICTs in Intelligent and Sustainable Manufacturing in a Developing Context", determines the essential elements required for Industry 4.0 success. Chapter 4, "Digital Twins and IoT for

Sustainable Manufacturing: A Survey of Current Practices and Future Directions",
discusses how the production process can be improved by using AI algorithms to
analyze data from DTs and IoT devices and find abnormalities. Chapter 5, "Overview
of Cyber Security in Intelligent and Sustainable Manufacturing", presents a detailed
discussion of cyber threats and incidents in the intelligent manufacturing industry,
along with acceptable available mitigation strategies. Chapter 6, "Service Quality
Improvement Using Fuzzy Inference Systems and Genetic Algorithm", provides
a reliable and efficient means of evaluating service quality and enables organizations to identify areas that need improvement. Chapter 7, "A Consilient Conspectus
Framework for Identification and Prioritization of Quality 4.0 Practices in Sustainable
Manufacturing Using Rough Number–Based AHP-MABAC" contributes to the body
of knowledge by identifying and prioritizing appropriate Q 4.0 practices for achieving
sustainable operations in context of Indian manufacturing organizations. Chapter 8,
"Intelligent Manufacturing in the Context of Industry 4.0 in Belarus: Overview and
Perspectives", clarifies the measures for acceleration and successful implementation
of intelligent manufacturing in Belarus. Chapter 9, "Investigation of Chip Reduction
Coefficient of X-625 Using Coated Tools", analyzes variations of the chip reduction
coefficient in the context of rotational speed, feed, and depth of cut. Chapter 10,
"Effect of ECM, Hard Turning and Deep Cryogenic on Properties of AISI S-1 Tool
Steel", delivers a demanding review of tool steels till improved type utilizing welding
processes. The objective is to flourish knowledge of recent mutated tool steel learning. Chapter 11, "A Study on Adoption of Information and Communication Tools in
Emergency Disaster Management", aims to conduct a literature study on available
disaster management solutions putting information and communication tools into
practice. Chapter 12, "Optimization of Process Parameters of Counter Bore Hole
Made on Work Piece of Al-6061 Using DOE And MCDM Technique", discusses
the relation of parameters by experimentation technique. Chapter 13, "Application
of IoT and Artificial Intelligence in Smart Manufacturing towards Industry 4.0",
discusses the importance of IoT and AI in smart manufacturing, which can intercon-
nect equipment and processes for increasing production efficiency and provides a
new level of manufacturing. Chapter 14, "Intelligent Manufacturing in the Financial
Sector: Application in Asset Management and Trading", shows that intelligent manufacturing in the financial sector develops advanced finance, capable of predicting
and analyzing future data trends, providing insights to maximize investment returns,
and managing emotional aspects, in compliance with ethics and regulatory standards. Chapter 15, "Machine Learning Applications in Industry 4.0: Opportunities
and Challenges", provides a review and discusses the importance of machine learning for manufacturing. Chapter 16, "Intelligent and Sustainable Manufacturing
Applications in the Automotive Industry", emphasizes the importance of collaboration among academic institutions, government bodies, and industry stakeholders
to advance and implement smart and sustainable manufacturing technologies in the
automotive sector.

This book is intended for both academia and industry. Postgraduate students,
PhD students, and researchers in universities and institutions who are involved in

the areas of intelligent and sustainable manufacturing will find this compilation useful.

The editors acknowledge the professional support received from CRC Press and express their gratitude for this opportunity.

Reader's observations, suggestions, and queries are welcome,

Editors
Ajay Kumar
Parveen Kumar
Yang Liu
Rakesh Kumar

Acknowledgments

The editors are grateful to CRC Press for showing their interest to publish this book in the buzz area of intelligent and sustainable manufacturing. The editors express their personal adulation and gratitude to Ms. Cindy Renee Carelli (executive editor) of CRC Press for giving consent to publish our work. She undoubtedly imparted great and adept experience in terms of systematic and methodical staff who have helped the editors compile and finalize the manuscript. The editors also extend their gratitude to Ms. Kaitlyn Fisher, CRC Press, for support during her tenure.

The editors wish to thank all the chapter authors who contributed their valuable research and experience to compile this volume. The chapter authors, corresponding authors in particular, deserve special acknowledgment for bearing with the editors, who persistently kept bothering them for deadlines and with their remarks.

Dr. Ajay also wishes to express his gratitude to his parents, Sh. Jagdish and Smt. Kamla, and his loving brother Sh. Parveen for their true and endless support. They have made him able to walk tall before the world regardless of sacrificing their happiness and living in a small village. He cannot close these prefatory remarks without expressing his deep sense of gratitude and reverence to his life partner, Mrs. Sarita Rathee, for her understanding, care, support, and encouragement to keep his morale high all the time. No magnitude of words can ever quantify the love and gratitude the he feels in thanking his daughters, Sejal Rathee and Mahi Rathee, and son, Kushal Rathee, who are the world's best children.

Finally, the editors dedicate this work to the divine creator and express their indebtedness to the "ALMIGHTY" for gifting them the power to yield their ideas and concepts into substantial manifestation. The editors believe that this book will enlighten readers about each feature and characteristic of intelligent and sustainable manufacturing.

Ajay Kumar
Parveen Kumar
Yang Liu
Rakesh Kumar

Editors

Dr. Ajay Kumar is currently serving as professor in the Mechanical Engineering Department, School of Engineering and Technology, JECRC University, Jaipur, Rajasthan, India. He received his PhD in the field of advanced manufacturing from Guru Jambheshwar University of Science & Technology, Hisar, India, after his BTech (Hons.) in mechanical engineering and MTech (Distinction) in manufacturing and automation. His areas of research include incremental sheet forming, artificial intelligence, sustainable materials, additive manufacturing, mechatronics, smart manufacturing, Industry 4.0, waste management, and optimization techniques. He has over 60 publications in international journals of repute, including SCOPUS, Web of Science, and SCI indexed databases and refereed international conferences. He has also co-authored and co-edited many books and proceedings.

He has organized various national and international events, including an international conference on mechatronics and artificial intelligence (ICMAI-2021) as conference chair. He has more than 20 national and international patents to his credit. He has supervised more than eight MTech and PhD scholars and numerous undergraduate projects/theses. He has a total of 15 years of experience in teaching and research. He is Guest Editor and Review Editor of reputed journals, including *Frontiers in Sustainability*. He has contributed to many international conferences/symposia as a session chair, expert speaker, and member of the editorial board. He has won several proficiency awards during the course of his career, including merit awards, best teacher awards, and so on.

He is an adviser of QCFI, Delhi Chapter, student cell at JECRC University and has also authored many in-house course notes, lab manuals, monographs, and invited chapters in books. He has organized a series of faculty development programs, international conferences, workshops, and seminars for researchers and PhD-, UG-, and PG-level students. He is associated with many research, academic, and professional societies in various capacities.

Mr. Parveen Kumar is currently serving as an assistant professor and head in the Department of Mechanical Engineering at Rawal Institute of Engineering and Technology, Faridabad, Haryana, India. Currently, he is pursuing PhD from the National Institute of Technology, Kurukshetra, Haryana, India. He completed his BTech (Hons.) from Kurukshetra University, Kurukshetra, India, and M. Tech (Distinction) in manufacturing and automation from Maharshi Dayanand University, Rohtak, India. His areas of research include intelligent manufacturing systems, materials, die-less forming, design of automotive systems, additive manufacturing, CAD/CAM and artificial intelligence, machine learning and Internet of Things in manufacturing, and multi-objective optimization techniques. He has over 30 publications in international journals of repute including SCOPUS, Web of Science, and SCI indexed databases and refereed international conferences. He has eight national and international patents in his credit. He has supervised more than ten M. Tech scholars and numerous undergraduate projects/theses. He has a total of 13 years of experience in teaching and research. He has co-authored/co-edited several books. He has organized a series of faculty development programs, workshops, and seminars for researchers and UG-level students.

Dr. Yang Liu Biography received a MSc (Tech.) degree in telecommunication engineering and DSc (Tech.) degree in industrial management from the University of Vaasa, Finland, in 2005 and 2010, respectively. He is currently a tenured associate professor and doctoral supervisor with the Department of Management and Engineering, Linköping University, Sweden; a visiting faculty with the Department of Production, University of Vaasa, Finland; and a chair professor with Jinan University, China. Meanwhile, he is an adjunct/visiting professor at multiple other universities. His research interests include sustainable smart manufacturing, product service innovation, decision support systems, competitive advantage, control systems, autonomous robots, signal processing, and pattern recognition.

Prof. Liu has authored or co-authored more than 130 peer-reviewed scientific articles and is ranked No. 1 among the top authors on "big data analytics in manufacturing". His publications have appeared in multiple distinguished journals, and some ranked as top 0.1% ESI Hot Papers and top 1% ESI Highly Cited Papers. He serves as an associate editor of the prestigious *Journal of Cleaner Production* and *Journal of Intelligent Manufacturing*, referees in over 60 SCI leading journals, and is an external reviewer for NSERC of Canada and CONICYT of Chile.

Rakesh Kumar is currently serving as the vice president of Operations, Sigma Electric Manufacturing Corporation Pvt Ltd, Sigma Engineered Solutions, Jaipur, Rajasthan, India. He received his master of science (MS) in quality management from BITS-Pilani, India, after receiving a BE from same institute. His areas of expertise are manufacturing excellence, Industry 4.0, waste management, and optimization techniques. He is also Six Sigma Black Belt certified from ASQ. He has also worked in multinational companies like Asahi Glass, Bawal, India, and Rieter Automotive, India. At his current company (Sigma), he is leading the operations of five world-class manufacturing plants. He has worked on improving human efficiency, robotics, advanced machining, automatic loading and unloading, and human-less material transferring. He has a total of 20 years of experience in manufacturing industries. He is a member of the Indian Institute of Foundry (IFEX) and Confederation of Indian Industries (CII). He has contributed to many conferences as a session chair and expert speaker. He has won several proficiency awards during the course of his career, including Operations Excellence awards, best Quality award, Zero PPM award. He is an enterprising leader with a strong record of contributions in streamlining operations and procedures; invigorating businesses; and aiming for senior-level assignments in strategic planning, operational excellence, plant operations, business management, stakeholder management, and team management with high repute.

He has core competencies of strategic planning, business transformation, business excellence, cost optimization, OPEX/CAPEX reduction, continuous process improvements, plant operations, and TPM/TQM/Six Sigma/Lean Manufacturing. He has organized a series of management development programs and workshops in industries. His areas of interest include intelligent manufacturing systems. He also teaches Six Sigma techniques to industrial plant teams and leads Six Sigma projects in manufacturing processes and smart manufacturing techniques.

Contributors

Rosa Adamo
Department of Business and Law
University of Calabria
Arcavacata of Rende (CS), Italy

Maria Anastasia Arcuri
University of Salerno
Salerno, Italy

Rana Basu
School of Management Sciences
Indian Institute of Engineering Science
    and Technology
Shibpur, Howrah, India

G. Bhavani
Department of Computer science and
    Engineering
Thiagarajar College of Engineering
Madurai, India

Manjeet Bohat
Department of Mechanical Engineering,
    Maharishi Markandeshwar
    Engineering College
Maharishi Markandeshwar (Deemed to
    be University)
Mullana, India

Manik Chandra Das
Department of Industrial Engineering
    and Management
Maulana Abul Kalam Azad University
    of Technology
West Bengal, India

Sandeep Chhillar
Department of Mechanical
    Engineering
JECRC University
Jaipur, India

Rajesh Choudhary
Department of Mechanical Engineering
JECRC University
Jaipur, Rajasthan, India

Shirasthi Choudhary
Banasthali Vidyapith (Deemed to be
    University)
Rajasthan, India

Kamil Dimililer
Department of Electrical and Electronic
    Engineering
and
Applied Artificial Intelligence Research
    Centre (AAIRC)
Near East University
Nicosia, North Cyprus, Turkey

Yogesh Dubey
Department of Mechanical Engineering
JECRC University
Jaipur, India

Domenica Federico
Department of Economics
eCampus University
Novedrate (CO), Italy

Taha-Hossein Hejazi
Department of Industrial Engineering
Amirkabir University of Technology
    (Tehran Polytechnic)
Garmsar Campus
Garmsar, Iran

R. K. Jain
AcSIR, DMSE Group/Micro Robotics La
CSIR-Central Mechanical Engineering
    Research Institute (CMERI)
Durgapur, West Bengal, India

Manish Katariya
Multidisciplinary Engineering
  Department (Mechanical
  Engineering)
The NorthCap University
Gurugram, India

Bhargavi Krishnamurthy
Department of Computer science and
  Engineering
Siddaganga Institute of Technology
Bengaluru, Karnataka

Ajay Kumar
Department of Mechanical
  Engineering
School of Engineering and Technology
JECRC University
Jaipur, Rajasthan, India

Lavish Kumar
Department of Mechanical Engineering
Sharda University
Greater Noida, India

Parveen Kumar
Department of Mechanical Engineering
Rawal Institute of Engineering and
  Technology
Faridabad, Haryana, India

Pradeep Kumar
Department of Applied Science, BPIT
  College
IP University
Delhi, India

Raja Lavanya
Department of CSE
Thiagarajar College of Engineering
Madurai, India

Sheltar Marambi
Department of Information Systems
Midlands State University
Gweru, Midlands Province, Zimbabwe

Akanksha Mathur
Multidisciplinary Engineering
  Department (Mechanical
  Engineering)
The NorthCap University
Gurugram, India

Natraj Mishra
Department of Mechanical Engineering
Adamas University
Kolkata, W.B, India

Saumya Mishra
Banasthali Vidyapith (Deemed to be
  University)
Rajasthan, India

Samuel Musungwini
Department of Information and
  Marketing Sciences, Department of
  Business Sciences
Midlands State University
Gweru, Midlands Province,
  Zimbabwe

Antonella Notte
Department of Economics
eCampus University
Novedrate (CO), Italy

Ezekiel Tijesunimi Ogidan
Department of Computer Engineering
Near East University
Nicosia, North Cyprus, via Mersin 10,
  Turkey

Oluwaseun Priscilla Olawale
Department of Software Engineering
Near East University
Nicosia, North Cyprus, via Mersin 10,
  Turkey

Aditi Paul
Banasthali Vidyapith (Deemed to be
  University)
Rajasthan, India

Yuliya Pranuza
Francisk Skorina Gomel State University
Gomel, Belarus

S. K. Samanta
AcSIR, Head Foundry Group
CSIR-Central Mechanical Engineering
    Research Institute (CMERI)
Durgapur-713209, West Bengal, India

Krishna Samdani
Banasthali Vidyapith (Deemed to be
    University)
Rajasthan, India

C. Santhiya
Department of CSE
Thiagarajar College of Engineering
Madurai, India

Bijan Sarkar
Department of Production Engineering
Jadavpur University
Kolkata, India

Neeraj Sharma
Department of Mechanical Engineering
Maharishi Markandeshwar Engineering
    College, Maharishi Markandeshwar
    (Deemed to be University)
Mullana, India

Pankaj Sharma
Department of Mechanical Engineering,
    JECRC University
Jaipur, India

Mahendra Pratap Singh
Department of Mechanical Engineering,
    JECRC Foundation
Jaipur, India

Ranbir Singh
Department of Mechanical
    Engineering
BML Munjal University
Sidhrawali, Gurgaon, Haryana

Satnam Singh
Multidisciplinary Engineering
    Department (Mechanical
    Engineering)
The NorthCap University
Sector 23A, Gurugram, India

Somnath Sinha
CHRIST (Deemed to be
    University)
Bengaluru, Karnataka, India

Saurabh Tege
Mechanical Engineering,
    Geetanjali Institute of Techncial
    Studies
Udaipur, Rajasthan, India

Gaurav Tiwari
Department of Mechanical
    Engineering
JECRC University
Jaipur, Rajasthan, India

G. Vinoth Chakkaravarthy
Department of CSE
Velammal College of Engineering and
    Technology
Madurai, India

Minoo Zarghami
Department of Industrial
    Management
Caspian Higher Education Institute
Valiasr Choka Town, Iran

1 Human–Robot/Machine Interaction for Sustainable Manufacturing
Industry 5.0 Perspectives

Manish Katariya[1], Akanksha Mathur[1], and Satnam Singh[2]

1 Multidisciplinary Engineering Department (Mechanical Engineering), The NorthCap University, Gurugram, India

2 Multidisciplinary Engineering Department (Mechanical Engineering), The NorthCap University Sector 23A, Gurugram, India

1.1 INTRODUCTION

The application of automation using robotics, sensors, and smart grids in the manufacturing process is one of the main development drivers for industries. However, the effective application of these technologies requires the peaceful coexistence of people and robots/machines. As the manufacturing industry continues to shift towards the fifth industrial revolution [1], with its focus on sustainability, productivity, and human-centered design, human–robot/machine interaction (HRMI) [2–9] is becoming more significant. Sustainable manufacturing [10–12] techniques are becoming increasingly necessary as the industrial sector moves closer to Industry 5.0 [13–16]. A key component of Industry 5.0 [17–20], human–robot interaction has the capability to increase the sustainability of production procedures. HRMI involves a common workplace where people and robots interact. The advantages of HRMI are coming into more focus as manufacturing industries have included more robots recently in the production line. The development of a mutually beneficial partnership between human workers and robots is essential for a sustainable manufacturing future. HRMI can raise shop floor safety, decrease waste, and increase productivity by combining the best qualities of humans and robots. Additionally, HRMI can open new chances for skilled employment because people can shift their focus to supervisory roles, specialized programming, and maintenance tasks that require complex understanding and problem-solving skills beyond the capabilities of robots.

DOI: 10.1201/9781003405870-1

1

Although HRMI in sustainable manufacturing provides numerous advantages, there are some drawbacks as well. Ensuring robots perform safely and do not endanger human workers is one of the primary challenges. Making sure that employees are professionally trained to collaborate with robots is another difficulty. To guarantee a smooth transition, it is necessary and important to carefully manage the changes required such that the integration of HRMI facilitates growth and revolutionizes the industrial sector.

In this chapter, we'll examine HRMI's function in sustainable manufacturing from an Industry 5.0 viewpoint. We will examine the advantages and difficulties of HRMI and talk about how it might enhance manufacturing sustainability. We'll also look at some of the most important tools and techniques employed in manufacturing to support HRMI, as well as the implications of HRMI for how manufacturing jobs will be performed in the future. AI/ML and similar innovative technologies are used in intelligent manufacturing to streamline and automate production procedures [21–25]. To make real-time analysis, forecasting, and control of the manufacturing process possible, it needs the integration of data-driven decision-making, innovative automation, and networked systems. Industry 4.0, in which digitization, automation, and smart technologies are employed to improve productivity, quality, and sustainability, is made possible through intelligent manufacturing. Numerous industries, including aerospace, automotive, healthcare, and consumer goods, can benefit from intelligent manufacturing. It can be used to increase customer satisfaction, lower expenses, improve product quality, and optimize manufacturing processes. Intelligent manufacturing can help producers create more individualized, environmentally friendly, and productive products. The capacity of monitoring and managing the production plant and processes in a real-time environment is one of the main advantages of intelligent manufacturing. Manufacturers may identify and address production difficulties in real time by combining sensors, data analytics, and self/machine learning codes/algorithms. This can reduce downtime, help raise product quality, and boost productivity in manufacturing plants.

The ability to conduct predictive maintenance [11, 26] is another advantage of intelligent manufacturing. Manufacturers can predict future equipment failures by examining data from on board/off grid sensors and other data sources. This can enhance and increase the overall efficacy of the equipment while lowering maintenance expenses and downtime. Additionally, firms can enhance product quality through intelligent manufacturing. Manufacturers can spot patterns and trends in the manufacturing process that might have an impact on product quality by accessing data from sensors and other sources. This can assist in finding and fixing quality problems before they arise, which will result in fewer defects and greater customer satisfaction. Using intelligent manufacturing methods can also increase sustainability [20, 27, 28]. A more sustainable manufacturing process will arise from manufacturers optimizing their processes to utilize less energy and cause less waste. Predictive maintenance can also assist in reducing the environmental effects of production by reducing the frequency of equipment replacement. The adoption of intelligent manufacturing is, however, not without its difficulties. The integration of innovative technology into current systems is one of the major obstacles. A considerable investment in hardware and software may also be necessary, along with adjustments to company culture and procedures. The timely collection of the right data at the right time and its analysis in real time might pose security issues; therefore, industries may be worried about their data privacy and security. The industrial sector could undergo the next revolution with the integration of intelligent manufacturing systems. Manufacturers may optimize manufacturing processes, cut costs, raise product quality, and improve sustainability by incorporating various artificial intelligence techniques, self/machine learning methods, and other smart/intelligent technologies. Although there are certain obstacles to the adoption of such intelligent manufacturing, the advantages are substantial where these technologies can play a more predominant role in manufacturing/production plants in the future.

Additionally, another important pillar of Industry 5.0 is *sustainable manufacturing*, which seeks to maximize positive economic and social effects while minimizing harmful environmental effects. It is a method of production that values the preservation of Earth's resources and the reduction of waste, along with the lowering of greenhouse gas emissions. To maximize resource utilization and reduce waste, a circular economy must include sustainable production. The food and beverage, textile, construction, and electronics industries are just a few that can use sustainable manufacturing techniques. Utilizing sustainable resources, methods, and technology can lower waste and emissions, boost energy effectiveness, and enhance product quality. The ability of sustainable manufacturing to lessen negative environmental effects is one of its main advantages. Manufacturing firms can opt for sustainable techniques which can significantly reduce the amount of waste and pollution produced during manufacturing. These may include utilizing sustainable materials, processing methods, and technologies which may involve material handling and transportation. These steps tend to abolish the negative effects of manufacturing and processing engineering on the environment, pollution, and communities. The overall impact may be much larger in term of cost effectiveness and market capture in the long term. Further, the advantage of sustainable manufacturing can be related to production methods and their efficiency and effectiveness. Sustainable technologies can help manufacturers to significantly save more energy and cut power expenses and can boost production by streamlining operations and eliminating wastes. Such methods can positively impact environmental policies as laid out by governments and can improve internal production costs while simultaneously enhancing economic viability. The key benefits of utilizing sustainable manufacturing technologies are that they can help firms compete in the market and become global leaders by adapting sustainable methodologies. By using sustainable materials, processes, and technologies, manufacturers can minimize the amount of waste and pollution generated during production. The previously mentioned steps can help to restrict/reduce the environmental impact of manufacturing/production, which can benefit both the environment and the communities where manufacturing takes place.

The quality and quantity of products can also be improved through sustainable manufacturing technologies, which may give a competitive edge over other players. Producing goods with greater quality and longevity is possible for manufacturers by utilizing sustainable materials and methods. Customers may be more satisfied as a result, and regular replacements or repairs may be less necessary. The implementation of sustainable manufacturing does face certain difficulties, though. The excessive cost of integrating sustainable technologies and practices is one of the major obstacles. This could necessitate substantial investments in new machinery, instruction, and certification, which not all enterprises may be able to afford. The adoption of sustainable manufacturing techniques may also be impeded by legal and regulatory restrictions, which can present extra difficulties for producers.

In summary, sustainable manufacturing is essential to a circular economy because it may increase economic and social advantages while minimizing harmful environmental effects. Manufacturers may cut waste, boost productivity, and enhance product quality by utilizing sustainable materials, methods, and technology. Sustainable manufacturing processes have their limitations, but they also have many advantages that can contribute to a more robust and sustainable manufacturing sector. The major findings from the literature are shown in Table 1.1.

TABLE 1.1
Major Findings and Outcomes of the Literature

Reference(s)	References	Major Findings/Outcomes
[1]	Adel, A., 2022	Focus on sustainability, productivity, and human-centered design.
[2–4, 6]	Kanda, T. and Ishiguro, H., 2017; Kidd, C.D. and Breazeal, C., 2008 Goodrich, M.A. and Schultz, A.C., 2008; Inkulu, A.K., Bahubalendruni, M.R. and Dara, A., 2022; Hu, W., Lim, K.Y.H. and Cai, Y., 2022	Human–robot teaming and collaboration: As robots and humans work side by side, it becomes crucial to understand the dynamics of human–robot teams and how to optimize collaboration. Research could focus on areas such as shared decision-making, communication strategies, trust building, and task allocation, balancing and managing between humans and robots in a manufacturing setting.
[10–12]	Chourasia, S., Pandey, S.M., Gupta, K., Murtaza, Q. and Walia, R.S., 2023; Lindström, J., Larsson, H., Jonsson, M. and Lejon, E., 2017; Machado, C.G., Winroth, M.P. and Ribeiro da Silva, E.H.D., 2020	Sustainability considerations: Industry 5.0 emphasizes the importance of sustainability in manufacturing processes. Research gaps may exist in understanding how human–robot interaction can contribute to sustainable manufacturing practices, including energy efficiency, waste reduction, resource optimization, and the environmental impact of increased automation.
[13, 14, 16]	Broo, D.G., Kaynak, O. and Sait, S.M., 2022; Maddikunta, P.K.R., Pham, Q.V., Prabadevi, B., Deepa, N., Dev, K., Gadekallu, T.R., Ruby, R. and Liyanage, M., 2022; Güğerçïn, S., 2021; Xu, X., Lu, Y., Vogel-Heuser, B. and Wang, L., 2021	Human-centered design and user experience: Industry 5.0 emphasizes the collaboration and coexistence of humans and machines/robots in the manufacturing environment. Research could explore how to enhance the user experience, design intuitive interfaces, and develop effective training methods to ensure that humans can easily and safely interact with machines.

TABLE 1.1 *(Continued)*
Major Findings and Outcomes of the Literature

Reference(s)	References	Major Findings/Outcomes
[2, 18, 20]	Akundi, A., Euresti, D., Luna, S., Ankobiah, W., Lopes, A. and Edinbarough, I., 2022; Grabowska, S., Saniuk, S. and Gajdzik, B., 2022; Nirmala, P., Ramesh, S., Tamilselvi, M., Ramkumar, G. and Anitha, G., 2022	Adaptive automation and cognitive capabilities: Industry 5.0 envisions the utilization of progressive and advanced technologies, such as artificial intelligence and machine learning algorithms, to create adaptive and intelligent manufacturing systems. There is a need for research on how to design robots and machines that can learn from and adapt to human input, making them more flexible, efficient, and responsive to changing production demands.
[3, 19, 23, 25]	Wang, W. and Siau, K., 2019; Krarti, M., 2003; Rizvi, A.T., Haleem, A., Bahl, S. and Javaid, M., 2021; Kidd, C.D. and Breazeal, C., 2008	The need for research on how to design cobots and machines that can learn from and adapt to human input, making them more flexible and efficient, to reduce waste and optimize production capabilities. Standardization and interoperability: As the number of robots and machines in manufacturing environments increases, there is a need for research on developing standards and protocols that enable interoperability between different systems. This can include research on communication protocols, data exchange formats, and compatibility between various hardware and software platforms.

1.2 INDUSTRIAL REVOLUTIONS

Developments and industrial revolutions took place over the past centuries and are represented in a timeline in Figure 1.1. During the eighteenth and nineteenth centuries, a time of transition known as the Industrial Revolution, mechanized and improved industrial processes took the place of antiquated pre-Industrial Age production methods [29, 30]. During this time, innovative technologies and equipment were created, enabling far more widespread and effective production of goods and services. As a result, there was a substantial movement in the global economy toward industrialization, which had a tremendous effect on people's lives all over the world. The various Industrial Revolutions will be briefly discussed in the coming sections, focusing on what they were and how they affected the world's economies and societies. This will be followed by various technologies created over time and how they

2023

1750-1850	1850-1915	1970-2010	2011-2025	2025 -
1.0	**2.0**	**3.0**	**4.0**	**5.0**
First Industrial Revolution	Second Industrial Revolution	Third Industrial Revolution	Fourth Industrial Revolution	Fifth Industrial Revolution

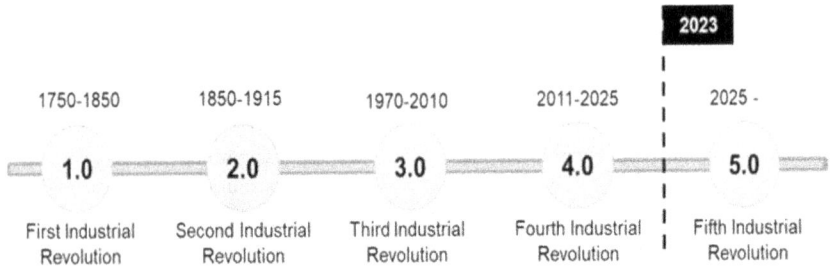

FIGURE 1.1   Timeline representing the evolution of industrial revolution eras.

enhanced production methods and industrial output. Finally, this chapter will discuss how the Industrial Revolution 4.0 is transitioning into 5.0 with integration of technological advancements while focusing on their strengths and challenges.

1.2.1 INDUSTRY 1.0—MECHANIZATION ERA

The use of new production techniques utilizing water and steam can be dated to roughly 1760, marking the beginning of the first industrial revolution. This had enormous advantages for increasing the variety of items produced and raising some people's quality of life. Industrialization changed both the transportation and textile industries. The use of machines became more applied and real, because fuel sources like steam and coal started commercial success in many applications, and the concept of machine manufacturing increased. The progress in machine development facilitated all kinds of technological advancements and made production quicker and easier. Industry 1.0 enabled production by using mechanization, as shown in Figure 1.2.

1.2.2 INDUSTRY 2.0—ELECTRIFICATION ERA

The years from about 1760 to 1840 are referred to as the first industrial revolution. Thereafter, the second industrial revolution took place. This period, which predominantly occurred in Britain, Germany, and the United States, is sometimes referred to as the "Technical Revolution" by historians. New technical systems and improved electrical technologies were developed at this time, which led to increased production rates and more sophisticated machines. At its inception, electricity and assembly line manufacturing were introduced. This revolutionizing enabler is represented in Figure 1.3. In comparison to machinery powered by water and steam, operations were made more efficient. Mass production and assembly lines were common.

1.2.3 INDUSTRY 3.0—AUTOMATION

Automation started with the introduction of machine linkages with computers in the early computer era. These early computers were frequently overly simplistic, bulky,

FIGURE 1.2 Representation of Industry 1.0 machine (start of mechanization era) [31].

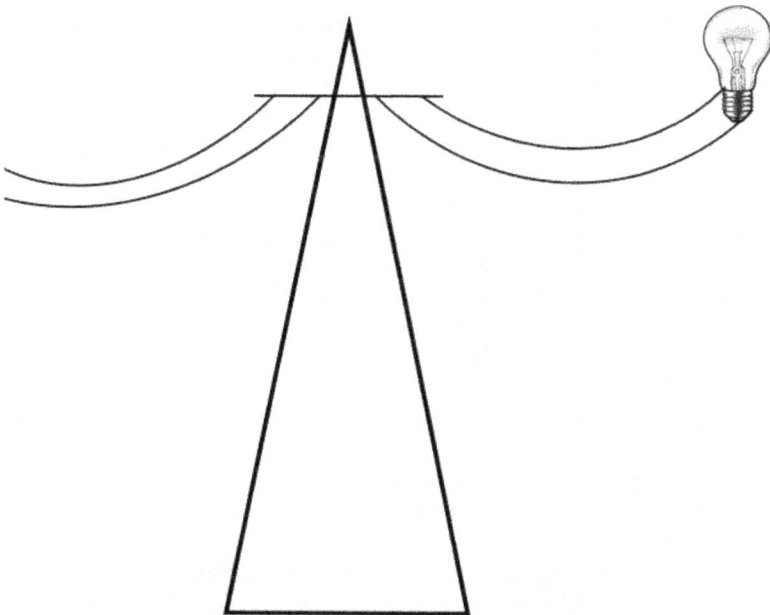

FIGURE 1.3 Important enablers for Industry 2.0 (birth of electricity).

and huge for the required computational power. However, they played a significant role and set the stage for the world to innovate and grow in the future. The third industrial revolution, which began around 1970 and involved the use of various electronics and computing technologies (information technologies using computers) to further automate industries. Due to the expansion in the domain of the availability of the internet, connectivity, and energy, automation in manufacturing sectors substantially improved. Due to the integration of robotics and automation, product variety was also enhanced with respect to productivity. In the manufacturing industry, added terms like "just-in-time" and "lean manufacturing" were frequently employed. Figure 1.4 shows pick and drop robots used for initial automation in industries under Industry 3.0.

The introduction of advanced assembly lines in Industry 3.0 induced more automated technologies, specifically the usage of programmable logic controllers (PLCs) to conduct human activities. Although automated systems were present, they still required input and human involvement. This revolution got its start with the development and production of electronic devices, including computers, memory-programmable controllers, and transistors. When procedures were mechanized, human assistance was minimized or eliminated. Examples include robots that followed orders to do a variety of tasks. Among the developed software systems and procedures are enterprise resource planning (ERP), product flow scheduling, and shipping and logistics. The concept of supply chain management was officially established at this time.

1.2.4 INDUSTRY 4.0—DIGITALIZATION

The onset of the fourth industrial revolution focused on the integration of progressive advanced technologies and is the period of self-aware/automated decision-making

FIGURE 1.4 Representation of Industry 3.0 (robotics and automation).

machines, storage systems, and manufacturing services that can communicate, initiate operations, and operate on their own without human input. Currently, we are at this point in industrial evolution and slowly transiting to Industry 5.0. As a result, we are changing the ways in which we share information and communicate. Processes have become more automated because of the development of linked devices, that is, Internet of Things (IoT), cloud computing, additive manufacturing, robots, augmented reality, smart factories, data analytics, and artificial intelligence technologies. Information and communication technology as well as the use of various cyber-physical systems are some of the major enablers for the overall architecture of modern manufacturing companies under Industry 4.0. Figure 1.5 shows the pillars required to support the architecture of Industry 4.0.

This is currently known as the Industrial Internet of Things (IIoT), which lets information interchange occur at high speeds and with ease. The key elements used in Industry 4.0 include:

- **Cyber-physical systems**: a type of mechanical/electrical/pneumatic device that can be controlled through a computer algorithm and can perform per instruction sets.
- **Internet of Things (IoT)** [32–41]: an interconnected system of various sensors controlling vehicles and machines through automated scanning, monitoring, sensing, and feedback analysis.
- **Cloud Computing**: off-site web hosting and data backup.
- **Cognitive Computing**: a technology platform which uses artificial intelligence to simulate human thought and approaches and process signals as required.

The tailoring of customer demands through flexible systems by manufactures led to the ongoing renovation of Industry 4.0 into Industry 5.0. A computerized smart

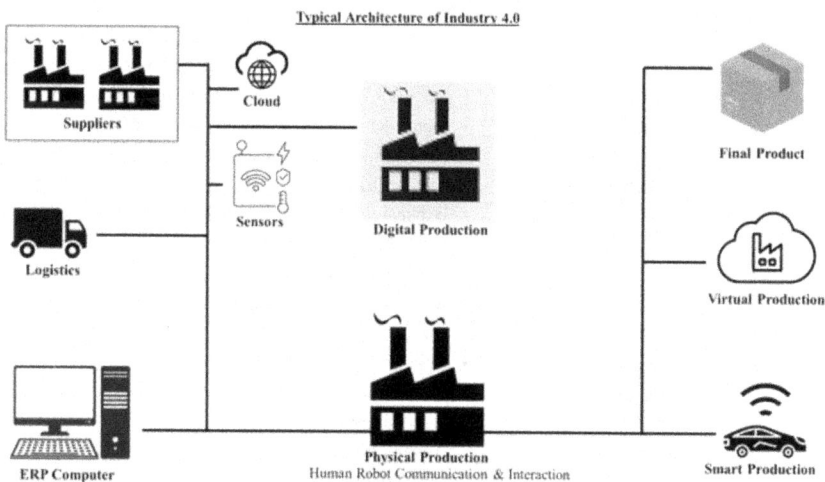

FIGURE 1.5 Typical architecture of Industry 4.0 of manufacturing company.

manufacturing process is necessary for this transformation. In the industries, mass production is currently making way for personalized items. Because of the industries' quick advancements in manufacturing technology and applications, productivity has improved. The developments in Industry 4.0 led to new levels of organization, which started controlling the whole value chain of a product's life cycle. These developments were defined by Industry 4.0 with a focus on the more specific needs of the consumer and tailoring these needs to the supply. The concept of Industry 4.0, which unites the IoT, the Industrial Internet (Big Data, analytical tools and Wi-Fi), smart manufacturing/production plants, and cloud-based manufacturing technologies, is still in the growing phase and is necessary for transitioning to Industry 5.0. However, Industry 4.0 was highly concerned with the integration of humans in the industrial processes to have continuous improvement and to concentrate on various value-adding activities.

Table 1.2 presents the key features and outcomes of all the industrial revolutions to date starting from the mechanization era to the intelligent manufacturing era.

TABLE 1.2

Outcomes of Various Industrial Revolutions

Industrial Revolution	Outcomes	Key Features
Industry 1.0	Mechanization and the advent of steam power revolutionized production	• Transition from manual labor to machine-based production. • Introduction of steam engines and water-powered mechanization. • Development of new manufacturing processes, such as textile mills. • Growth of industries like textiles, iron, and coal mining. • Increased productivity and efficiency in manufacturing.
Industry 2.0	Mass production and electrification transformed industries	• Adoption of electric power and the assembly line. • Increased specialization and division of labor. • Introduction of interchangeable parts and standardization. • Expansion of industries like automotive, steel, and consumer goods. • Rise of industrial giants and large-scale production. • Improvement in transportation and communication networks.
Industry 3.0	Automation and computerization revolutionized production and services	• Emergence of computers, robotics, and automation. • Integration of digital technologies into manufacturing and operations. • Advancements in information technology and telecommunications. • Introduction of computer-controlled machines and industrial robots.

TABLE 1.2 *(Continued)*
Outcomes of Various Industrial Revolutions

Industrial Revolution	Outcomes	Key Features
		• Globalization of supply chains and increased outsourcing.
		• Transition from analog to digital systems.
		• Expansion of service-based industries and the rise of the internet.
Industry 4.0–based manufacturing	Smart/intelligent manufacturing and the integration of cyber-physical systems	• Interconnectivity and data exchange between machines and systems (Internet of Things).
		• Integration of artificial intelligence and machine learning into manufacturing processes/plants.
		• Use of big data analytics for predictive maintenance and optimization.
		• Cyber-physical systems and smart factories with autonomous decision-making capabilities.
		• Customization and personalization of products through digital technologies.
		• Increased emphasis on sustainability and resource efficiency.
		• Augmented reality (AR) and virtual reality (VR) applications in industrial settings/manufacturing simulations, etc.

To support the growth of Industry 4.0, many tools, processes, and methods were developed, evolved, and implemented. Some of the key management tools implemented in the Industry 4.0 framework include manufacturing execution systems (MESs) and manufacturing operations management (MOM). These are discussed briefly with their roles in the next sections.

1.2.5 MANUFACTURING EXECUTION SYSTEMS

The developed systems for scheduling and planning shop floor activities are known as manufacturing execution systems, and they also provide information about the manufacturing process to enterprise resource planning stakeholders [8, 42, 43]. Manufacturing execution systems are a type of manufacturing-oriented software, which was developed explicitly for the goal of improving overall efficiency and controlling manufacturing operations. These systems methodically and proactively incorporate various work orders, history sheets, sigma levels, control limits, and job activities into the various manufacturing processes. Though MESs are used in many industries as top-notch software programs, they lack the intelligence which can continuously and automatically improve performance through controlled decision making. Manufacturing execution systems are utilized in production systems for tracking and documenting manufacturing plans, starting from the conversion of available virgin materials into semifinished/finished goods. The data from MESs can be used by decision-makers in production plants and industries to better understand the current state of production systems and their environments. This can be utilized on the factory floor to make appropriate changes to boost output with optimized conditions. The spectrum of the manufacturing process, such as materials, energy input, workers, machines, maintenance schedules, and various support services, can be controlled using a real-time monitoring system known as MES.

To achieve overall equipment effectiveness (OEE), factors like quality of product and various material tracking, tracing, and controlling, advanced MESs may perform additional functions in the product's life cycle, such as planning, humanpower scheduling and handling, order control and execution, outbound dispatch management, production monitoring, analysis and optimization, and downtime management. The MES, which logs the details, actions, and results of the manufacturing process, produces the "as-built" record. This is important in industries with regulations, such the food and beverage and pharmaceutical industries, where it may be necessary to document and certify events, processes, and actions. The notion of MES can be viewed as an interlinkage between the available enterprise resource planning (ERP) systems and a supervisory control and data acquisition system (SCADA) or process control system, even though historical definitions have evolved. In the early 1990s, trade groups like Manufacturing Enterprise Solutions Association (MESA) International were founded to understand and address the complexity of industries and to offer support and advice on the adoption of MESs controlling production and other processes. The numerous benefits of MESs are:

- Production execution systems "assist" in creating ideal manufacturing processes by providing real-time feedback on changes in demand and information from a sole source
- Faster setup times
- Less waste with minimum rework and scrap
- More accurate cost data capture (e.g., labor, scrap, downtime, and tooling)
- Longer uptime
- Include processes that use paperless workflow
- Traceability of manufacturing operations
- Reduce downtime and make fault finding simple.
- Elimination of just-in-case inventories to reduce inventory

1.2.6 Manufacturing Operations Management

Manufacturing operations management is a term used to describe the broad management of production activities. It is a comprehensive solution that makes use of innovative technology like the Internet of Things to provide detailed real-time visibility into the many operations and activities. Manufacturing execution systems with usable intelligence to enhance real-time execution are regarded to have evolved into manufacturing operations management. They are recognized as a solution which unifies the complete manufacturing processes via manufacturing intelligence. This helps in the improvement of quality management, manufacturing execution systems, and advanced planning and scheduling mechanisms through enterprise resource planning, as well as effective management of research and development (R&D).

1.3 INTRODUCTION TO INDUSTRY 5.0

Industry 4.0 only entered the production sphere a little over ten years ago, but visionaries already foresee Industry 5.0 as the following revolution. Industry 5.0 will emphasize bringing human hands and brains back into an industrial setting, in contrast to the current revolution, which focuses on the transformation of industries into IoT-enabled smart facilities/production centers that use cognitive computers and are interconnected via cloud servers. Industry 5.0, also known as the Industrial Internet of Things, is the next major step in the evolution of industrial automation and robotics. It is a unification of the latest technologies and processes like cloud computing, artificial intelligence, analytics, and the Internet of Things to create a seamless, interconnected system that can revolutionize the way industrial manufacturers and other organizations operate. It concerns customization. After manufacturing processes are automated and human interaction is eliminated, manufacturing will once again have a human touch. To offer a tailored experience, it would be in step with technological development and human cognitive and thinking abilities. As shown in Figure 1.6, when Industry 4.0 is integrated with sustainable manufacturing, personalization, and intelligent manufacturing, it leads to Industry 5.0.

The five pillars of Industry 5.0 are:

- **Cloud Computing** [44]: Industry 5.0 relies heavily on cloud computing since it enables data to be shared, stored, and accessed from any location in the world. As a result, operating costs are decreased, and data-driven decision-making is made possible.

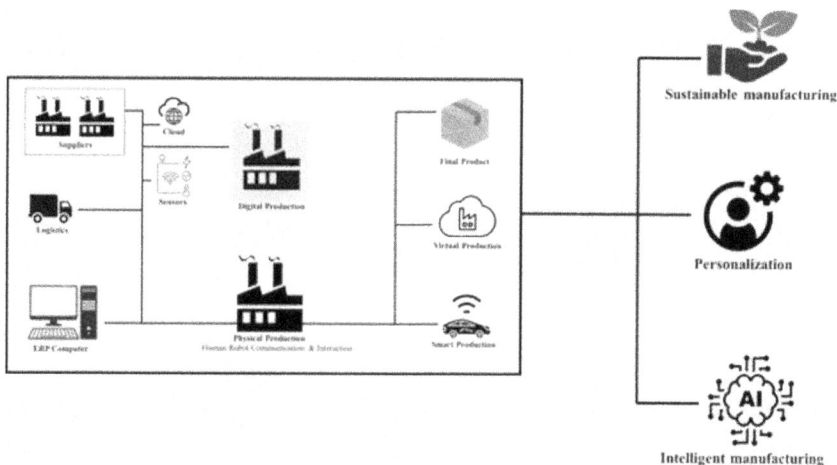

FIGURE 1.6 Typical architecture of Industry 5.0 as a new era.

- **Analytics**: Manufacturers may use data to spot trends, improve decision-making, and optimize processes with the aid of analytics. They can eliminate waste, enhance the quality of their products, and optimize their processes thanks to this.
- **Artificial Intelligence**: AI enables machines to learn and adapt to their environments, allowing them to take on tasks that would otherwise be too difficult or time consuming for humans. AI can also help to automate processes, identify patterns in data, and improve product quality.
- **Internet of Things**: A network of sensors known as the Internet of Things enables real-time data collection, monitoring, and analysis. Overall efficiency is increased, and improved process optimization is made possible.
- **Human–Machine Interaction:** Industry 5.0 relies heavily on human–machine interaction because it enables both parties to cooperate toward a common objective. Errors can be decreased, productivity can go up, and safety can get better.

These five pillars—which form the basis of Industry 5.0—are crucial for businesses that want to benefit from the most recent developments in automation and technology. Organizations may transform how they conduct business and establish a more profitable environment by utilizing these technologies and procedures. Industry 5.0 mainly involves the three major parameters, as shown in Figure 1.7. These three pillars are discussed in detail.

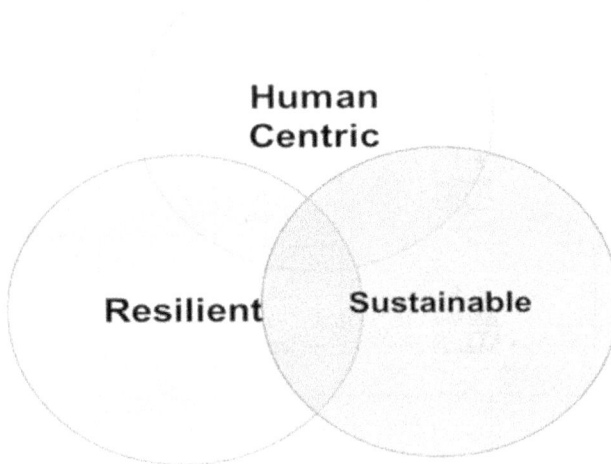

FIGURE 1.7 Representation of Industry 5.0 as an intersection of pillars in a Venn diagram.

- **A human-centered approach**

  Industry 5.0 places it all in the context of a larger human-centered strategy, whereas Industry 4.0 was focused on the use of networked automated systems, machines, and robots for optimum performance optimization and efficiency. Due to this, talent plays a bigger part and is more crucial to the success of operations. Industry 5.0 stresses cooperation and interaction between humans and machines through implementing robots and other technology.

  Because it puts fundamental human wants and interests at the center of the production process and makes it easier to adapt industrial automation technology to the needs of industrial employees, this method is frequently referred to as "socio-centric". Industry 5.0 understands the need to establish a work environment that is safe and inclusive and prioritizes the health and well-being of all employees in addition to achieving high productivity.

- **Resilience and sustainability**

  In addition to empowering individuals, Industry 5.0 emphasizes resilience and sustainability in a big way. The idea of Industry 5.0 offers a framework that must combine the original vision's competitiveness and commercial efficiency with a focus on sustainability, a careful consideration of the environment, and the effects of industrial automation on our world. Industry 5.0 wants to be resilient, and one way to do this is by relying on technology that is flexible, agile, and adaptable.

- **Human–Machine Collaboration**

  The emphasis has changed to cooperative interactions between humans and machines, even if Industry 5.0 does not undervalue the crucial role that robots and automated machines play in the new industrial revolution. Industry 5.0 acknowledges all the drawbacks of over-automation in addition to the well-known advantages and benefits of robotic automation, such as the fact that their labor is more accurate, reliable, and effective than that of human workers. For instance, highly automated solutions are not flexible or easily adapted to changing demands.

In the Industry 5.0 revolution, humans and machines collaborate and create new ways to produce goods more effectively. It's interesting to note that businesses that have only recently embraced Industry 4.0 ideas may already be experiencing the fifth revolution. Even if businesses adopt innovative technology, they won't quickly eliminate most of their labor and go fully computerized.

1.3.1 COMPARISON OF INDUSTRY 4.0 AND INDUSTRY 5.0 VISIONS

There are major differences between the Industry 4.0 and Industry 5.0 visions, and they are presented in Table 1.3 [5].

1.3.2 DIFFERENCE BETWEEN INDUSTRY 4.0 AND INDUSTRY 5.0 APPROACHES

The major differences between the approaches of Industry 4.0 and Industry 5.0 are presented in Table 1.4.

TABLE 1.3

Comparison between Visions of Industrial Revolutions (Demir, Döven, and Sezen 2019)

	Industry 4.0	Industry 5.0 (Vision 1)	Industry 5.0 (Vision 2)
Key Feature	Integration of smart manufacturing technologies	Human–robot co-working	Bioeconomy
Driver	Mass production	Smart society	Sustainability
Power Source	• Electrical power • Fossil-based fuels • Renewable power sources	• Electrical power • Renewable power sources	• Electrical power • Renewable power sources
Involved Technologies	• Internet of Things • Cloud computing • Big Data • Robotics and artificial intelligence	• Human–robot collaboration • Renewable resources	• Sustainable agriculture • Production • Bionics • Renewable resources
Involved Research Areas	• Organizational research • Process improvement and innovation • Business administration	• Smart environments • Organizational research • Process improvement and innovation • Business administration	• Agriculture • Biology • Waste prevention • Process improvement and innovation • Business administration • Economy

TABLE 1.4

Differences between Industry 4.0 and 5.0

Industry 4.0 Approaches	Industry 5.0 Approaches
• Enhanced efficiency	• Enhanced efficiency with competitiveness and sustainability
• Digital connectivity	• Digital connectivity with self learning
• Technology limited to cyber-physical objectives	• Emphasis on governance of sustainability and resilience
• Goal for cost minimization and profit maximization	• Human-centric approach to technology
• Aligned with optimization of business models	• Emphasizes alternate models of technology
• No focus on design and material use	• Ensures design and material use considers environmental aspects and sustainability

1.3.3   SWOT ANALYSIS OF INDUSTRY 5.0

The quick transition of Industry 4.0 to Industry 5.0 is triggered by numerous factors like sustainability, customization, cyber risks, and cost effectiveness. A detailed strengths, weaknesses, opportunities, and threats (SWOT) analysis of Industry 5.0 from the Industry 4.0 point of view is carried out and is presented in Figure 1.8. The major strengths of Industry 5.0 mainly include a human-centric approach with a focus on customization of products, sustainability, and new technological advancements. The opportunities lie in the domain of efficiency and mass as well as enhanced individual customer experience. Some of the identified weaknesses are high implementation costs and the requirement of a skilled human workforce. Cyber security risks and human–robot/machine collaborative interaction are major threats in the implementation of Industry 5.0.

The key challenges associated with the successful implementation of Industry 5.0 in the manufacturing sector are human–robot/machine collaborative protocols. This is a global challenge which needs to be addressed for the success of the fifth industrial revolution. The challenges from an Indian perspective of human–robot collaboration in manufacturing industries are presented in the next section.

1.4   CHALLENGES ASSOCIATED WITH ROBOT INTEGRATION IN ORGANIZATIONS

In the Indian context, the integration of advanced automation, including robot integration, involves challenges. These challenges/roadblocks are due to the following factors:

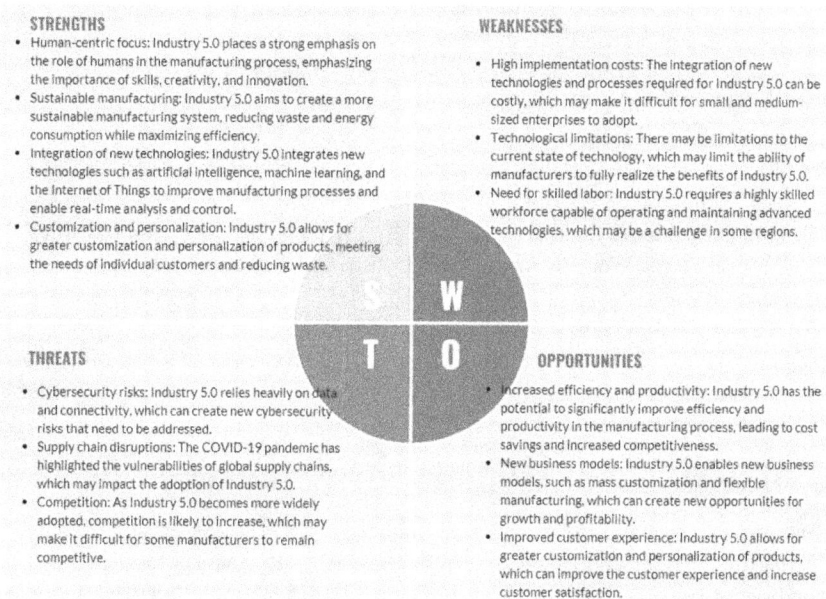

STRENGTHS
- Human-centric focus: Industry 5.0 places a strong emphasis on the role of humans in the manufacturing process, emphasizing the importance of skills, creativity, and innovation.
- Sustainable manufacturing: Industry 5.0 aims to create a more sustainable manufacturing system, reducing waste and energy consumption while maximizing efficiency.
- Integration of new technologies: Industry 5.0 integrates new technologies such as artificial intelligence, machine learning, and the Internet of Things to improve manufacturing processes and enable real-time analysis and control.
- Customization and personalization: Industry 5.0 allows for greater customization and personalization of products, meeting the needs of individual customers and reducing waste.

WEAKNESSES
- High implementation costs: The integration of new technologies and processes required for Industry 5.0 can be costly, which may make it difficult for small and medium-sized enterprises to adopt.
- Technological limitations: There may be limitations to the current state of technology, which may limit the ability of manufacturers to fully realize the benefits of Industry 5.0.
- Need for skilled labor: Industry 5.0 requires a highly skilled workforce capable of operating and maintaining advanced technologies, which may be a challenge in some regions.

THREATS
- Cybersecurity risks: Industry 5.0 relies heavily on data and connectivity, which can create new cybersecurity risks that need to be addressed.
- Supply chain disruptions: The COVID-19 pandemic has highlighted the vulnerabilities of global supply chains, which may impact the adoption of Industry 5.0.
- Competition: As Industry 5.0 becomes more widely adopted, competition is likely to increase, which may make it difficult for some manufacturers to remain competitive.

OPPORTUNITIES
- Increased efficiency and productivity: Industry 5.0 has the potential to significantly improve efficiency and productivity in the manufacturing process, leading to cost savings and increased competitiveness.
- New business models: Industry 5.0 enables new business models, such as mass customization and flexible manufacturing, which can create new opportunities for growth and profitability.
- Improved customer experience: Industry 5.0 allows for greater customization and personalization of products, which can improve the customer experience and increase customer satisfaction.

FIGURE 1.8   SWOT analysis representing aspects and impact of the fifth industrial revolution.

1. Organizational Behavior Change
2. Workplace Acceptance of Robots
3. Changes in Workflows and Organizational Structures
4. Changes in Work Ethics
5. Discrimination against Either People or Robots
6. In a Human–Robot Collaborative Work Environment, Privacy and Trust
7. Training and Education Robotics Workplace Redesign

The biggest challenge is workplace acceptance of robots, which threatens the job security of workers. It is exceedingly difficult to make workers aware of the benefits of integration; rather they look at this as negative change. The mindset of workers needs to be changed through proper training and education. Further, the costs involved in initial integration and maintenance costs are remarkably high. However, developments are going on in India where people are getting their skills upgraded and accepting robots as a collaborative work force.

1.5　CHALLENGES OF HUMAN–ROBOT COLLABORATION

The integration of only robots in various manufacturing industries was the major challenge for Industry 4.0; however, human–robot collaboration is a much more complicated task and needs further attention [45]. The various issues and challenges involved in human–robot collaboration in the production industries are:

1. Issues with Laws and Regulations
2. Personal Preference for Using Robots at Work
3. Psychological Problems Caused by Human–Robot Collaboration
4. Social Effects of Human–Robot Collaboration
5. The Shifting Functions of Human Resources Divisions
6. The Evolution of Robotics Departments and the Changing Role of Information Technology Departments
7. Ethical Problems Caused by Human–Robot Collaboration—Robot Ethics
8. Preference for Working with Specific Types of Robots (Learning or Rule-Based Robots)
9. Learning Robotics at Work
10. Due to a Declining Human Workforce, a Negative Attitude toward Robots
11. Robots and Humans Competing or Humans and Robots Working Together

1.6　HUMAN–ROBOT/MACHINE INTERACTION IN INDUSTRY 5.0

In Industry 5.0, human–robot/machine interaction (HRMI) is essential to achieve the objective of efficient and sustainable manufacturing. By incorporating HRMI into the industrial process, humans and robots/machines may work together more effectively, boosting productivity and efficiency. Advanced technologies like augmented reality and virtual reality can be used in production processes thanks to HRMI, making them more engaging and human centered.

1.6.1 HRMI TECHNIQUES

There are several techniques used in HRMI to improve the interaction between humans and robots/machines. These techniques include:

- **Natural language processing:** This entails conversing with robots and machines using natural language, which makes the encounter more natural and human-like.
- **Gesture recognition**: This entails using gestures to engage and operate robots and machines, making interaction easier and requiring less manual input.
- **Augmented and virtual reality**: This entails utilizing augmented and virtual reality technology to enhance human–robot interaction, resulting in a more immersive and human-centered experience [26, 46–49].

Augmented reality is a technology that continuously superimposes digital information over the physical world. By incorporating digital components like images, text, or sounds into the real-world environment, augmented reality improves the user's experience of reality. Through gadgets like smartphones, tablets, or head-mounted displays, AR can be experienced. Users can engage with a simulated environment thanks to the technology known as virtual reality. VR can be experienced using head-mounted displays that completely immerse the user in a virtual environment. Users can interact with digital content in ways that aren't feasible in the real world because of the unique and immersive experience that virtual reality offers.

1.6.2 BENEFITS OF HRMI IN MANUFACTURING

The integration of HRMI into the manufacturing process provides several benefits, including:

- **Improved efficiency and productivity:** HRMI allows for more collaborative and efficient interaction between humans and robots/machines, resulting in increased productivity and efficiency.
- **Improved safety:** HRMI enables the implementation of advanced safety features, reducing the risk of accidents and injuries in the workplace.
- **Improved sustainability:** HRMI can help to reduce waste and minimize energy consumption, making the manufacturing process more sustainable. The detailed parameters affecting the sustainability of any manufacturing process are explained by the authors [50].
- **Improved human experience:** HRMI provides a more human-centered and immersive experience, improving the satisfaction and well-being of workers in the manufacturing process.

1.7 CONCLUSION

Industry 5.0's sustainable manufacturing has the potential to greatly benefit from HRMI. By facilitating seamless collaboration between humans, robots, and machines, HRMI can improve workplace safety, minimize environmental impact, create new job opportunities, and improve overall efficiency. However, it is important that these benefits be balanced against potential challenges such as data privacy and security concerns and that HRMI be designed and implemented in a way that respects human dignity and autonomy. The major conclusions are as follows:

- Industry 5.0 with the appropriate incorporation of HRMI can achieve sustainability in manufacturing systems by enhancing environmental, social, and economic performances.
- The goal of HRMI is to enhance the efficiency and effectiveness of human, robot, and machine collaboration to increase the sustainability of industrial processes.
- HRMI can lower workplace accidents and raise workplace safety by automating some operations and letting robots and machines conduct hazardous or repetitive duties.
- The environmental effect of industrial processes can be reduced with the use of HRMI by streamlining procedures and cutting waste.
- HRMI enables people to learn new skills and acquire experience with innovative technologies by allowing them to collaborate with robots and machines. This can support Sustainable Development Goal (SDG) 4 (Quality Education) and SDG 8 by ensuring that workers are prepared to adapt to changing manufacturing processes and emerging technology (Decent Work and Economic Growth).
- HRMI can enhance the general effectiveness of production procedures, resulting in lower costs and higher productivity. As a result, SDG 9 (Industry, Innovation, and Infrastructure) may benefit from more economically viable sustainable manufacturing.

The major concerns for successful implementation of HRMI are issues about data security, privacy, and traceability.

1.8 FUTURE SCOPE OF WORK

HRMI-enabled Industry 5.0 has the highest potential to impart sustainability in future manufacturing technologies. With the advancements and integration of artificial intelligence techniques equipped with machine and deep learning algorithms, future processes can be monitored, traced, and rectified on a real-time basis, which can lead to optimized production and sustainable manufacturing. However, there are many concerns requiring thorough testing and validation studies before successful implementation of HRMI. The major concern in Industry 5.0 and integration of HRMI is the breach of data and cyber security–related issues. Defining and standardizing global safety and communication protocols in HRMI-enabled manufacturing systems would be a challenging task. Optimization of energy consumption in HRMI-enabled Industry 5.0 is another concern which needs to be addressed.

REFERENCES

[1] Adel, A. "Future of industry 5.0 in society: Human-centric solutions, challenges, and prospective research areas." *Journal of Cloud Computing* 11, no. 1 (2022): 1–15. https://doi.org/10.1186/s13677-022-00314-5.

[2] Kanda, Takayuki, and Hiroshi Ishiguro. *Human–robot Interaction in Social Robotics.* CRC Press, 2017, ISBN: 9781138071698.

[3] Kidd, Cory D., and Cynthia Breazeal. "Robots at home: Understanding long-term human–robot interaction." In *2008 IEEE/RSJ International Conference on Intelligent Robots and Systems*, pp. 3230–3235. IEEE, 2008. https://doi.org/10.1109/IROS.2008.4651113.

[4] Goodrich, Michael A., and Alan C. Schultz. "Human–robot interaction: A survey." *Foundations and Trends® in Human–Computer Interaction* 1, no. 3 (2008): 203–275. http://dx.doi.org/10.1561/1100000005

[5] Demir, Kadir Alpaslan, Gözde Döven, and Bülent Sezen. "Industry 5.0 and human–robot co-working." *Procedia Computer Science* 158 (2019): 688–695. https://doi.org/10.1016/j.procs.2019.09.104

[6] Inkulu, Anil Kumar, M. V. A. Raju Bahubalendruni, Ashok Dara, and K. Sankaranarayana Samy. "Challenges and opportunities in human robot collaboration context of industry 4.0-a state of the art review." *Industrial Robot: The international Journal of Robotics Research and Application* 49, no. 2 (2021): 226–239. https://doi.org/10.1108/IR-04-2021-0077

[7] Hu, Wei, Kendrik Yan Hong Lim, and Yiyu Cai. "Digital twin and industry 4.0 enablers in building and construction: A survey." *Buildings* 12, no. 11 (2022): 2004. https://doi.org/10.3390/buildings12112004

[8] Ajoudani, Arash, Andrea Maria Zanchettin, Serena Ivaldi, Alin Albu-Schäffer, Kazuhiro Kosuge, and Oussama Khatib. "Progress and prospects of the human–robot collaboration." *Autonomous Robots* 42 (2018): 957–975. https://doi.org/10.1007/s10514-017-9677-2

[9] Ranavolo, Alberto, Giorgia Chini, Francesco Draicchio, Alessio Silvetti, Tiwana Varrecchia, Lorenzo Fiori, Antonella Tatarelli et al. "Human–robot collaboration (HRC) technologies for reducing work-related musculoskeletal diseases in industry 4.0." In *Proceedings of the 21st Congress of the International Ergonomics Association (IEA 2021) Volume V: Methods & Approaches* 21, pp. 335–342. Springer International Publishing, 2022. https://doi.org/10.1007/978-3-030-74614-8_40

[10] Chourasia, Shubhangi, Shailesh Mani Pandey, Kalpana Gupta, Qasim Murtaza, and R. S. Walia. "Industry 5.0 for sustainable manufacturing: New product, services, organizational and social information." In *Surface Engineering*, pp. 243–255. CRC Press, 2023.

[11] Lindström, John, Hans Larsson, Martin Jonsson, and Erik Lejon. "Towards intelligent and sustainable production: Combining and integrating online predictive maintenance and continuous quality control." *Procedia CIRp* 63 (2017): 443–448. https://doi.org/10.1016/j.procir.2017.03.099

[12] Machado, Carla Gonçalves, Mats Peter Winroth, and Elias Hans Dener Ribeiro da Silva. "Sustainable manufacturing in Industry 4.0: An emerging research agenda." *International Journal of Production Research* 58, no. 5 (2020): 1462–1484. https://doi.org/10.1080/00207543.2019.1652777

[13] Broo, Didem Gürdür, Okyay Kaynak, and Sadiq M. Sait. "Rethinking engineering education at the age of industry 5.0." *Journal of Industrial Information Integration* 25 (2022): 100311. https://doi.org/10.1016/j.jii.2021.100311

[14] Maddikunta, Praveen Kumar Reddy, Quoc-Viet Pham, B. Prabadevi, Natarajan Deepa, Kapal Dev, Thippa Reddy Gadekallu, Rukhsana Ruby, and Madhusanka Liyanage. "Industry 5.0: A survey on enabling technologies and potential applications." *Journal of Industrial Information Integration* 26 (2022): 100257. https://doi.org/10.1016/j.jii.2021.100257

[15] Rahardjo, Benedictus, Fu-Kwun Wang, Shih-Che Lo, and Tzu-Hsien Chu. "A sustainable innovation framework based on Lean Six Sigma and Industry 5.0." *Arabian Journal for Science and Engineering* (2023): 1–18.

[16] Güğerçin, S. "How employees survive in the industry 5.0 era: In-demand skills of the near future." *International Journal of Disciplines in Economics and Administrative Sciences Studies (IDEAstudies)* 7, no. 31 (2021): 524–533.

[17] Xu, Xun, Yuqian Lu, Birgit Vogel-Heuser, and Lihui Wang. "Industry 4.0 and Industry 5.0—Inception, conception and perception." *Journal of Manufacturing Systems* 61 (2021): 530–535. https://doi.org/10.1016/j.jmsy.2021.10.006

[18] Akundi, Aditya, Daniel Euresti, Sergio Luna, Wilma Ankobiah, Amit Lopes, and Immanuel Edinbarough. "State of industry 5.0—Analysis and identification of current research trends." *Applied System Innovation* 5, no. 1 (2022): 27. https://doi.org/10.3390/asi5010027

[19] Wang, Weiyu, and Keng Siau. "Artificial intelligence, machine learning, automation, robotics, future of work and future of humanity: A review and research agenda." *Journal of Database Management (JDM)* 30, no. 1 (2019): 61–79.

[20] Grabowska, Sandra, Sebastian Saniuk, and Bożena Gajdzik. "Industry 5.0: Improving humanization and sustainability of Industry 4.0." *Scientometrics* 127, no. 6 (2022): 3117–3144. https://doi.org/10.1007/s11192-022-04370-1

[21] Levitt, Jonathan, and Mike Thelwall. "Does the higher citation of collaborative research differ from region to region? A case study of economics." *Scientometrics* 85, no. 1 (2010): 171–183. https://doi.org/10.1007/s11192-010-0197-5

[22] Nirmala, P., S. Ramesh, M. Tamilselvi, G. Ramkumar, and G. Anitha. "An artificial intelligence enabled smart industrial automation system based on internet of things assistance." In *2022 International Conference on Advances in Computing, Communication and Applied Informatics (ACCAI),* pp. 1–6. IEEE, 2022. https://doi.org/10.1109/ACCAI53970.2022.9752651.

[23] Krarti, Moncef. "An overview of artificial intelligence-based methods for building energy systems." *Journal of Solar Energy Engineering* 125, no. 3 (2003): 331–342. https://doi.org/10.1115/1.1592186.

[24] Kumar, Ajay, Hari Singh, Parveen Kumar, and Bandar AlMangour, eds. *Handbook of Smart Manufacturing: Forecasting the Future of Industry 4.0.* CRC Press, 2023. https://doi.org/10.1201/9781003333760

[25] Rizvi, Ali Tarab, Abid Haleem, Shashi Bahl, and Mohd Javaid. "Artificial intelligence (AI) and its applications in Indian manufacturing: A review." *Current Advances in Mechanical Engineering: Select Proceedings of ICRAMERD 2020* (2021): 825–835. https://doi.org/10.1007/978-981-33-4795-3_76

[26] Pech, Martin, Jaroslav Vrchota, and Jiří Bednář. "Predictive maintenance and intelligent sensors in smart factory." *Sensors* 21, no. 4 (2021): 1470. https://doi.org/10.3390/s21041470

[27] Moldavska, Anastasiia, and Torgeir Welo. "A Holistic approach to corporate sustainability assessment: Incorporating sustainable development goals into sustainable manufacturing performance evaluation." *Journal of Manufacturing Systems* 50 (2019): 53–68. https://doi.org/10.1016/j.jmsy.2018.11.004

[28] Malek, Javed, and Tushar N. Desai. "A systematic literature review to map literature focus of sustainable manufacturing." *Journal of Cleaner Production* 256 (2020): 120345. https://doi.org/10.1016/j.jclepro.2020.120345

[29] Ruiz-De-La-Torre, Aitor, Rosa M. Rio-Belver, Wilmer Guevara-Ramirez, and Christophe Merlo. "Industry 5.0 and human-centered approach. Bibliometric review." In *The International Conference on Industrial Engineering and Industrial Management,* pp. 402–408. Springer International Publishing, 2022. https://doi.org/10.1007/978-3-031-27915-7_71

[30] Shahbakhsh, Mehrangiz, Gholam Reza Emad, and Stephen Cahoon. "Industrial revolutions and transition of the maritime industry: The case of Seafarer's role in autonomous shipping." *The Asian Journal of Shipping and Logistics* 38, no. 1 (2022): 10–18. https://doi.org/10.1016/j.ajsl.2021.11.004

[31] Brown, Michael. "Making sense of modernity's maladies: Health and disease in the Industrial Revolution." *Endeavour* 30, no. 3 (2006): 108–112. https://doi.org/10.1016/j.endeavour.2006.08.001

[32] Vermesan, Ovidiu, and Peter Friess. *Building the Hyperconnected Society-Internet of Things Research and Innovation Value Chains, Ecosystems and Markets.* Taylor & Francis, 2015. https://doi.org/10.1201/9781003337454

[33] Friess, Peter, and Francisco Ibanez. "Putting the internet of things forward to the next Nevel." In *Internet of Things Applications-From Research and Innovation to Market Deployment*, pp. 3–6. River Publishers, 2022.

[34] Sadhu, Pintu Kumar, Venkata P. Yanambaka, and Ahmed Abdelgawad. "Internet of things: Security and solutions survey." *Sensors* 22, no. 19 (2022): 7433. https://doi.org/10.3390/s22197433

[35] Dhanaraju, Muthumanickam, Poongodi Chenniappan, Kumaraperumal Ramalingam, Sellaperumal Pazhanivelan, and Ragunath Kaliaperumal. "Smart farming: Internet of things (IoT)-based sustainable agriculture." *Agriculture* 12, no. 10 (2022): 1745. https://doi.org/10.3390/agriculture12101745.

[36] Andersson, Per, and Lars-Gunnar Mattsson. "Service innovations enabled by the 'internet of things'." *IMP Journal* 9, no. 1 (2015): 85–106. https://doi.org/10.1108/IMP-01-2015-0002

[37] Gubbi, Jayavardhana, Rajkumar Buyya, Slaven Marusic, and Marimuthu Palaniswami. "Internet of Things (IoT): A vision, architectural elements, and future directions." *Future Generation Computer Systems* 29, no. 7 (2013): 1645–1660. https://doi.org/10.1016/j.future.2013.01.010

[38] Alavi, Amir H., Pengcheng Jiao, William G. Buttlar, and Nizar Lajnef. "Internet of Things-enabled smart cities: State-of-the-art and future trends." *Measurement* 129 (2018): 589–606. https://doi.org/10.1016/j.measurement.2018.07.067

[39] Rejeb, Abderahman, Steve Simske, Karim Rejeb, Horst Treiblmaier, and Suhaiza Zailani. "Internet of Things research in supply chain management and logistics: A bibliometric analysis." *Internet of Things* 12 (2020): 100318. https://doi.org/10.1016/j.iot.2020.100318

[40] Haghnegahdar, Lida, Sameehan S. Joshi, and Narendra B. Dahotre. "From IoT-based cloud manufacturing approach to intelligent additive manufacturing: Industrial Internet of Things—An overview." *The International Journal of Advanced Manufacturing Technology* (2022): 1–18. https://doi.org/10.1007/s00170-021-08436-x

[41] Blazek, Roman, Lenka Hrosova, and Janet Collier. "Internet of medical things-based clinical decision support systems, smart healthcare wearable devices, and machine learning algorithms in COVID-19 prevention, screening, detection, diagnosis, and treatment." *American Journal of Medical Research* 9, no. 1 (2022): 65–80. https://doi.org/10.1007/s00170-021-08436-x

[42] Lattanzi, Luca, Roberto Raffaeli, Margherita Peruzzini, and Marcello Pellicciari. "Digital twin for smart manufacturing: A review of concepts towards a practical industrial implementation." *International Journal of Computer Integrated Manufacturing* 34, no. 6 (2021): 567–597. https://doi.org/10.1080/0951192X.2021.1911003

[43] Novák, Petr, Petr Douda, Petr Kadera, and Jiří Vyskočil. "PyMES: Distributed manufacturing execution system for flexible industry 4.0 cyber-physical production systems." In *2022 IEEE International Conference on Systems, Man, and Cybernetics (SMC)*, pp. 235–241. IEEE, 2022. https://doi.org/10.1109/SMC53654.2022.9945350

[44] Pham, Quoc-Viet, Kapal Dev, Praveen Kumar Reddy Maddikunta, Thippa Reddy Gadekallu, and Thien Huynh-The. "Fusion of federated learning and industrial internet of things: A survey." *arXiv preprint*, arXiv:2101.00798 (2021). https://doi.org/10.48550/arXiv.2101.00798

[45] Mahiri, Fadwa, Aouatif Najoua, Souad Ben Souda, and Najia Amini. "From industry 4.0 to industry 5.0: The transition to human centricity and collaborative hybrid intelligence." *Journal of Hunan University Natural Sciences* 50, no. 4 (2023): 84–94. https://doi.org/10.55463/issn.1674-2974.50.4.8

[46] Xiong, Jianghao, En-Lin Hsiang, Ziqian He, Tao Zhan, and Shin-Tson Wu. "Augmented reality and virtual reality displays: Emerging technologies and future perspectives." *Light: Science & Applications* 10, no. 1 (2021): 216. https://doi.org/10.1038/s41377-021-00658-8

[47] Kumar, Love, and Rajiv Kumar Sharma. "Smart manufacturing and industry 4.0: State-of-the-art review." In Ajay, Hari Singh, Parveen, Bandar AlMangour (Eds), *Handbook of Smart Manufacturing*, pp. 1–28. CRC Press, 2023. https://doi.org/10.1201/9781003333760-1

[48] Arena, Fabio, Mario Collotta, Giovanni Pau, and Francesco Termine. "An overview of augmented reality." *Computers* 11, no. 2 (2022): 28. https://doi.org/10.3390/computers11020028

[49] Zhao, Ziyue, Zhiting Jia, and Yan Li. "Introduction to the application of augmented reality technology in the digital manufacturing process." In *Ninth Symposium on Novel Photoelectronic Detection Technology and Applications,* vol. 12617, pp. 391–397. SPIE, 2023. https://doi.org/10.1117/12.2664386

[50] Leng, Jiewu, Weinan Sha, Baicun Wang, Pai Zheng, Cunbo Zhuang, Qiang Liu, Thorsten Wuest, Dimitris Mourtzis, and Lihui Wang. "Industry 5.0: Prospect and retrospect." *Journal of Manufacturing Systems* 65 (2022): 279–295. https://doi.org/10.1016/j.jmsy.2022.09.017.

2 Nature-Inspired Optimization Techniques of Industry 4.0 for Sustainable Manufacturing

Bhargavi Krishnamurthy[1] and Parveen Kumar[2]
1 Department of Computer science and Engineering,
  Siddaganga Institute of Technology, Bengaluru, Karnataka
2 Department of Mechanical Engineering, Rawal Institute of
  Engineering and Technology, Faridabad, Haryana, India

2.1 INTRODUCTION

One of the preliminary challenge faced by the globalization is to satisfy the rapidly growing demand for consumer goods for ensuring the sustainable evolution of the human race in economic dimensions. The adoption of sustainable manufacturing of sustainable supply chain management, sustainable industries, and sustainable smart cities. In order to meet this challenge, Industry 4.0 has evolved. The manufacturing of fourth industrial revolution as revolutionized the manufacturing procedure followed by the companies. The practical implementation of Industry 4.0 poses several new challenges with respect to economic investment and adopting to new technologies [1, 2].

Nature-inspired optimization algorithms are a collection of algorithms inspired by the behavior of animals, chemical reactions, and biological processes to solve complex engineering and medical problems. They use a group of innovative mathematical procedures for nonlinear optimization of various species for their survival. These algorithms are used in various applications like vehicle routing, smart transportation, supply chain management, industrial manufacturing, data mining, image processing, and medical solutions development. Some widely used optimization algorithms are ant colony optimization, particle swarm optimization, genetic algorithms, and many more. But the old optimization algorithms are not suitable for addressing the current issues of sustainable manufacturing due to their inherent limitations in terms of convergence rate, scalability, tuning of parameters, and stability attainment [3, 4].

However, recent nature-inspired optimization algorithms are capable of overcoming the inherent limitations of traditional algorithms. Some of the potential recent

DOI: 10.1201/9781003405870-2

25

nature-inspired optimization algorithms are the Adam optimizer, stochastic gradient descent (SGD), water wave optimization, seed-based plant propagation algorithm, bumblebee mating optimization (BBMO), and collective animal behavior (CAB) algorithm. The advantages offered by these algorithms are general-purpose problem solving, providing new insights for complex problems, better parameter tuning capability, and better exploration of huge search spaces [5, 6].

In this chapter, these recent nature-inspired algorithms are applied to address the sustainable manufacturing problem for Industry 4.0. Each of these algorithms is discussed briefly along with its pros and cons for the proper management and distribution of consumer goods. Further, the performance analysis of these algorithms towards sustainable manufacturing in Industry 4.0, with respect to performance metrics like social indicator, governance indicator, industry productivity, is discussed.

2.2 ADAM OPTIMIZER

The improved version of the stochastic gradient descent algorithm is referred to as the Adam optimizer. It is widely used in several applications, which include computer vision, natural language processing, and sustainable manufacturing. It is adaptive in nature and computes individual learning rates by considering different sets of parameters. A generic representation of the Adam optimizer is given in Algorithm 2.1.

Algorithm 2.1. Generic Representation of Adam Optimizer

Step 1: **Begin**
Step 2: Initialize the stepsize α
Step 3: Exponential decay rates for moment estimates $\beta_1, \beta_1 \in [0,1]$
Step 4: Assign stochastic objective function $f(\theta)$ with θ
Step 5: Initialize parameters i.e., $\theta_0 \leftarrow 0$, $m_0 \leftarrow 0$, $v_0 \leftarrow 0$, and t
Step 6: **While** θ_0 not converged **do**
Step 7:   Increase the step size $t = t+1$
Step 8:   Update gradient with respect to stochastic objectives $g_t \leftarrow \nabla_\theta f_t(\theta_{t-1})$
Step 9:   Update biased first moment estimate $m_t \leftarrow \beta_1 * m_{t-1} + (1-\beta_1) * g_t$
Step 10:   Update biased second moment estimate $v_t \leftarrow \beta_2 * v_{t-1} + (1-\beta_2) * g_t^2$
Step 11:   Calculate bias-corrected first moment $\hat{m}_t \leftarrow m_t / (1-\beta_1^t)$
Step 12:   Calculate bias-corrected second raw moment $\hat{v}_t \leftarrow v_t / (1-\beta_2^t)$
Step 13:   Update the initial parameters $\theta_t \leftarrow \theta_{t-1} - \alpha * \hat{m}_t / \sqrt{\hat{v}_t} + \in$
Step 14: **End While**
Step 15: Output the resulting initial parameters θ_t
Step 16: **End**

The advantages of the Adam optimizer for sustainable manufacturing are as follows: It incorporates the best features of the adaptive gradient algorithm (AdaGrad) and rate-monotonic scheduling (RMS) algorithms which makes it eligible to perform a sparse gradient on noisy problems. It attains an accuracy of 76% compared to stochastic gradient descent. It is suitable to solve problems which involve large amounts of data with many functioning parameters. Efficient computation is performed with lower memory requirements.

However, some of the limitations of the Adam optimizer for sustainable manufacturing are as follows:

- It is not able to converge to an optimal solution in computation-intensive areas
- Weight decay problem is also surrounded around the Adam optimizer.
- Initial version of Adam are found to operate at less speed.

The learning rate is found to be very small and the tendency towards converging to suboptimal solutions is very high, it doesn't work properly under longer training phases, and it is subject to an early stopping mechanism [7, 8].

2.3 STOCHASTIC GRADIENT DESCENT

Stochastic gradient descent is a very old machine learning algorithm used to determine the machine learning model parameters which fit perfectly between predicted output and actual output. A key idea used by SGD is that by observing only one training example at a single time step, inference about the direction of decreasing slope level is determined. The technique selects any data point randomly from the existing set of training samples at every iteration, which reduces the amount of computation involved in the operation. A generic representation of SGD is given in Algorithm 2.2.

Algorithm 2.2. Generic Representation of Stochastic Gradient Descent

Step 1: **Begin**
Step 2: Start with random initial values for the parameters
Step 3: Extract the feature and target variable for each of the random index
Step 4: **While** predict the target variable value for every random selected point
      by choosing current parameters
Step 5:   Determine the cost involved in the prediction
Step 6:   **If** the cost is minimized
Step 7:   Update the parameters
Step 8:   **Else**
Step 9:   Update the parameters using gradient descent technique
Step 10:   **End If**
Step 11: **End While**
Step 12: Output the resulting stochastic gradient descent parameters
Step 13: **End**

The advantages of SGD for sustainable manufacturing are as follows: It accurately determines the optimal set of configuration parameters for the kind of machine learning algorithm in less time. The error involved in computation is less, as it makes minor adjustments to the network configuration in every iteration. Whenever larger data samples are considered, its convergence rate is faster, as it frequently updates the training parameters. The effort involved in computation drastically decreases as it achieves better tradeoff with lowered convergence rate. It achieves higher efficiency as its efficiency is linearly proportional to number of training examples considered for training the algorithm.

However, some of the limitations of SGD for sustainable manufacturing are as follows: Compared to the original gradient descent method, SGD's convergence rate is slow, as its performance is poor when handling noisy data. It exhibits high variance in the direction of learning and learning values. The process followed to update the weights is noisy, and denoising the gradient value is difficult. The learning rate achieved is less as it reaches the minima because of more oscillation. The SGD model lacks interpretability due to a poor explanation factor. A prediction done over imbalanced data samples is poor as it fails to precisely predict rare outcomes. If the data samples don't fit in memory, then the algorithm becomes intractable. The SGD model is not able to be updated in online mode [9, 10].

2.4 WATER WAVE OPTIMIZATION

The water wave optimization algorithm is inspired by shallow water wave theory. It uses three operators which are inspired by water wave movements: water propagation, water refraction, and breaking. These three operators make it able to explore high-dimensional problem spaces by providing optimization solutions. The structure of the WWO algorithm is very simple with good realization, and its performance is found to be good enough over small population problems. The core concept of WWO is that it allocates a wavelength number to every solution, and the wavelength is proportional to fitness of the solution. The higher the water wave, the larger the height of the water wave. After sufficient iterations, if the efficiency of the solution doesn't improve, once the wave height becomes zero, a random solution between the current solution and optimal solution is chosen to avoid search stalling. A generic representation of water wave optimization is given in Algorithm 2.3.

Algorithm 2.3. Generic Representation of Water Wave Optimization

Step 1: **Begin**
Step 2: Input the given combinatorial optimization problem
Step 3: **While** concrete WWO solution is obtained for the problem
Step 4:   Define the representation of solution along with the neighbor structure
Step 5:   Redefine the propagation operator
Step 6:   Redefine the wavelength calculation
Step 7:   Tune the optional aspects of the metaheuristic water wave optimization
Step 8: **End While** concrete WWO solution is obtained
Step 9: Output the resulting initial parameters θ_t
Step 16: **End**

The advantages of WWO for sustainable manufacturing are as follows: It achieves better balance between exploration and exploitation activities. It also exhibits a sufficient amount of ability to prevent the weeding-out effect caused by the water refraction operator. The computation effort involved in implementation is much less and is capable of achieving multiple objectives. The chances of premature convergence are less, as the water refraction operator escapes from stagnation of the search space, which improves the diversity of the solution developed. By using the water breaking operator, a global optimal solution is guaranteed, as it enables an intensive search operation around the premises of promising areas of solution.

However, some of the limitations of WWO for sustainable manufacturing are as follows: The optimal behavior of WWO is subjective in nature and depends on the nature of the problem. It is sometimes time consuming when exposed to high-dimensional problems. The solutions achieved are far from optimal solutions for uncertainty-prone large-scale problems [11, 12].

2.5   SEED-BASED PLANT PROPAGATION

Seed based plant propagation is inspired by taking the benefits of all different types of intelligent agents like animals, birds, and water movement. It is inspired by the way a strawberry plant propagates. Plants propagate at a border range to determine the proper estimation to mark boundaries. Two basic methods of SPP are wind-based and ballistic-based methods. Even animals and birds play a vital role in seed dispersion. In particular, the strawberry plant attracts birds and spreads seeds across many areas located far away. But the way strawberry plant does spatial distribution of seeds on the presence of strawberries in the plant is a tedious activity. The rate at which the seeds propagate depends on the number of times different agents like animals/birds come to eat the strawberry fruits. The abundant availability of seeds to spread locally and globally across the region of plant areas is determined via a seed shadow. One of the promising SSPs is the strawberry feeding station model, which operates based on two factors: the amount of seeds collected by each intelligent agent and the service rate of intelligent agents for parent plants. A generic representation of seed-based plant propagation is given in Algorithm 2.4.

Algorithm 2.4. Generic Representation of Seed-Based Plant Propagation

Step 1: **Begin**
Step 2: Initialize the seed population
Step 3: Start with the seed production
Step 4: Perform abiotic dispersal of seeds
Step 5: Perform seed removal operation
Step 6: **While** seed present **do**
Step 7:   Remove the seeds from the existing set of available seeds
Step 8:   Perform seeding establishment
Step 9:   perform survival of seeding
Step 10:   Reject the poor quality seeds
Step 11: **End While**
Step 15: Output the resulting high quality seeds as solution
Step 16: **End**

The advantages of SPP for sustainable manufacturing are as follows: The performance of SPP is found to be better than all kinds of intelligent agents like wind, water, or even animals, as it combines the advantages of all the intelligent agents. From an implementation point of view, it's easy, as it consists of fewer parameters for learning purposes. Over a span of time, the generation of output produced is of higher quality, as to produce high-quality outputs, high-quality seeds will be considered. The solution generated through the seed propagation mechanism exhibits long-lasting stability, as it exhibits higher resistance to all kinds of attacks. Genetic uniformity is sustained via consistent seed propagation. Hybrid seed propagation helps in delivering solutions which are capable of adapting to uncertainties and the dynamically varying nature of computing environment.

However, some of the limitations of SPP for sustainable manufacturing are as follows: There are chances of seedling solutions generating inferior-quality solutions. Some kinds of seeds (algorithms) are difficult to germinate. Because of genetic variation among the seeds, it is very difficult to produce and retain superior-quality solutions. If the propagation of seeds is carried out using a vegetative mechanism, then the solutions proposed by them are short lived. Precise identification of the stage of the solution is difficult, especially at the inoculation time of the solution [13, 14].

2.6 BUMBLEBEE MATING OPTIMIZATION

Bumblebees mating optimization is a recent nature-inspired algorithm which mimics the mating strategy of bumblebees. The working of BBMO is broadly classified into three different modes of operation: queen mode, worker mode, and drone mode. The working of the algorithm begins with the initialization of bumblebee parameters; then, the initial set of the population of solutions is generated. The fitness value of each of the bumblebees is computed. Among all bumblebees, the bumblebee with the highest fitness shall be elected as queen. The rest of the bumblebees are elected as drones. The existing drones are sorted based on their respective fitness values. The best drone will be selected for mating with the queen to generate new possible offspring. A generic representation of bumblebee mating optimization is given in Algorithm 2.5.

Algorithm 2.5. Generic Representation of Bumblebee Mating Optimization

Step 1: **Begin**
Step 2: Generate the initial population of solution
Step 3: Randomly initialize the parameters
Step 4: Determine the fitness value of each of the bumblebee
Step 5: Select the best fitness value bee as queen
Step 6: Sort the remaining bees according to their fitness value for mating with the queen
Step 7: **Do While** maximum number of iterations are reached
Step 8:   Create the broods using crossover operator
Step 9:   Determine the fitness of each of the brood
Step 10:   Selection of best brood as the new queens

Step 11:   Selection of rest of the brood as workers
Step 12:   **Do While** maximum number of meetings for each new queen has been reached
Step 13:   Select the drone for mating
Step 14:   Store the drone genotypes
Step 15:   Survival of new queens for next generations
Step 16:   Dying of all other members of the population
Step 17:   Update the parameters of the algorithm
Step 18:   **End Do While**
Step 19: **End Do While**
Step 20: Output the best solution found as queen
Step 21: **End**

The advantages of BBMO for sustainable manufacturing are as follows: The cost of operation involved in the implementation of BBMO is very low, which leads to wider applicability. Bumblebees exhibit high flexibility by arriving at optimal results. The offspring solutions produced by bumblebees are of high quality as it saves considerable amount of time during mating. Through a controlled mating strategy, genetic diversity is achieved among the newer offspring. Usually, male and female mates once, which helps in maintaining lifetime fitness of the bees. If the queen bees mate with more than two male bees, then they will be executed by worker bees, which yields increased productivity.

However, some of the limitations of BBMO for sustainable manufacturing are as follows: BBMO suffers from global-level unconstrained optimization problems. Feature selection becomes difficult while dealing with a variety of database applications. Finding suitable parameters for tuning the BBMO algorithm is troublesome [15, 16].

Table 2.1 provides a comparison between the selected potential nature-inspired optimization techniques for sustainable manufacturing applications. From the

TABLE 2.1

Comparison of Efficiency Achieved by Nature-Inspired Optimization Techniques for Sustainable Manufacturing

Nature-Inspired Optimization Types	Learning Parameters	Error Rate	Depth of Learning	Precision Achieved	Potential Features
Adam optimizer [7]	More	10.0%	Low	Medium	Operates well under uncertainty
Stochastic gradient descent [9]	More	4.7%	Low	Low	Convergence rate is faster
Water wave optimization [11]	Less	5.0%	High	Medium	Achieves better balance between exploration and exploitation activities
Seed-based plant propagation [13]	Less	6.5%	Medium	Medium	Produces high-quality outputs in fewer iterations of learning
Bumblebee mating optimization [15]	Less	1.0 %	High	High	Cost of operation is minimum

analysis, it is found that the performance of BBMO is the best compared to all other algorithms. Similarly, the performance of Adam optimizer is the worst, as the learning rate is slow, and the precision achieved is also lower.

2.7   CONCLUSION

This chapter provides a brief introduction to potential nature-inspired optimization techniques like the Adam optimizer, stochastic gradient descent, water wave optimization, seed-based plant propagation algorithm, and bumblebee mating optimization. These algorithms are discussed, along with their advantages and disadvantages. From the performance analysis, it is found that the performance of BBMO is the best compared to all other algorithms, as it achieves higher performance pertaining to metrics like learning parameters, error rate, depth of learning, and precision achieved.

REFERENCES

[1] Kumar, R., Singh, R. K., & Dwivedi, Y. K. (2020). Application of industry 4.0 technologies in SMEs for ethical and sustainable operations: Analysis of challenges. *Journal of Cleaner Production*, Vol. 275, p. 24063. https://doi.org/10.1016/j.jclepro.2020.124063

[2] Kumar, A., Singh, H., Kumar, P., & AlMangour, B. (Eds.). (2023). *Handbook of Smart Manufacturing: Forecasting the Future of Industry 4.0*. CRC Press. https://doi.org/10.1201/9781003333760

[3] Abdel-Basset, M., Abdel-Fatah, L., & Sangaiah, A. K. (2018). Metaheuristic algorithms: A comprehensive review. *Computational Intelligence for Multimedia Big Data on the Cloud with Engineering Applications*, pp. 185–231. https://doi.org/10.1016/B978-0-12-813314-9.00010-4

[4] Dokeroglu, T., Sevinc, E., Kucukyilmaz, T., & Cosar, A. (2019). A survey on new generation metaheuristic algorithms. *Computers & Industrial Engineering*, Vol. 137, p. 106040. https://doi.org/10.1016/j.cie.2019.106040

[5] Sharma, V., & Tripathi, A. K. (2022). A systematic review of meta-heuristic algorithms in IoT based application. *Array*, p. 100164. https://doi.org/10.1016/j.array.2022.100164

[6] Khanduja, N., & Bhushan, B. (2021). Recent advances and application of metaheuristic algorithms: A survey. *Metaheuristic and Evolutionary Computation: Algorithms and Applications*, pp. 207–228. https://doi.org/10.1007/978-981-15-7571-6_10

[7] Zhang, Z. (2018, June). Improved Adam optimizer for deep neural networks. In *2018 IEEE/ACM 26th International Symposium on Quality of Service (IWQoS)*. IEEE, pp. 1, 2. https://doi.org/10.1109/IWQoS.2018.8624183

[8] Bock, S., & WeiB, M. (2019, July). A proof of local convergence for the Adam optimizer. In *2019 International Joint Conference on Neural Networks (IJCNN)*. IEEE, pp. 1–8. https://doi.org/10.1109/IJCNN.2019.8852239.

[9] Liu, Y., Gao, Y., & Yin, W. (2020). An improved analysis of stochastic gradient descent with momentum. *Advances in Neural Information Processing Systems*, Vol. 33, pp. 18261–18271.

[10] Sebbouh, O., Gower, R. M., & Defazio, A. (2021, July). Almost sure convergence rates for stochastic gradient descent and stochastic heavy ball. In *Conference on Learning Theory*, pp. 3935–3971.

[11] Zhou, X. H., Zhang, M. X., Xu, Z. G., Cai, C. Y., Huang, Y. J., & Zheng, Y. J. (2019). Shallow and deep neural network training by water wave optimization. *Swarm and Evolutionary Computation*, Vol. 50, p. 100561. https://doi.org/10.1016/j.swevo.2019.100561.

[12] Zhao, F., Shao, D., Wang, L., Xu, T., & Zhu, N. (2022). An effective water wave optimization algorithm with problem-specific knowledge for the distributed assembly blocking flow-shop scheduling problem. *Knowledge-Based Systems*, Vol. 243, p. 108471.

[13] Hazubska-Przybył, T. (2019). Propagation of Juniper species by plant tissue culture: A mini-review. *Forests,* Vol. 10, Issue. 11, p. 1028. https://doi.org/10.3390/f10111028

[14] Nath, S., Ghosh, N., Ansari, T. A., Mundhra, A., Patil, M. T., Mane, A., . . . & Dey, A. (2022). Genetic diversity assessment and biotechnological aspects in Aristolochia spp. *Applied Microbiology and Biotechnology*, Vol. 10, pp. 6397–6412. https://doi.org/10.1007/s00253-022-12152-1

[15] Marinakis, Y., Marinaki, M., & Migdalas, A. (2017). An adaptive bumble bees mating optimization algorithm. *Applied Soft Computing*, Vol. 55, pp. 13–30. https://doi.org/10.1016/j.asoc.2017.01.032

[16] Alotaibi, M. T., Almalag, M. S., & Werntz, K. (2020, December). Task scheduling in cloud computing environment using bumble bee mating algorithm. In *2020 IEEE Global Conference on Artificial Intelligence and Internet of Things (GCAIoT)*. IEEE, pp. 01–06. https://doi.org/10.1109/GCAIoT51063.2020.9345824

3 Framework for Implementation of Disruptive Emerging ICTs in Intelligent and Sustainable Manufacturing in a Developing Context

Samuel Musungwini[1] *and Sheltar Marambi*[2]

1 Department of Information and Marketing Sciences, Department of Business Sciences, Midlands State University, Gweru, Midlands Province, Zimbabwe

2 Department of Information Systems, Midlands State University, Gweru, Midlands Province, Zimbabwe

3.1 INTRODUCTION

The evolution of information and communication technology (ICT) and continuous revolution have changed the way we do business. So rapidly advancing and multi-faceted is ICT that it has evolved into several technologies which are disruptive in nature. These ICT shoots encompass but are not limited to artificial intelligence, cloud computing solutions, virtual reality, big data, and the Internet of Things (IoT), just to name a few, as posited by Tsanousa et al. (2022). This new automation is pervasive and straddles many spheres of economic activities, such that it has been felt across different economic sectors like tourism (Romão, 2020), banking (Li et al., 2021), agriculture (Musungwini et al., 2022), education (Brunetti et al., 2020), and health (Furusa & Coleman, 2018) over the whole world, but the application in the manufacturing industry (Moeuf et al., 2018) has been limited to the developed world. However, the advent of Industry 4.0, also known as the fourth wave of industrialisation, is marked by the convergence of next-generation systems. These emerging smart systems are popularly known as cyber-physical systems (CPSs). CPSs integrate the real world with the virtual world to create seamless innovative possibilities. The term "Industry 4.0" is often used to refer to a broad array of ICT technological tools

DOI: 10.1201/9781003405870-3

which are used in the manufacturing sector, including supply chain management (SCM) and product design, as asserted by Stentoft et al., 2021). This prospect presents an opportunity for developing countries like Zimbabwe to harness these emerging cutting-edge technologies to resuscitate the manufacturing industry to boost the economy.

This chapter examines the distinctive fourth industrial revolution technologies used in intelligent production processes to advance scholarship on Industry 4.0. First, a comprehensive overview of potential Industry 4.0 implementations for smart enterprise applications is provided. Furthermore, through various illustrative instances, CPS-enabled smart design is introduced. Finally, new developments and their possible application through smart manufacturing in Zimbabwe's manufacturing industries are examined based on these example cases. To help industrial sectors grow stronger, an array of Industry 4.0 challenges and opportunities are identified and examined. The chapter culminates with the development of a framework for the implementation of emerging ICTs in intelligent and sustainable manufacturing in a developing context like Zimbabwe.

3.2   DEVELOPING CONTEXT: CASE OF ZIMBABWE

Due to several circumstances, Zimbabwe's manufacturing sector experienced a downturn that was analogous to the country's overall economic activity over the last 20 years. One of these is the state of hyperinflation, which peaked in 2008 at an official rate of 231% (Damiyano et al., 2012). This, coupled with several negative factors, including a lack of working capital support; a contraction in the domestic market; costly utility rates; tax structures that were significantly high when compared to the region; high wages; a credit and liquidity crunch; and several supply-side impediments, including those involving fuel, electricity supply, imported inputs, and skills, were the main features of the corporate environment during this time (Mhaka et al., n.d.). A change in consumer habits in favour of cheaper imported goods has lowered the appetite for locally produced goods. According to industry officials, over 400 businesses have shut down since the year 2000, and this rendered more than 90% of employees jobless (Gadzikwa, 2013). That is why the Zimbabwean manufacturing sector needs a plan for the future to boost productivity and regain the industrial competitiveness clout it once had. There is a need to acknowledge that the level of competition has drastically shifted in addition to the new economic climate bringing new difficulties.

As globalisation and trade liberalisation are now facts of life in Zimbabwe, they present both opportunities and risks to the local economy. As a result, Zimbabwean businesses are finding it difficult to keep up with the competition and manage both imports and exports in a global economic environment, as asserted by Kwenda (2014). Companies are increasingly under unprecedented pressure to increase operational effectiveness for increased competitiveness and better business performance as a whole. These challenges include foreign product competition, competitor new product introductions, quick technological innovations and reduced product lifespan, unpredictable customer changes, and innovations in manufacturing and information technology (Damiyano et al., 2012). Therefore, in the context of the present economic climate, the business must satisfy client expectations by providing a dependable product or service on schedule. Thus, there is a need for a radical shift which can usher in change (Musungwini & Mondo, 2019).

That is why Zimbabwean manufacturing organisations must improve their operational efficiency and management practises, which can boost production which meets the local requirements at competitive prices and compete competitively in the global arena in a bid to overcome these difficulties and make the most of the possibilities that come with globalisation. We believe that Zimbabwe's domestic policies must foster an environment that enables local companies to expand and raise their level of manufacturing supremacy to compete internationally.

This suggests that the Zimbabwe economy is at a defining moment where it may potentially rebound from more than two decades of recession and achieve its former level of international competitiveness (Gadzikwa, 2013). Therefore, productivity, profitability, and growth of the local manufacturing and services sectors will depend on its ability to compete in the evolving economic climate that presents a more liberal national and global setting. Yet the country boasts some of the world's most precious mineral resources like lithium, gold, diamonds, platinum, and natural gas, among other precious mineral resources. Thus, there is a need to tap into these resources, and ultimately the internationally competitive manufacturing industry in Zimbabwe must generate more wealth and make a greater contribution to the development of a more sustainable and stable economy, which will in turn promote both foreign and domestic investment. But for this to happen, there is a need to harness emerging cutting-edge ICT, as suggested by Mahlangu et al. (2018), using innovation hubs. That is why we propose fashioning a framework for the implementation of emerging ICTs in intelligent and sustainable manufacturing in a developing context which is in sync with the Smart Zimbabwe 2030 master plan, which is anticipated to result in the development of a digital economy in the nation.

This is in line with the Sustainable Development Goals (SDGs) framework, which is based on the idea of using ICTs to create the world we want, as asserted by Sununtar Setboonsarng and Elsbeth E. Gregorio (2017). The economic activity that arises from billions of daily online interactions between individuals, corporations, devices, and data is known as the "digital economy". Research papers, dissertations, and policy documents on the adoption, application, and use of ICTs in many fields, such as banking and finance (Akinyemi & Mushunje, 2020), health (Zhou et al., 2015), tourism (Njerekai, 2020), agriculture (Musungwini et al., 2023), education (Magura et al., 2022), and manufacturing have been published in the past ten years. Therefore, manufacturing companies in Zimbabwe will have new potential for digital transformations as a result of technological breakthroughs like those connected to Industry 4.0, enabling them to offer competitively priced goods and services to both established and emerging markets. A qualitative approach is used to explore the factors that influence Industry 4.0 readiness and adoption among Zimbabwe's manufacturers, who are primarily small and medium-sized businesses at the moment.

The authors engaged with literature on the manufacturing industry and emerging cutting-edge ICTs like IoT, cloud computing, big data, artificial intelligence, digital robots, virtual and augmented reality, and digital twins, among other evolving digital technologies. The empirical data was gathered using in-depth interviews with key informants who are key industry captains who were purposively sampled with the use of a snowball effect to contextualise this work. The authors think that drawing on the knowledge of these professionals allowed for more nuanced assessments of the relevance of the Industry 4.0 problems encountered by businesses in the manufacturing sector in a changing environment.

3.3 METHODOLOGY

This work follows a qualitative approach which is premised on a literature review and in-depth interviews with key informants made up of industry and technology experts in Zimbabwe in line with the dictates of Patton and Cochran (2002). The researchers conducted a comprehensive systematic literature review to identify the existing research and knowledge on the subject domain of disruptive innovative emerging ICTs that are being used in intelligent and sustainable manufacturing in the developed world and have the potential to be used in a developing context. This provided our work an understanding of the state of what is known and not known, thereby revealing the existing gaps that needed to be addressed through further research. This also informed the formulation of our eight semi-structured interviews that we conducted with experts from various backgrounds, specifically industry, academia, and government, who had intimate experience with the implementation of emerging ICTs in intelligent and sustainable manufacturing in developing contexts. The interviews focused on identifying key challenges and opportunities that are associated with the harnessing of disruptive emerging Industry 4.0 ICTs and also fashioned potential solutions for overcoming these challenges.

To analyse the data collected from both the literature review and interviews, we used thematic analysis to identify patterns and themes related to the implementation of emerging ICTs in intelligent and sustainable manufacturing in general and developing contexts specifically (Thomas, 2003). After the data analysis, there was a need for discussion of the findings, thereby synthesising the findings from both the literature review and interviews into a cohesive narrative so that we project the key insights of the work and implications for academia, practice, and policy formulation.

3.4 LITERATURE REVIEW

3.4.1 INTELLIGENT AND SUSTAINABLE MANUFACTURING IN DEVELOPING CONTEXT

In developing countries, the manufacturing industry is seen as a crucial sector for economic growth and development of those countries to catch up with the developed world. Yet these countries have traditional manufacturing processes that are typically associated with significant amounts of waste and pollution, both of which have severe consequences for the environment and human health. The concept of intelligent manufacturing (IM) is a human–machine integrated intelligent system that can perform intelligent manufacturing tasks such as analysis, reasoning, judgement, conception, and decision-making, as reported by Qin et al. (2022). The concept of sustainable manufacturing is gaining popularity, as it tries to reduce the environmental and social effects of manufacturing processes and activities while maintaining or improving economic performance (Aidonis et al., 2019). Concurrently, improvements in information and communication technologies (ICTs) have created opportunities for developing countries to integrate intelligent and sustainable manufacturing methods.

The Fourth Industrial Revolution (4IR) is a new era of technical advancement characterised by the adoption of advanced digital technologies in all aspects of the industry. Industry 4.0 requires sustainability, which necessitates the use of ICTs for sustainable manufacturing (Hansen & Lilja, 2020). Cimini et al. (2020) agree with the use of ICTs for sustainable development, as they discuss several technologies that underpin the concept of sustainable manufacturing, such as the Industrial Internet of Things, cloud computing, and big data analytics, among others. The application of modern technologies such as artificial intelligence (AI), machine learning, and the Internet of Things (IoT) to improve manufacturing processes and products is referred to as intelligent manufacturing (Kumar et al., 2023). These technologies have the potential to minimise waste, increase efficiency, and optimise industrial processes, resulting in better economic and environmental consequences.

Numerous studies have been conducted to demonstrate the potential benefits of intelligent and sustainable manufacturing in developing countries. Implementing intelligent manufacturing methods in a Mexican manufacturing plant resulted in considerable increases in energy efficiency, waste reduction, and productivity (Sanchez-Planelles et al., 2022). However, the adoption of intelligent and sustainable manufacturing practices in developing countries is not without its challenges. Lack of technical expertise and infrastructure and limited access to financing and regulatory policies are some of the challenges that hinder the adoption of advanced manufacturing technologies (Gunasekaran et al., 2018). Additionally, cultural factors and resistance to change may also impede the adoption of sustainable manufacturing practices (Aidonis et al., 2019).

Despite these challenges, there is growing recognition of the importance of intelligent and sustainable manufacturing in developing countries, and efforts are being made to promote the adoption of these practices. For example, initiatives such as the United Nations Industrial Development Organization's (UNIDO) Global Cleantech Innovation Programme for SMEs in developing countries aim to support the adoption of sustainable manufacturing technologies by small and medium-sized enterprises in developing countries. The integration of intelligent and sustainable manufacturing practices in developing countries has the potential to improve economic and environmental outcomes but also presents significant challenges. A review of the different ICTs that can be used for intelligent and sustainable manufacturing, including their benefits and challenges, is explained further in the next sections.

3.4.2 General Overview of ICT

A wide range of technologies used to access, process, and communicate information are known as ICTs and include hardware, such as computers, mobile devices, and servers, as well as software, such as operating systems, productivity tools, and applications (Chu et al., 2018). ICT has become an integral part of modern society, and its applications are widespread across many sectors, including education, healthcare, finance, and manufacturing. Advancement in ICTs has been driven by several factors, including advances in computer hardware and software, the growth of the internet, and the increasing demand for data-driven decision-making (W. Lu et al., 2021). The use of ICT has been shown to have numerous benefits, including improved efficiency, increased productivity, and enhanced communication and collaboration (Chu et al., 2018). Additionally, the use of ICT has been linked to improvements in education and healthcare outcomes and the ability to promote economic growth and development.

ICT has also enabled the development of new technologies, such as AI, machine learning, and the Internet of Things. These technologies are increasingly being integrated into a wide range of applications, including manufacturing, transportation, and energy management (Y. Lu et al., 2020). The integration of these technologies has the potential to revolutionise traditional industries and create new economic opportunities. ICT has become an integral part of modern society and has the potential to drive innovation and economic growth. The integration of new technologies, such as AI and IoT, is likely to continue to transform traditional industries and create new opportunities. However, there are also several challenges associated with ICT, including the digital divide and concerns around data privacy and security. The digital divide refers to the gap in access to ICT and the Internet between different regions and socioeconomic groups (Musungwini et al., 2022). The digital divide can exacerbate existing social and economic inequalities and limit access to critical services and opportunities. Other challenges associated with ICT include concerns around data privacy and security, as well as the potential for job displacement as automation and AI technologies become more prevalent (W. Lu et al., 2021). Additionally, the rapid pace of technological change and the need for ongoing training and education can create barriers to adoption and limit the benefits of ICT for some individuals and organisations.

There are also technologies like augmented reality (AR) and virtual reality (VR), as well as mixed reality (MR), which have gained increasing attention in recent years due to their potential to transform many industries, including healthcare, education, and entertainment. AR refers to technology that overlays digital information onto the physical world, while VR creates an entirely immersive digital environment, and MR simply fuses the two (AR and VR), as posited by Gattullo et al. (2022). Both technologies have potential for the manufacturing sector to boost new product offerings and experimentation. The development of AR and VR has been driven by several factors, including advances in computer hardware, software, and mobile devices (Lacoche et al., 2019). In recent years, AR and VR have become more accessible and affordable and their applications have expanded to include areas such as remote work and telemedicine. AR and VR have been used in a variety of industries, including healthcare, where they have been used for medical training, patient education, and surgery simulation (Abdel-Basset et al., 2021). In education, AR and VR have been used to enhance learning and engagement by providing immersive experiences and interactive simulations. In entertainment, AR and VR have been used to create new forms of gaming, storytelling, and immersive experiences.

In the manufacturing industry, AR and VR have gained attention, as they have the potential to revolutionise the way manufacturing processes are executed, improving productivity and reducing errors. Mixed reality, a seamless combination of physical and virtual objects, is another technology that is being used to achieve a sustainable environment in the manufacturing sector (Sullivan et al., 2022). Hence, we think that developing countries like Zimbabwe can also take advantage of the wave of transformation and overhaul their manufacturing industry.

Despite the many benefits of AR and VR, there are also several challenges associated with their use. One of the most significant challenges is the high cost of hardware and software, which can limit access for some individuals and organisations (Bhat et al., 2022). Additionally, the development of AR and VR content requires specialised skills and expertise, which can create barriers to adoption. Other challenges associated with AR and VR include concerns around user safety and discomfort, as well as the potential for addiction and overuse. There are also concerns about data privacy and security, particularly in the case of AR, which relies on accessing and analysing real-world data.

3.4.3 ROLE OF ICT IN INTELLIGENT AND SUSTAINABLE MANUFACTURING

The integration of ICT in manufacturing processes has significantly changed the way industries operate. The use of ICT in manufacturing has enabled intelligent and sustainable manufacturing practices, leading to increased efficiency, reduced waste, and improved sustainability. ICTs are not only tools but driving factors for sustainable manufacturing practices, as they drive the economic, social, environmental and technological aspects of an organisation (Alhassan & Scholtz, 2019). ICT is being used in various applications in intelligent and sustainable manufacturing, such as predictive maintenance, real-time monitoring, and supply chain management.

Predictive maintenance enables manufacturers to identify potential equipment failures before they occur, reducing downtime and improving productivity, whilst real-time monitoring enables manufacturers to monitor production processes in real time, enabling them to optimise processes and reduce waste (Zain et al., 2020). On the other hand, supply chain management gives manufacturers the ability to monitor and trace their products through the supply chain, boosting the effectiveness of the supply chain and transparency. The integration of ICT in manufacturing processes has significant implications for sustainability. ICT-enabled sustainable manufacturing practices can reduce energy consumption, waste, and emissions and enable manufacturers to implement circular economy principles such as product reuse and recycling. For example, the use of data analytics in manufacturing processes can enable manufacturers to identify waste-reduction opportunities and implement strategies to reduce waste (Zheng et al., 2021).

Additionally, the use of ICT can enable manufacturers to implement closed-loop supply chain systems, enabling them to recover and reuse materials throughout the supply chain. AR and VR technologies are being used in various applications in the manufacturing industry, such as assembly, maintenance, training, and quality control. AR and VR technologies can be used to create immersive training experiences for workers, simulate production processes, and provide remote support to technicians. For example, Boeing uses AR technology to provide its technicians with real-time guidance and instructions during aircraft assembly (Urban & Łukaszewicz, 2020).

AR is being used in manufacturing to improve training, reduce errors, and enhance the overall manufacturing process. The technology can provide workers with real-time information and guidance, reducing the risk of mistakes and improving efficiency. It can also be used to simulate and test products before they are manufactured, reducing the time and cost associated with prototyping. AR technology has been used to guide workers during assembly processes, providing them with step-by-step instructions and real-time feedback. This technology has been shown to improve assembly speed and accuracy while reducing errors (Burova et al., 2020). VR technology, on the other hand, is being used for training and simulation purposes, allowing workers to practice their skills in a safe and controlled environment. VR technology is also being used for quality control purposes, allowing manufacturers to detect defects and errors in products before they are released to the market. The integration of AR and VR technologies in manufacturing systems has brought about many benefits in intelligent and sustainable manufacturing. The following are some of the emerging ICT technologies for intelligent and sustainable manufacturing:

3.4.3.1 Internet of Things

IoT involves the integration of sensors, devices, and systems that can collect and share data in real-time. In manufacturing, IoT is being used to enable real-time monitoring and control of the production process, reducing waste and improving efficiency. IoT is also being used to create smart factories, where machines can communicate with each other, predict maintenance needs, and optimise production processes (L. Da Xu et al., 2018). IoT technology has the potential to revolutionise the manufacturing industry by connecting machines, sensors, and devices, and enabling real-time data exchange. This can help manufacturers to monitor their operations, optimise production processes, and reduce costs.

3.4.3.2 Artificial Intelligence

Artificial intelligence is being used in manufacturing to automate processes, increase efficiency, and reduce costs. AI-powered machines can learn from data and make decisions based on that data, reducing the need for human intervention. In the field of predictive maintenance, whereby machines can identify and address problems before they result in downtime, AI is also applied (Ikumapayi et al., 2023). This technology has the potential to automate and optimise many aspects of the manufacturing process. For example, AI-powered robots can be used for tasks such as assembly, quality control, and material handling. Additionally, AI can be used to analyse large amounts of data to identify patterns and insights that can help manufacturers to make better decisions.

3.4.3.3 Blockchain

Blockchain technology is being used in manufacturing to improve supply chain transparency, traceability, and security. Blockchain can create a secure and transparent record of transactions between suppliers, manufacturers, and customers, reducing the risk of fraud and improving trust (W. Lu et al., 2021). It can also be used to create smart contracts that automatically trigger actions based on predefined rules, reducing the need for intermediaries. Through creating a decentralised ledger of transactions, manufacturers can track their products from raw materials to finished goods and ensure that their suppliers are adhering to ethical and sustainable practices. For example, IBM is working with Walmart to use blockchain technology to improve the traceability of food products (Westerlund et al., 2021).

3.4.3.4   Cloud Computing

Cloud computing is a technology that enables manufacturers to access and store data and applications on remote servers. This allows manufacturers to reduce their IT infrastructure costs and improve their scalability and flexibility. Additionally, cloud computing can provide manufacturers with real-time data analysis and collaboration capabilities.

3.4.3.5   Digital Twins

A digital twin is a virtual model of a physical product, process, or system. It allows manufacturers to simulate and analyse the performance of their products in real time, and optimise their production processes accordingly. Digital twins can also be used for predictive maintenance, quality control, and product design (Friederich et al., 2022).

3.4.3.6   Big Data

Big data refers to the large quantities of structured and unstructured data that are generated by manufacturing operations. By analysing this data, manufacturers can identify trends, patterns, and insights that can help them to advance their operations, reduce costs, and increase customer satisfaction (Gunasekaran et al., 2018).

3.4.4   BENEFITS OF ICT FOR INTELLIGENT AND SUSTAINABLE MANUFACTURING

ICTs have transformed the manufacturing industry in numerous ways. By providing manufacturers with advanced tools and systems, ICT enables them to improve efficiency, reduce costs, and enhance sustainability. The benefits of ICT for intelligent and sustainable manufacturing can be seen as follows.

3.4.4.1   Increased Productivity

One of the most significant benefits of ICT for manufacturing is increased productivity. Advanced sensors and data analytics systems enable manufacturers to collect real-time data on their production processes, which they can use to optimise their operations. For example, in a study by Lee and Lee (2015), they found that the use of ICT tools, including big data analytics and the IoT, can significantly improve manufacturing efficiency, leading to higher productivity. Wuest et al. (2016) also concluded that the integration of ICT in manufacturing systems has resulted in the development of smart factories. These factories use advanced technologies such as big data analytics, the IoT, and cloud computing to optimise manufacturing processes, minimise downtime, and reduce energy consumption.

3.4.4.2   Reduced Costs

The integration of ICT in manufacturing processes can also reduce costs. By providing manufacturers with real-time data on production processes, ICT enables them to identify inefficiencies and areas where costs can be reduced. The use of ICT tools such as cloud computing and data analytics can significantly reduce manufacturing costs.

3.4.4.3   Improving Worker Training

According to a study by Pan and Zhang (2021), AR/VR technologies can be used to improve worker training in manufacturing systems and can provide realistic and interactive training simulations, which can improve worker skills and reduce the risk of accidents. A/VR simulations can afford a secure and coordinated working environment for workers to learn new skills and techniques, reducing the risk of accidents and errors.

3.4.4.4   Enhancing Collaboration

AR/VR technologies are a means to enhance collaboration in manufacturing systems. These technologies can facilitate communication and collaboration between workers, leading to improved productivity and reduced downtime. Collaboration between different stakeholders in the manufacturing process, such as suppliers, designers, and customers, can be enhanced by enabling real-time communication and information sharing. The use of ICT in manufacturing can improve collaboration by up to 60%, as established by Xu et al. (2020).

3.4.4.5   Improving Environmental Sustainability

AR/VR technologies can be used to improve environmental sustainability in manufacturing systems and can help optimise manufacturing processes, leading to reduced waste and energy consumption (Lo et al., 2021). AR/VR simulations can help workers to identify and eliminate waste, reducing the environmental impact of the manufacturing process (Yao et al., 2021). The integration of AR/VR technologies in manufacturing systems can help improve worker training, enhance collaboration, improve environmental sustainability, and enhance maintenance and repair activities. These technologies have the potential to improve the efficiency and sustainability of manufacturing systems, leading to a more intelligent and sustainable manufacturing industry.

3.4.4.6   Enhancing Product Design and Prototyping

AR/VR simulations can help designers and engineers to visualise and test products in a virtual environment, reducing the need for physical prototypes and improving the sustainability of the manufacturing process. This can enhance product design and prototyping processes (Burova et al., 2020).

3.4.4.7   Enhancing Maintenance and Repair

AR/VR technologies can be used as a means to enhance maintenance and repair activities in manufacturing systems. They can provide workers with real-time information and guidance, leading to faster and more efficient maintenance and reducing downtime of equipment (Carvalho et al., 2019).

3.4.4.8 Improved Product Quality

The use of ICT in manufacturing processes can also lead to improved product quality. By providing manufacturers with real-time data on production processes, ICT enables them with real-time monitoring, tracking and control of production processes, leading to higher product quality. The integration of ICT in manufacturing systems has resulted in improved product quality (McFarland et al., 2017). They found that the use of intelligent sensors and data analytics can help detect defects early in the production process, resulting in higher-quality products and lower production costs. Implementation of automated quality control systems that can identify defects and ensure that products meet quality standards is enabled using ICTs. This reduces the cost of quality control and improves the overall quality of products (Sioma, 2018).

3.4.4.9 Process Optimisation and Predictive Maintenance

ICT allows manufacturers to optimise their production processes by providing real-time data about the manufacturing process. This enables manufacturers to make real-time decisions and improve the efficiency of their production process. ICT enables the use of predictive maintenance systems that can classify probable equipment breakdowns before they occur. This reduces downtime and maintenance costs and ensures that production runs smoothly.

3.4.4.10 Resource Efficiency

Manufacturers can optimise the use of resources such as energy, water, and materials through the use of ICTs. This reduces waste and helps manufacturers to achieve sustainability goals (Ikumapayi et al., 2023). ICT has the potential to increase efficiency in manufacturing by optimising production processes, reducing waste, and minimising downtime.

3.4.4.11 Customisation and Supply Chain Management

ICT can increase the flexibility of manufacturing by enabling real-time adjustments to production processes, reducing the need for human intervention. Customised products by using real-time data to modify production processes can also be achieved when ICTs are used in manufacturing processes. This allows manufacturers to meet the specific needs of their customers and improve customer satisfaction (Parrott et al., 2023). ICT enables manufacturers to manage their supply chain more efficiently by providing real-time data about the movement of goods and materials. This allows manufacturers to optimise their inventory levels, reduce waste, and improve the efficiency of their supply chain (NEAGU et al., 2019).

3.4.4.12 Reduced Environmental Impact

The environmental impact of manufacturing is reduced by optimising energy consumption, reducing waste, and minimising the use of hazardous materials. The use of ICT in manufacturing systems leads to a significant reduction in waste, as posited by Verdouw et al. (2019). They found that the use of real-time monitoring and control systems can help minimise waste by optimising production processes, reducing energy consumption, and improving material utilisation. The integration of ICT in manufacturing systems has led to improved environmental sustainability, as reported by Cimini et al. (2020). They noted that the use of IoT, big data analytics, and cloud computing can help reduce greenhouse gas emissions, improve energy efficiency, and promote sustainable manufacturing practices.

3.4.4.13 Increased Innovation

The use of ICT in manufacturing can enable new forms of innovation, such as the development of smart products and services, and the creation of new business models. It can be seen that the integration of ICT in manufacturing systems can result in more efficient production processes, improved product quality, reduced waste, and improved environmental sustainability. These benefits are supported by various studies in the literature, and ICT will continue to play a crucial role in the manufacturing industry's future.

3.4.5 CHALLENGES OF ICT FOR INTELLIGENT AND SUSTAINABLE MANUFACTURING

Despite the many benefits of using ICT in manufacturing, several challenges must be addressed. The following are some of the challenges that need addressing to enjoy the benefits of ICT usage for intelligent and sustainable manufacturing.

3.4.5.1 Cost

The high cost of implementing advanced ICT systems includes hardware, software, and training. The high cost of ICT systems is a significant challenge for the implementation of intelligent and sustainable manufacturing. It is found that the cost of hardware, software, and personnel can be a significant barrier to the adoption of ICT in manufacturing systems. Because of this, manufacturers need to carefully evaluate the costs and potential ROI before investing in these technologies. This involves assessing the potential benefits of AR/VR, such as improved efficiency and productivity, against the costs of implementation and maintenance.

3.4.5.2 Cybersecurity Risks

The integration of ICT in manufacturing systems has led to increased cybersecurity risks. It is noted that the use of IoT devices can increase the attack surface and make manufacturing systems vulnerable to cyber threats. There are cybersecurity concerns, as the integration of ICT in manufacturing processes increases the risk of cyber-attacks. Manufacturers need to implement robust security measures to protect their data and ensure the privacy of their employees and customers (Ahram et al., 2017). While AR/VR hold great promise for intelligent and sustainable manufacturing, they also present substantial challenges that need to be addressed to fully realise their potential. Manufacturers should cautiously reflect on these challenges and develop strategies to overcome them to successfully implement AR/VR in their manufacturing processes.

3.4.5.3 Data Quality and Privacy

Data quality and privacy are major challenges in the integration of ICT in manufacturing systems. AR/VR generates a massive amount of data, which can be overwhelming to manage and analyse. The use of big data analytics can improve production processes, but the quality of data must be ensured, and privacy concerns must be addressed. Data security is a significant challenge for the implementation of ICT in manufacturing systems and the integration of ICT increases the risk of cyber-attacks and data breaches, which could compromise the safety and security of production systems. To address this challenge, manufacturers need to adopt advanced analytics tools that can process and analyse large datasets in real time (Rodriguez-Lluesma et al., 2021).

3.4.5.4   Organisational and Cultural Challenges

The integration of ICTs in manufacturing systems can also present organisational and cultural challenges and their implementation requires changes in organisational structures, job roles, and work processes, which can be difficult to manage.

3.4.5.5   Skill Gap and Workforce Development

The skills gap and workforce development are significant challenges in the integration of ICT in manufacturing systems, and there is a need for specialised skills and knowledge to operate and maintain ICT-enabled manufacturing systems. A lack of skilled labour is also a significant challenge for the implementation of ICT in manufacturing systems, and the shortage of workers with the necessary ICT skills could impede the adoption of intelligent and sustainable manufacturing practices (Jiang & Dong, 2020). Implementation of AR/VR requires specialised skills and knowledge that are not commonly found in manufacturing organisations. Therefore, manufacturers need to invest in training and development programs to ensure that their employees have the necessary skills to use these technologies effectively (Viglialoro et al., 2021).

3.4.5.6   Lack of Standards

One of the challenges of implementing ICT in manufacturing systems is the lack of standards (Davies et al., 2017). The absence of common standards makes it difficult for different systems to communicate with each other, resulting in integration issues and increased costs. The lack of standardisation in ICT systems makes it difficult to integrate different systems and technologies (Sharma et al., 2020). While the integration of ICT in manufacturing systems brings many benefits, it also comes with challenges, including the lack of standards, data security concerns, cost, and a shortage of skilled labour. These challenges need to be addressed to ensure the successful implementation of intelligent and sustainable manufacturing practices.

3.5   CONTEXTUAL EMPIRICAL FINDINGS

This section presents the empirical data collected from key informants through in-depth interviews. The key informants were made up of three (3) industrial experts, three (3) ICT experts, and two (2) academic experts who possessed a wealth of experience on the subject matter under examination. To begin with, the key informants were asked to identify the key challenges faced by the manufacturing sector in a developing context like Zimbabwe and suggest how emerging ICTs like AI, AR,

VR, MR, big data, cloud computing solutions, and the IoT can help address these challenges.

According to the key informants, the Zimbabwe manufacturing industry resembles a caricature, as some likened it to a patient in the intensive care unit battling for survival. Five key themes were generated under this section, and they are shown in Table 3.1. As a result, participants proposed that disruptive emerging ICTs including cloud computing, big data, and the IoT, can be seen as solutions to the problem of Zimbabwean manufacturers' lack of access to financing. They think that small and medium-sized firms, which currently make up most businesses in Zimbabwe, may access finance through cloud-based financial services. They also made note of the fact that credit scoring models can be improved through the analysis of financial risk using big data. IoT-enabled equipment can assist in boosting productivity and cutting costs, facilitating easier capital investment on the part of firms. Experts proposed that artificial intelligence can be utilised to automate numerous production processes, thereby eliminating the requirement for skilled staff, in response to the challenge of a scarcity of skilled labour. AI can also be used to enhance training, bridging the gap between the skills of current employees and what the industry needs. It was claimed that VR may be used to simulate manufacturing processes to solve the problem of inadequate infrastructure.

This would enable manufacturers to experiment with and create new goods without needing to build physical prototypes. Because companies can access computing power without having to buy and maintain IT hardware, cloud computing solutions can also help to lower the cost and complexity of infrastructure. a lack of producing large volumes of data may be gathered and analysed using big data and IoT, which can be used to spot production trends and patterns. As a result, producers will be better able to spot inefficiencies and prescribe changes accordingly. To address the innovation gap identified, the experts asserted that AI can be utilised to find new opportunities and create novel goods. Manufacturers may analyse consumer behaviour and create goods and services that satisfy their demands by utilising data from IoT devices. VR may also be used to create immersive experiences that aid in the testing and development of new concepts by manufacturers. Therefore, harnessing these disruptive emerging technologies may assist the manufacturing sector in Zimbabwe by granting access to new markets, facilitating quicker decision-making, and lowering the cost of production. Hence, this can contribute to boosting production efficiency and enhancing product quality.

> The country is reeling under difficult circumstances . . . there is a need for proper understanding of the existing systems and a clear understanding of these new ICTs' pros and cons.

Inter-Expert-2

> To integrate disruptive technologies we need an all-encompassing plan for the implementation and operation required for the effective integration of emerging ICTs with existing manufacturing processes and systems in Zimbabwe.

Inter-Expert-8

TABLE 3.1
Showing Themes' Major Causes and Interview Excerpts Supporting the Themes

	Theme	Major Cause/S	Supporting Interview Excerpts	
1	There is a lack of access to capital in Zimbabwe.	A hostile economic environment which does not attract foreign direct investment. Corruption demands kickbacks from potential investors.	In my view the absence of access to necessary affordable capital.	Inter-Expert-3
2	There is a lack of skilled labour in Zimbabwe.	Poor remuneration continuously bleeds the country of its finest human resources to other powerful economies.	Attracting and retaining skilled workers is a huge challenge for the manufacturing sector.	Inter-Expert-4
3	There is poor infrastructure in Zimbabwe.	Lack of investments in capital infrastructure projects.	There is an absence of modern technology.	Inter-Expert-2
4	There is underproductivity in Zimbabwe.	The culmination of all challenges faced by the industry renders it uncompetitive and therefore produces poor-quality but expensive products.	The country is reeling under difficult circumstances . . . there is a need for a proper understanding of the existing systems and a clear understanding of these new ICTs' pros and cons.	Inter-Expert-1 Inter-Expert-5
5	There is a lack of innovation in Zimbabwe.	Technological experts lack motivation or rewards for innovators.	There is a need for investing in the development of the appropriate skills and capabilities . . . this is essential to ensuring the successful installation and operation of developing ICTs in the industrial sector in Zimbabwe.	Inter-Expert-7 Inter-Expert-8 Inter-Expert-6

I think it is critical to understand that the successful adoption and operation of new ICTs in Zimbabwe's manufacturing sector requires a continuous process of learning and development rather than a one-time activity. To guarantee that staff members can keep up with the newest advancements and stay ahead of the curve, it is crucial to give them the necessary assistance and tools. Employees can stay current by keeping in touch with industry experts and having access to the most recent research, among other things.

Inter-Expert-5

There is a need for investing in the development of the appropriate skills and capabilities . . . this is essential to ensuring the successful installation and operation of developing ICTs in the industrial sector in Zimbabwe.

Inter-Expert-7

3.5.1 THE VITAL DISRUPTIVE EMERGING ICTS THAT HAVE THE POTENTIAL TO TRANSFORM THE MANUFACTURING SECTOR IN ZIMBABWE

Participants proposed that by offering safe and transparent tracking of commodities whilst lowering the cost of manufacturing processes and boosting supply chain efficiency, blockchain is one technology that has the potential to revolutionise the manufacturing industry in Zimbabwe and sub-Saharan Africa. AI technology like 3D printing has the potential to completely transform the manufacturing industry in sub-Saharan Africa by making it possible to produce customised goods and parts quickly and affordably. The Internet of Things can enable effective monitoring and control of production processes as well as the gathering and analysis of data to increase production efficiency. IoT is another disruptive technology that has the potential to revolutionise the manufacturing sector in sub-Saharan Africa. By automating intricate activities and enabling effective data analysis for better decision-making, AI technology has the potential to revolutionise the manufacturing industry in sub-Saharan Africa. Augmented reality technological advances have the power to completely transform the industrial industry in sub-Saharan Africa by giving employees immediate access to information and improving the way that products and processes are visualised.

3.5.2 REQUIREMENTS FOR SUCCESSFULLY INTEGRATING EMERGING ICTS WITH EXISTING MANUFACTURING PROCESSES AND SYSTEMS IN THE DEVELOPING CONTEXT

Participants suggested that manufacturing businesses should engage in digital transformation projects to update outdated industrial systems and processes. This includes making investments in cutting-edge technology like cloud computing, 3D printing, automation, robots, and artificial intelligence. As a result, businesses will be able to work more effectively, spend less money, and produce better work overall. Businesses should use analytics and data to make better judgements. This involves making better use of data to locate bottlenecks, streamline procedures, and enhance the customer experience. To obtain a competitive edge, businesses should investigate and experiment with emerging ICTs like blockchain and virtual reality. Companies will be able to respond to customer needs more quickly and fluidly as a result. To enable the successful integration of the newest ICTs, businesses should invest in enhancing their internet access. This includes making investments in mobile networks and high-speed broadband. Businesses should spend money on imparting knowledge to their staff so they can use the most recent ICTs efficiently. Training in new systems and software is part of this revolution.

3.5.3 THE ESSENTIAL SKILLS AND KNOWLEDGE REQUIRED TO IMPLEMENT AND OPERATE EMERGING ICTS IN THE MANUFACTURING SECTOR IN A DEVELOPING CONTEXT LIKE ZIMBABWE

According to the research, participants' expertise in technology is a boon; hence they suggested that people must possess a solid foundation in ICTs, including understanding of the hardware, software, networking, and programming, as well as the ability to incorporate these technologies into industrial processes. To deploy ICTs successfully in the manufacturing sector, people need to be aware of the business models and procedures of the industry and how ICTs may be used to optimise them. People should be able to plan and think strategically to deploy ICTs as efficiently as possible. In project management, people must be capable of managing teams, budgets, timeframes, and risks in addition to having experience in project management. People need to have good interpersonal skills, such as communication and bargaining, to deploy and use ICTs in the industrial industry. These information and skill-building opportunities can be found in formal education, career advancement, and on-the-job training. Additionally, people should constantly work to stay informed of the newest developments in ICTs and the manufacturing industry, as well as the best integration techniques.

3.5.4 THE KEY BARRIERS TO THE ADOPTION AND IMPLEMENTATION OF EMERGING ICTS IN THE MANUFACTURING SECTOR IN ZIMBABWE

Participants pointed out that Zimbabwe lacks access to stable telecommunications, internet, and energy networks. This restricts how easily firms may implement and utilise ICTs. That is why the government should make investments in infrastructure that are both dependable and reasonably priced to address this.

Cost: Buying and implementing ICTs can be expensive for enterprises. The government can solve this by providing tax breaks and financial aid to companies that make ICT investments.

The shortage of skilled labour: Businesses in Zimbabwe struggle to deploy and make use of ICTs due to a shortage of skilled labour. The government ought to support training and education initiatives to boost the number of skilled workers to address this.

Lack of knowledge: Businesses in Zimbabwe do not fully grasp or are not aware of the power of ICTs. The government should fund awareness-raising initiatives to inform businesses about the possibilities of ICTs to remedy this. ICT use in Zimbabwe is hampered by security issues, which prevent firms from putting their full trust in ICTs. The government should make investments in security infrastructure and offer businesses cyber security education and training to address this.

> this entails giving staff members access to the most recent tools and resources, as well as the chance to attend workshops and conferences. Giving staff members access to industry experts and the most recent research can also help to ensure that they can stay on top of things.

Inter-Expert-6

3.5.5   THE KEY SOCIAL AND ENVIRONMENTAL CONSIDERATIONS THAT NEED TO BE CONSIDERED WHEN IMPLEMENTING EMERGING ICTS IN THE MANUFACTURING SECTOR IN A DEVELOPING CONTEXT

It is challenging for enterprises to have access to the necessary technology because many regions of the country have poor or no connection to the internet and other ICTs. ICT implementation can be prohibitively expensive in Zimbabwe's manufacturing sector. Businesses must take into account the cost of acquiring and maintaining the required machinery as well as the expense of instructing workers on how to use the technology. ICTs have the potential to significantly alter a nation's social structure. ICT adoption in the industrial industry may result in job losses as machines and computers begin to perform many tasks that once required human labour. Businesses must be conscious of the possible social effects of deploying ICTs and take appropriate action. ICTs' potential to significantly harm the environment depends on how they are used and managed. That is why businesses must be conscious of the potential environmental effects of using ICTs and make sure they are taking precautions to limit any unfavourable effects. When implementing ICTs in the manufacturing sector, security is a crucial factor. Therefore, to safeguard their data and to stop any unauthorised access or manipulation, businesses must make sure that their systems are secure.

3.5.6   THE BEST WAY TO EFFECTIVELY ENGAGE STAKEHOLDERS IN THE PROCESS OF IMPLEMENTING EMERGING ICTS IN THE MANUFACTURING SECTOR IN A DEVELOPING CONTEXT

There is a need to create an all-encompassing forum for industry stakeholders to discuss the opportunities and challenges of implementing cutting-edge ICTs in the manufacturing sector. Representatives from business, government, and academia should participate in this discussion. Create a roadmap with specific goals and deadlines for implementing new ICTs in the manufacturing sector. To guarantee their participation and support, this roadmap should be designed in conjunction with stakeholders. Hence, there is a need to establish the obligations of each participant in the implementation process. The ability of each stakeholder to contribute successfully will be ensured by doing this. Give stakeholders opportunities for training and capacity building to make sure they comprehend new technologies and can use them well.

There is a need to regularly check on and assess the progress to make sure the implementation is going according to schedule. This will make it possible for interested parties to spot any issues early and fix them.

> A clear grasp of the goals and objectives of the stakeholders, as well as an efficient communication strategy, are necessary for the effective engagement of stakeholders in the process of implementing new ICTs in Zimbabwe's manufacturing sector.
>
> **Inter-Expert-1**

> making sure that stakeholders are involved in decision-making and that their concerns are taken into account is crucial.
>
> **Inter-Expert-4**

3.5.7 MEASURING THE SUCCESS OF THE IMPLEMENTATION OF EMERGING ICTs IN THE MANUFACTURING SECTOR IN A DEVELOPING CONTEXT LIKE ZIMBABWE

Participants suggested that, in a developing country like Zimbabwe, the adoption of new ICTs in the manufacturing sector can be monitored and evaluated by keeping track of productivity changes. This can be accomplished by contrasting the output of the same kind of product before and after ICT adoption. Monitoring the cost savings related to the implementation of emerging ICTs is another way to assess the implementation's performance. This can include the cost reductions brought on by lower labour costs, more effective operations, and better inventory control. Monitoring customer satisfaction is another way to gauge how well new ICTs are used in the manufacturing sector in a developing country like Zimbabwe. To monitor customer happiness, surveys or consumer feedback may be used. Hence, it is imperative that in a developing country like Zimbabwe, monitoring employee satisfaction can help to determine how well emergent ICTs are implemented in the industrial sector. Monitoring changes in market share is another way to assess how well emergent ICTs are being implemented in Zimbabwe's industrial sector. This may involve monitoring shifts in market share for the good or service before and after ICT adoption.

> The potential effects on employment, the potential for increased resource consumption, and the potential for increased environmental pollution are the key social and environmental considerations that need to be taken into account when implementing emerging ICTs in the manufacturing sector in the developing context. . . . It is crucial to make sure that the technology is implemented in a way that does not negatively affect these regions and that all stakeholders benefit from it.
>
> **Inter-Expert-6**

3.6 DISCUSSION OF FINDINGS

To address this research, an extensive review was conducted on relevant literature related to disruptive technologies for smart factories and the Zimbabwean context. This included scholarly articles published within different journals such as the *Technology Innovation Management Review, Journal of Cleaner Production* (Elsevier and Taylor and Francis), but most of these are indexed in the Google Scholar, EBSCO, and Scopus databases. To situate the research in a developing context, the researchers gathered empirical data by conducting in-depth interviews with three academics from three universities specialising in technological fields in Zimbabwe. One is from the National University of Science and Technology, one from the Harare Institute of Technology, and one from Chinhoyi University of Technology. There were also three information technology professionals working at different companies implementing IoT solutions in Zimbabwe and two industry practitioners who specialise in new technological implementations in manufacturing firms in Zimbabwe.

The literature review conducted revealed that there are several key challenges faced by the manufacturing sector in the developing context like Zimbabwe, and emerging ICTs can help to address these challenges. These challenges include problems such as underproductivity, high cost of production, a lack of access to technology, a lack of infrastructure and resources, a lack of vital knowledge and key relevant skills, a lack of effective stakeholder participation, and social and environmental concerns. In addition, some ICT solutions have been found that could revolutionise the manufacturing industry in a developing country. These include blockchain solutions, Internet of Things (IoT), augmented and virtual reality, cloud computing solutions, robotic process automation (RPA), and other artificial intelligence technologies like machine learning and deep learning (DL). Interviews with three academic experts provided additional insights on how best to implement these disruptive ICTs into existing manufacturing processes in a developing context. The consensus was that effective implementation requires an understanding of both technical aspects such as hardware and software requirements but also non-technical factors such as organisational culture change management strategies; stakeholder engagement plans, resource planning and allocation, and user experience design, all tailored to suit specific local contexts.

The findings suggest that there exist various opportunities for utilising new technologies such as the Internet of Things, artificial intelligence, augmented reality, robotics, cloud computing, and digital twins which could be used to provide the necessary support for transforming manufacturing industries into smarter factories in developing countries like Zimbabwe as they have done in the developed world. However, the research equally identified that there exist certain challenges that militate against the adoption and implementation of these disruptive ICTs which have to be overcome. Such challenges include a lack of funding or financial resources that are needed for the implementation and operationalisation of these emerging ICTs. The research findings also pointed out that there is a need to overhaul the infrastructure currently in place, as it is difficult to adapt the current practices or infrastructure to the impending new normal. There is also bound to be a failure to properly plan out projects due to insufficient understanding or expertise around the technological applications being used. This is in line with the findings of the research by some scholars (El-Motasem et al., 2021; Factory, 2020; Gunal, 2019; Zheng et al., 2021).

The findings also suggest that there is a need for all stakeholders to be properly engaged throughout the entire process, especially those that are involved directly or indirectly through manufacturing and the ICT value chain so that they understand their roles and responsibilities before any manufacturing ICT automation project is conceptualised. Additionally, the findings suggest that social and environmental contextual factors should also be taken into consideration while planning any implementations of these disruptive emerging technologies in the manufacturing industries in developing countries since they could greatly influence the success rate, depending upon local laws and regulations present within the countries where operations take place. The three ICT experts interviewed further elaborated on how different types of emerging technologies could be effectively integrated into existing systems through careful consideration of system architecture designs and standards, data security protocols, and scalability and compatibility concerns, while the two industrial experts highlighted practical considerations around skill development needs for successful implementations, including training programs, both online and offline platforms, the need to provide career guidance services, and mentorship programs.

Finally, the findings highlighted that when the disruptive emerging ICTs have been implemented, there is a need for measuring success/failure, whichever the case would be. Hence there is a need to apply both qualitative and quantitative metrics based upon which the expected outcomes are set during initial stages like efficiency gains achieved after introducing newer elements into the production line or customer satisfaction scores attained after introducing new product offerings. Based on this, the researchers can conclude that although there might exist certain primary obstacles when implementing emerging ICTs in the manufacturing industries in a developing context, if this is planned purposefully within the confines of a guiding framework, then the outcomes of such an endeavour could prove very beneficial for the firms that implement the technologies, and subsequently customers can get better products faster while businesses increase efficiency, thereby reducing the costs associated with each operation cycle.

3.7 THE PROPOSED FRAMEWORK FOR IMPLEMENTATION OF EMERGING ICTs IN INTELLIGENT AND SUSTAINABLE MANUFACTURING IN A DEVELOPING CONTEXT

The essence of carrying out this research and writing this chapter was to establish the current status quo in the manufacturing industry in Zimbabwe, a developing country in SSA. Based on the literature review and the empirical contextual data, we then propose a guiding framework for the implementation of disruptive emerging ICTs in the intelligent and sustainable manufacturing industry in a developing context. Therefore, we propose a three-layer framework, which is described in detail as follows.

3.7.1 LAYER 1: DATA COLLECTION AND ANALYSIS = MIKMS

This is the first layer, labelled layer 1, and the framework requires that at this layer there be data collection and analysis of the contextual operational environment in each context to establish what is termed the manufacturing industry knowledge management system (MIKMS) of that operating environment. This entails establishing the current situation in the manufacturing industry to identify the relevant challenges therein. This process also encompasses the establishment of the current situation on the technological landscape that is the emerging ICT domain. In addition to that, the process involves conducting a human resource skills audit in the industries and the job market to establish the industry skills deficit. The outcomes from layer 1 are then used to inform the operations to be done on layer 2.

3.7.2   LAYER 2: ESTABLISH OPERATIONAL PROCEDURE STANDARDS AND REGULATION REQUIREMENTS

As alluded to, this layer is informed by layer 1 outcomes, and therefore it follows that in tandem with the findings in our study, there is a need to establish the operational procedures, standards, and regulatory requirements for implementing emerging ICTs. Our proposed framework requires the establishment of standards and regulations for the adoption, implementation, and usage of ICTs in intelligent and sustainable manufacturing. This includes setting guidelines for data protection, privacy, and security, as well as establishing protocols for the sharing of data between different organisations. There is also a need to develop sound digital infrastructure to provide the foundation on which to launch Industry 4.0 in Zimbabwe. Developing countries like Zimbabwe need to build a strong digital infrastructure to facilitate the widespread adoption of new ICTs. This includes the deployment of high-speed broadband networks and mobile internet access, as well as the development of cloud-based storage solutions.

This includes the need to obtain government support to have a solid regulatory framework. Governments of developing countries like Zimbabwe must provide the necessary support to promote the use of ICTs in intelligent and sustainable manufacturing. This includes providing financial incentives, offering tax breaks, and creating favourable regulations and policies. In addition to that, there is also a need to provide relevant education and training programs for the human resources of the country to support the implementation efforts to ensure successful implementation and use of ICTs in intelligent and sustainable manufacturing. It is essential to have an adequate number of skilled professionals, such as engineers, technicians, and data scientists. After this, we proceed to layer 3.

3.7.3   LAYER 3: INTEGRATE, IMPLEMENT, AND OPERATIONALISE INDUSTRY 4.0 TECHNOLOGIES IN SELECT INDUSTRIES FOR PIONEERING THE OPERATION

This is the final layer in our proposed framework, and at this layer, there is the integration, implementation, and operationalisation of Industry 4.0 technologies in select industries that would have been identified for pioneering the process. This would then require the monitoring of performance and evaluating the effect by performing data analytics on implemented systems in the pioneering manufacturing industries. This would result in the generation of data for decision-making and improving the layer 2 stages. The soundness of the framework requires that there be a continuous exchange of data between layer 2 and layer 3, as there is bound to be continuous learning and improvement to get to the point of perfecting this proposed framework.

We believe that manufacturing firms need to be willing to adopt disruptive ICTs and use them in their manufacturing processes; therefore this requires an intimate understanding of the benefits that these technologies can bring to their business operations, as well as the businesses' ability to identify and address potential risks associated with these technologies.

3.8   CONCLUSIONS AND RECOMMENDATIONS

This chapter presented the concept of disruptive emerging ICT technologies in the manufacturing industry, otherwise known as Industry 4.0. The authors conducted a systematic literature review analysis of empirical research articles on disruptive emerging technologies in the manufacturing industry, like Industry 4.0, robotics, artificial intelligence, and augmented and virtual reality. An analysis of the benefits and the associated challenges of these technologies was done but primacy was given to those articles conducted in developing contexts. This made it possible for the chapter to identify the crucial components needed for Industry 4.0 success. The chapter also outlined and addressed challenges and difficulties associated with the adoption, implementation, and operationalisation of emerging disruptive technologies. The exploration of the Zimbabwean context made it possible to contextualise Industry 4.0 technologies in the manufacturing industry following the book's thematic areas. The chapter's maturation of a framework for the implementation of emerging ICTs in intelligent and sustainable manufacturing in a developing context was its focus. Figure 3.1 shows the proposed framework for the implementation of disruptive emerging ICTs in sustainable manufacturing in a developing context developed by the authors.

This work adds to the body of knowledge and practice in different ways. The chapter begins by offering a thorough summary of recent studies on the evolution and implementation of disruptive emerging ICTs in Industry 4.0 but with a bias towards the

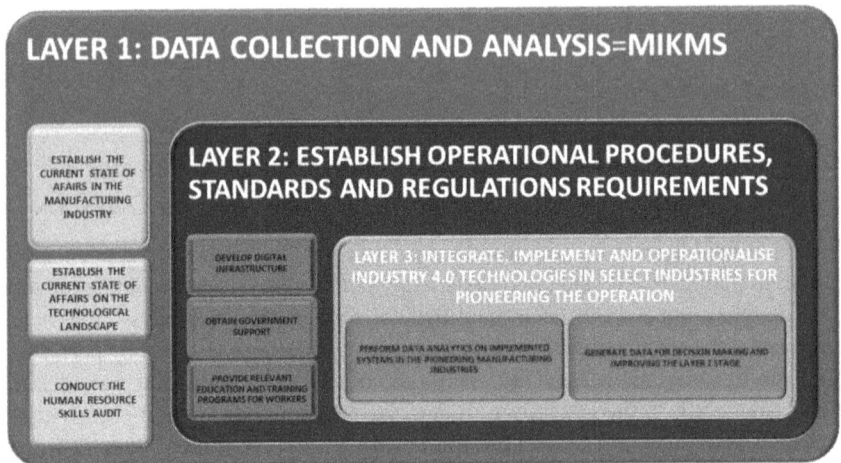

LAYER 1: DATA COLLECTION AND ANALYSIS=MIKMS

ESTABLISH THE CURRENT STATE OF AFAIRS IN THE MANUFACTURING INDUSTRY

LAYER 2: ESTABLISH OPERATIONAL PROCEDURES, STANDARDS AND REGULATIONS REQUIREMENTS

ESTABLISH THE CURRENT STATE OF AFFAIRS ON THE TECHNOLOGICAL LANDSCAPE

DEVELOP DIGITAL INFRASTRUCTURE

LAYER 3: INTEGRATE, IMPLEMENT AND OPERATIONALISE INDUSTRY 4.0 TECHNOLOGIES IN SELECT INDUSTRIES FOR PIONEERING THE OPERATION

OBTAIN GOVERNMENT SUPPORT

CONDUCT THE HUMAN RESOURCE SKILLS AUDIT

PROVIDE RELEVANT EDUCATION AND TRAINING PROGRAMS FOR WORKERS

PERFORM DATA ANALYTICS ON IMPLEMENTED SYSTEMS IN THE PIONEERING MANUFACTURING INDUSTRIES

GENERATE DATA FOR DECISION MAKING AND IMPROVING THE LAYER 2 STAGE

FIGURE 3.1   Proposed framework for the implementation of disruptive emerging ICTs in intelligent and sustainable manufacturing in a developing context (author's construction).

manufacturing industry, and primacy was given to literature on research carried out in developing countries in particular. In doing so, the chapter identified the key requirements for the successful implementation of Industry 4.0 digital technologies in the manufacturing industry in a developing context as well as their associated problems, controversies, and deleterious implications. The chapter proposes recommendations for the implementation of disruptive Industry 4.0 technologies in the manufacturing industry in a developing context, which adds to industrial practice. That is why we think that if developing country governments and policymakers want to improve their manufacturing industries, they would benefit from the proposed framework. Others in the academic community might review our proposed framework and empirically verify its applicability in different contexts; hence they may add any new elements they may discover in their empirical research or even disqualify it if that is the case.

To shed more light on the identified elements and the implications of implementing Industry 4.0 emerging ICT technologies in the manufacturing industry in a developing context, we suggest that subsequent studies can look at empirically evaluating our proposed framework in different contexts like other emerging technologies and underdeveloped countries. Therefore, to promote the endeavour of the industrial transformation of manufacturing industries in developing contexts using emerging Industry 4.0 ICTs, it is essential to have a strong and supportive business community, mainly the ICT, telecommunications, and banking sectors. This must be buttressed by a tranquil and supportive political environment that fosters stability and strategic thinking. The chapter contributes to the book, whose thematic emphasis is on the theme of intelligent and sustainable manufacturing: tools, principles, and strategies. It is, therefore, necessary to investigate the underlying causes of business organisations that are transforming, implementing Industry 4.0 ICT technologies, and integrating them into their manufacturing processes, as well as the inherent challenges they may encounter. This will help researchers better understand what is required to transform businesses into successful smart factories in developing contexts. This could provide more information about the endogenous and exogenous elements that influence the transformation of manufacturing industries in a developing context, as well as any positive or negative effects that may result from such a shift. Additionally, academia and practices must understand the precise impacts of Industry 4.0 transformation on management and human resource practises. For this reason, we believe that future studies on the implementation and transformation of industry in developing contexts and even different contexts could benefit from including such a perspective.

REFERENCES

Abdel-Basset, M., Chang, V., & Nabeeh, N. A. (2021). An intelligent framework using disruptive technologies for COVID-19 analysis. *Technological Forecasting and Social Change, 163*(July 2020), 120431. https://doi.org/10.1016/j.techfore.2020.120431

Ahram, T., Sargolzaei, A., Sargolzaei, S., Daniels, J., & Amaba, B. (2017). Innovaciones de la tecnología Blockchain. *2017 IEEE Technology and Engineering Management Society Conference, TEMSCON 2017, 2016*, 137–141.

Aidonis, D., Achillas, C., Folinas, D., Keramydas, C., & Tsolakis, N. (2019). Decision support model for evaluating alternative waste electrical and electronic equipment management schemes-A case study. *Sustainability (Switzerland), 11*(12), 1–13. https://doi.org/10.3390/su10023364

Akinyemi, B. E., & Mushunje, A. (2020). Determinants of mobile money technology adoption in rural areas of Africa. *Cogent Social Sciences*, *6*(1). https://doi.org/10.1080/2331188 6.2020.1815963

Alhassan, M., & Scholtz, B. (2019). Understanding the role of ICT in South African sustainable manufacturing practice. *Proceedings of 4th International Conference on the Internet, Cyber Security and Information Systems 2019*, Vol. 12, 32–42. https://doi.org/10.29007/7dtj

Bhat, S. A., Huang, N. F., Sofi, I. B., & Sultan, M. (2022). Agriculture-food supply chain management based on blockchain and IoT: A narrative on enterprise blockchain interoperability. *Agriculture (Switzerland)*, *12*(1). https://doi.org/10.3390/agriculture12010040

Brunetti, F., Matt, D. T., Bonfanti, A., De Longhi, A., Pedrini, G., & Orzes, G. (2020). Digital transformation challenges: Strategies emerging from a multi-stakeholder approach. *TQM Journal*, *32*(4), 697–724. https://doi.org/10.1108/TQM-12-2019-0309

Burova, A., Mäkelä, J., Hakulinen, J., Keskinen, T., Heinonen, H., Siltanen, S., & Turunen, M. (2020). Utilizing VR and Gaze tracking to develop AR solutions for industrial maintenance. *Proceedings of the 2020 CHI Conference on Human Factors in Computing Systems*, 1–13, Association for Computing Machinery New York, NY, United States.

Carvalho, T. P., Soares, F. A. A. D. M. N., Vita, R., Francisco, P., Basto, J. P., & Alcalá, S. G. S. (2019). Maintenance maintenance. *Computers & Industrial Engineering*, 106024. https://doi.org/10.1016/j.cie.2019.106024

Chu, M., Matthews, J., & Love, P. E. D. (2018). Integrating mobile Building Information Modelling and Augmented Reality systems: An experimental study. *Automation in Construction*, *85*(February 2017), 305–316. https://doi.org/10.1016/j.autcon.2017.10.032

Cimini, C., Lagorio, A., Romero, D., Cavalieri, S., & Stahre, J. (2020). Smart logistics and the logistics operator 4.0. *IFAC-PapersOnLine*, *53*(2), 10615–10620. https://doi.org/10.1016/j.ifacol.2020.12.2818

Damiyano, D., Muchabaiwa, L., & Mushanyuri, B. E. (2012). An investigation of Zimbabwe s manufacturing sector competitiveness. *International Journal of Development and Sustainability*, *1*(2), 581–598.

Davies, R., Coole, T., & Smith, A. (2017). Review of socio-technical considerations to ensure successful implementation of Industry 4.0. *Procedia Manufacturing*, *11*(June), 1288–1295. https://doi.org/10.1016/j.promfg.2017.07.256

El-Motasem, S., Khodeir, L. M., & Fathy Eid, A. (2021). Analysis of challenges facing smart building projects in Egypt. *Ain Shams Engineering Journal*, *12*(3), 3317–3329. https://doi.org/10.1016/j.asej.2020.09.028

Friederich, J., Francis, D. P., Lazarova-Molnar, S., & Mohamed, N. (2022). A framework for data-driven digital twins for smart manufacturing. *Computers in Industry*, *136*, 103586. https://doi.org/10.1016/j.compind.2021.103586

Furusa, S. S., & Coleman, A. (2018). A strategic framework for effective utilisation of eHealth tools by medical doctors in Zimbabwe's public hospitals. *Indian Journal of Public Health Research and Development*, *9*(8), 284–288. https://doi.org/10.5958/0976-5506.2018.00734.9

Gadzikwa, E. C. (2013, July). The future of the manufacturing sector in Zimbabwe. *Institute of Chartered Accountants of Zimbabwe Congress*. https://www.icaz.org.zw/iMISDocs/manufacture.pdf

Gattullo, M., Laviola, E., Boccaccio, A., Evangelista, A., Fiorentino, M., Manghisi, V. M., & Uva, A. E. (2022). Education laboratories. *Computers*, *11*(50), 1–24.

Gunal, M. M. (2019). Simulation for the better: The future in industry 4.0. In *Simulation for Industry 4.0: Past, Present, and Future*, 275–283. Springer.

Gunasekaran, A., Yusuf, Y. Y., Adeleye, E. O., & Papadopoulos, T. (2018). Agile manufacturing practices: The role of big data and business analytics with multiple case studies. *International Journal of Production Research*, *56*(1–2), 385–397. https://doi.org/10.108 0/00207543.2017.1395488

Hansen, D., & Lilja, J. (2020). Leading quality management transformation in complexity through adaptive space and metaphors. *Key Challenges and Opportunities for Quality, Sustainability and Innovation in the Fourth Industrial Revolution: Quality and Service Management in the Fourth Industrial Revolution—Sustainability and Value Co-Creation*, 203–230. https://doi.org/10.1142/9789811230356_0011

Ikumapayi, O. M., Afolalu, S. A., Ogedengbe, T. S., Kazeem, R. A., & Akinlabi, E. T. (2023). Human–robot co-working improvement via revolutionary automation and robotic technologies—An overview. *Procedia Computer Science*, *217*, 1345–1353. https://doi.org/10.1016/j.procs.2022.12.332

Jiang, L., & Dong, K. (2020). Artificial intelligence-based learning behavior data mining and network teaching quality monitoring mechanism. *Journal of Physics: Conference Series*, *1533*(3). https://doi.org/10.1088/1742-6596/1533/3/032058

Kumar, A., Singh, H., Kumar, P., & AlMangour, B. (Eds.). (2023). *Handbook of Smart Manufacturing: Forecasting the Future of Industry 4.0*. CRC Press. https://doi.org/10.1201/9781003333760

Kwenda, F. (2014). Trade credit in Zimbabwe's economic recovery. *Mediterranean Journal of Social Sciences*, *5*(2), 431–439. https://doi.org/10.5901/mjss.2014.v5n2p431

Lacoche, J., Le Chenechal, M., Villain, E., & Foulonneau, A. (2019). Model and tools for integrating IoT into mixed reality environments: Towards a virtual-real seamless continuum. *ICAT-EGVE 2019–29th International Conference on Artificial Reality and Telexistence and 24th Eurographics Symposium on Virtual Environments*, 97–104. https://doi.org/10.2312/egve.20191286

Lee, I., & Lee, K. (2015). The Internet of Things (IoT): Applications, investments, and challenges for enterprises. *Business Horizons*, *58*(4), 431–440. https://doi.org/10.1016/j.bushor.2015.03.008

Li, F., Lu, H., Hou, M., Cui, K., & Darbandi, M. (2021). Customer satisfaction with bank services: The role of cloud services, security, e-learning and service quality. *Technology in Society*, *64*(October 2020), 101487. https://doi.org/10.1016/j.techsoc.2020.101487

Lo, C. K., Chen, C. H., & Zhong, R. Y. (2021, March). A review of digital twin in product design and development. *Advanced Engineering Informatics*, *48*, 101297. https://doi.org/10.1016/j.aei.2021.101297

Lu, W., Li, X., Xue, F., Zhao, R., Wu, L., & Yeh, A. G. O. (2021). Exploring smart construction objects as blockchain oracles in construction supply chain management. *Automation in Construction*, *129*(November 2020), 103816. https://doi.org/10.1016/j.autcon.2021.103816

Lu, Y., Xu, X., & Wang, L. (2020). The smart manufacturing process and system automation—A critical review of the standards and envisioned scenarios. *Journal of Manufacturing Systems*, *56*(June), 312–325. https://doi.org/10.1016/j.jmsy.2020.06.010

Magura, Z., Zhou, T. G., & Musungwini, S. (2022). A guiding framework for enhancing database security in state-owned universities in Zimbabwe. *African Journal of Science, Technology, Innovation and Development*, *14*(7), 1761–1775. https://doi.org/10.1080/20421338.2021.1984010

Mahlangu, G., Musungwini, S., & Sibanda, M. (2018). A framework for creating an ICT knowledge hub in Zimbabwe : A holistic approach in fostering economic growth. *Journal of Systems Integration*, *2004*, 32–41. https://doi.org/10.20470/jsi.v9i1.327

McFarland, J., Hussar, B., de Brey, C., Snyder, T., Wang, X., Wilkinson-Flicker, S., Gebrekristos, S., Zhang, J., Rathbun, A., Barmer, A., Bullock Mann, F., & Hinz, S. (2017). The condition of education 2017. *National Center for Educational Statistics, NCES 2017-*, 1–133. https://nces.ed.gov/pubs2017/2017144.pdf

Mhaka, S., Runganga, R., Nyagweta, D. T., Kaseke, N., & Mishi, S. (n.d.). Impact of rural and urban electricity access on economic growth in Zimbabwe. *International Journal of Energy Economics and Policy*, *10*, 2020. https://doi.org/10.32479/ijeep.10141

Moeuf, A., Pellerin, R., Lamouri, S., Tamayo-Giraldo, S., & Barbaray, R. (2018). The industrial management of SMEs in the era of Industry 4.0. *International Journal of Production Research*, *56*(3), 1118–1136. https://doi.org/10.1080/00207543.2017.1372647

Musungwini, S., Furusa, S. S., Gavai, P. V., & Gumbo, R. (2022). Inclusive digital transformation for the marginalized communities in a developing context. *IGI Global*, 95–122. https://doi.org/10.4018/978-1-6684-3901-2.ch005

Musungwini, S., Gavai, P. V., Munyoro, B., & Chare, A. (2023). Emerging ICT technologies for agriculture, training, and capacity building for farmers in developing countries: A case study in Zimbabwe. In *Applying Drone Technologies and Robotics for Agricultural Sustainability*, 12–30. IGI Global. https://doi.org/10.4018/978-1-6684-6413-7.ch002

Musungwini, S., & Mondo, L. (2019). Developing a change management model for managing information systems initiated organisational change : A case of the banking sector in Zimbabwe. *Journal of System Integration.*, *10*(1), 49–61. https://doi.org/10.20470/jsi.v10i1.362

Neagu, G., Petre, I., Boncea, R., Barbu, D. C., & Dumitrache, M. (2019). Building a business model for service offer integrator in case of cloud-Iot based monitoring. *Proceedings of the 18th International Conference on INFORMATICS in ECONOMY Education, Research and Business Technologies*, 61–66. https://doi.org/10.12948/ie2019.02.05

Njerekai, C. (2020). An application of the virtual reality 360° concept to the Great Zimbabwe monument. *Journal of Heritage Tourism*, *15*(5), 567–579. https://doi.org/10.1080/1743873X.2019.1696808

Parrott, W. G., Bouchard, C., & Davies, C. T. H. (2023). B → k and D → k form factors from fully relativistic lattice QCD. *Physical Review D*, *107*(1), 14510. https://doi.org/10.1103/PhysRevD.107.014510

Patton, M. Q., & Cochran, M. (2002). *A Guide to Using: Qualitative Research Methodology*. Medecins Sans Frontieres.

Qin, T., Wang, L., Zhou, Y., Guo, L., Jiang, G., & Zhang, L. (2022). Digital technology-and-services-driven sustainable transformation of agriculture: Cases of China and the EU. *Agriculture (Switzerland)*, *12*(2), 1–16. https://doi.org/10.3390/agriculture12020297

Rodriguez-Lluesma, C., García-Ruiz, P., & Pinto-Garay, J. (2021). The digital transformation of work: A relational view. *Business Ethics, the Environment & Responsibility*, *30*(1), 157–167. https://doi.org/10.1111/beer.12323

Romão, J. (2020). Tourism, smart specialisation, growth, and resilience. *Annals of Tourism Research*, *84*. https://doi.org/10.1016/j.annals.2020.102995

Sanchez-Planelles, J., Segarra-Oña, M., & Peiro-Signes, A. (2022). Identifying different sustainable practices to help companies to contribute to the sustainable development: Holistic sustainability, sustainable business and operations models. *Corporate Social Responsibility and Environmental Management*, *29*(4), 904–917. https://doi.org/10.1002/CSR.2243

Setboonsarng, S., & Gregorio, E. E. (2017). *Achieving Sustainable Development Goals through Organic Agriculture: Empowering Poor Women to Build the Future*. (No. 15; ADB Southeast Asia, Issue 15).

Sharma, M., Fellow, S. R., Kamble, S., Venkatesh, M., Sehrawat, R., Belhadi, A., & Sharma, V. (2020). Industry 4.0 adoption for sustainability in multi-tier manufacturing supply chain in emerging economies. *Journal of Cleaner Production*, 125013. https://doi.org/10.1016/j.jclepro.2020.125013

Sioma, A. (2018). Applied sciences. *Early Writings on India*, 124–134. https://doi.org/10.4324/9781315232140-14

Stentoft, J., Adsbøll Wickstrøm, K., Philipsen, K., & Haug, A. (2021). Drivers and barriers for Industry 4.0 readiness and practice: Empirical evidence from small and medium-sized manufacturers. *Production Planning and Control*, *32*(10), 811–828. https://doi.org/10.1080/09537287.2020.1768318

Sullivan, B. P., Yazdi, P. G., Suresh, A., & Thiede, S. (2022). Digital value stream mapping: Application of UWB real time location systems. *Procedia CIRP*, *107*(March), 1186–1191. https://doi.org/10.1016/j.procir.2022.05.129

Thomas, D. R. (2003). A general inductive approach for qualitative data analysis. *Population English Edition*, *27*(2), 237–246. https://doi.org/10.1177/1098214005283748

Tsanousa, A., Bektsis, E., Kyriakopoulos, C., González, A. G., Leturiondo, U., Gialampoukidis, I., Karakostas, A., Vrochidis, S., & Kompatsiaris, I. (2022). A Review of Multisensor Data Fusion Solutions in Smart Manufacturing: Systems and Trends. *Sensors*, *22*(5), 1–27. https://doi.org/10.3390/s22051734

Verdouw, C., Sundmaeker, H., Tekinerdogan, B., Conzon, D., & Montanaro, T. (2019). Architecture framework of IoT-based food and farm systems: A multiple case study. *Computers and Electronics in Agriculture*, *165*(April), 104939. https://doi.org/10.1016/j.compag.2019.104939

Viglialoro, R. M., Condino, S., Turini, G., Carbone, M., Ferrari, V., & Gesi, M. (2021). Augmented reality, mixed reality, and hybrid approach in healthcare simulation: A systematic review. *Applied Sciences (Switzerland)*, *11*(5), 1–20. https://doi.org/10.3390/app11052338

Westerlund, M., Nene, S., Leminen, S., Rajahonka, M., & Kennedy, A. (2021). An exploration of blockchain-based traceability in food supply chains : On the benefits of distributed digital records from farm to fork. *Technology InnovationManagement Review*, *11*(6), 6–18. https://doi.org/http://doi.org/10.22215/timreview/1446.

Wuest, T., Weimer, D., Irgens, C., & Thoben, K. D. (2016). Machine learning in manufacturing: Advantages, challenges, and applications. *Production and Manufacturing Research*, *4*(1), 23–45. https://doi.org/10.1080/21693277.2016.1192517

Xu, L. Da, Xu, E. L., & Li, L. (2018). Industry 4.0: State of the art and future trends. *International Journal of Production Research*, *56*(8), 2941–2962. https://doi.org/10.1080/00207543.2018.1444806

Xu, S., Zhang, X., Feng, L., & Yang, W. (2020). Disruption risks in supply chain management: A literature review based on bibliometric analysis. *International Journal of Production Research*, *58*(11), 3508–3526. https://doi.org/10.1080/00207543.2020.1717011

Yao, L., Zhang, H., Zhang, M., Chen, X., Zhang, J., Huang, J., & Zhang, L. (2021). Application of artificial intelligence in renal disease. *Clinical EHealth*, *4*, 54–61. https://doi.org/10.1016/j.ceh.2021.11.003

Zain, M., Habib, M. M., Burbules, N. C., Luz Yolanda Toro Suarez, Peng, Y., Ilham, I., Post-modernisme, P., Untuk, S., Sosial, K., Us, A., Us, J., Us, C., Inglehart, R., Akhmad Hasan Saleh, Us, A., Us, J., Us, C., Siswiyanti, Y., Stefanov, S., . . . Power, M. (2020). No 主観的健康感を中心とした在宅高齢者における 健康関連指標に関する共分散構造分析 Title. *The Sociological Review*, *1*(1), 1–8. http://ezproxy.lib.uconn.edu/login?url=https://search.ebscohost.com/login.aspx?direct=true&db=eric&AN=EJ1143816&site=ehost-live%0Ahttp://eprints.utm.my/id/eprint/78124/%0Awww.researchgate.net/publication/328414890_TEROPONG_PENDIDIKAN_MARXISME%0Aht

Zheng, T., Ardolino, M., Bacchetti, A., & Perona, M. (2021). The applications of Industry 4.0 technologies in manufacturing context: A systematic literature review. *International Journal of Production Research*, *59*(6), 1922–1954. https://doi.org/10.1080/00207543.2020.1824085

Zhou, M., Herselman, M., & Coleman, A. (2015). USSD technology is a low-cost asset in complementing public health workers' work processes. *Lecture Notes in Computer Science (Including Subseries Lecture Notes in Artificial Intelligence and Lecture Notes in Bioinformatics)*, *9044*, 57–64. https://doi.org/10.1007/978-3-319-16480-9_6

4 Digital Twins and IoT for Sustainable Manufacturing
A Survey of Current Practices and Future Directions

Raja Lavanya[1], G. Bhavani[1], C. Santhiya[1], G. Vinoth Chakkaravarthy[2], and Parveen Kumar[3]

1 Department of Computer Science and Engineering, Thiagarajar College of Engineering, Madurai, India
2 Department of Computer Science and Engineering, Velammal College of Engineering and Technology, Madurai, India
3 Department of Mechanical Engineering, Rawal Institute of Engineering and Technology, Faridabad, Haryana, India

4.1 INTRODUCTION

Sustainable manufacturing entails the conscious use of manufacturing methods and products in a way that promotes long-term sustainability while minimising adverse effects on the environment, society, and economy [1]. It entails considering a product's complete life cycle, from design and raw material sourcing to production, usage, and end-of-life management, with the goal of lowering resource consumption, creating less waste and pollution, and enhancing social and economic well-being. In order to ensure that production practises are both now and, in the future, environmentally responsible, socially just, and economically feasible, sustainable production attempts to find a balance between these three factors. This entails implementing policies, procedures, and techniques that maximise resource efficiency, promote the consumption of green energy, reduce the release of greenhouse gases, increase product durability, give recycling and remanufacturing top priority, and advance social and economic equity throughout the supply chain. Achieving sustainable development goals and resolving global concerns like climate change, resource depletion, and social injustice require a critical approach that emphasises sustainable production.

DOI: 10.1201/9781003405870-4

4.1.1 IMPORTANCE OF SUSTAINABLE PRODUCTION

The following factors highlight the importance of sustainable production.

Resource efficiency: Sustainable manufacturing places a strong emphasis on maximising the use of resources, including raw materials, energy, water, and other inputs, to minimise waste and lower the manufacturing process' total environmental effect.

Holistic life cycle approach: Sustainable manufacturing is taking into account every stage of a product's life cycle, from design through end-of-life management, and making choices that minimise unfavourable effects and advance sustainability at every stage.

Pollution prevention: is one of the main objectives of sustainable production. This is done by implementing procedures, technologies, and practises that safeguard both the environment and public health.

Adoption of power from renewable sources: Sustainable manufacturing promotes the use of energy generated from renewable sources, such as solar, wind, and hydro-power, to lower greenhouse gas emissions and dependency on fossil fuels.

Durability and upgradeability of items are key components of sustainable manufacturing since these features help things last longer and require fewer replacements, which encourages resource conservation.

Recycling and refurbishment are key components of sustainable manufacturing because they encourage a circular economy and reduce waste by recovering and reusing materials from end-of-life products or garbage produced during the manufacturing process.

Social and economic justice: Through fair labour practises, safe working conditions, competitive pay, and ethical sourcing methods, sustainable production promotes a just distribution of social and economic benefits among stakeholders, including employees, communities, and supply chain partners.

Continuous innovation and improvement: By routinely assessing and enhancing production methods, technologies, and procedures, sustainable manufacturing strives to minimise negative effects on the environment, society, and the economy.

4.1.2 DIFFICULTIES IDENTIFIED IN SUSTAINABLE PRODUCTION

There are some difficulties that businesses using sustainable manufacturing methods may encounter, according to various research articles. The following are a few of these difficulties.

Financial sustainability for manufacturers, particularly with selected inadequate properties, the early costs of implementing sustainable manufacturing practises, such as investing in renewable energy systems and energy-efficient technologies, can be a financial barrier.

Education and awareness: The adoption and implementation of sustainable production practises may be hampered by worker ignorance, lack of expertise, and other factors.

Regulatory and policy framework: Manufacturers may experience difficulty adhering to and navigating a variety of requirements due to diverse sustainability-related legislation and policies that vary among regions and nations.

Supply chain complexity: Due to a lack of transparency, traceability, and accountability, managing sustainability risks and guaranteeing ethical raw material procurement can be difficult in complicated multi-supplier supply chains.

Technological constraints can make it challenging for manufacturers to identify appropriate technologies that meet their production processes and product requirements because of the sustainability of manufacturing technologies' cost, scalability, and availability.

Changing organisational culture, attitudes, and behaviour in the direction of sustainability may necessitate measures to overcome employee reluctance to change, a lack of motivation, and ingrained habits.

Measurement and reporting: Due to a lack of standardised procedures, data gathering, and reporting systems, it can be difficult to accurately measure and report the environmental, social, and economic implications of sustainable production practises.

Market demand and customer awareness: It can be difficult for producers to keep up with changing consumer preferences, market demand for sustainable products, and successfully communicating product sustainability qualities.

Balancing short-term company aims with long-term sustainability goals can be difficult since pressing financial objectives and shareholder expectations sometimes take precedence over sustainability considerations.

Stakeholder collaboration and engagement: It can be difficult to engage and collaborate for sustainable production when different stakeholders, such as suppliers, customers, regulators, and local communities, do not have mutually beneficial relationships or aligned interests.

To overcome obstacles and incorporate sustainable manufacturing practises into the core operations and culture of organisations, proactive strategies, teamwork, and continual improvement efforts are required. A digital representation of a real-time linked physical object, or process is known as a "digital twin". It functions as a dynamic and interactive digital model that, utilising information from sensors, Internet of Things (IoT) devices, and other sources, mimics the actions, traits, and results of a physical counterpart. A wide range of industries, including manufacturing, construction, transportation, healthcare, and energy, use digital twins [2]. The usage of digital twins helps to mimic, improve, and keep an eye on complicated industrial processes, goods, and machinery. They make it possible for manufacturers to monitor, analyse, and visualise sensor data and other data in real-time, gaining useful insights, spotting issues, and making wise decisions for process improvement, preventive maintenance, and product optimisation.

4.2   RELATED WORK

A contribution to the field of sustainability assessment in the context of smart manufacturing can be found in [1]. The authors outline a brand-new approach for assessing the sustainability effects of smart manufacturing processes using the methodology of digital twins. The framework offers a thorough method for evaluating the sustainability performance of smart manufacturing systems by considering a wide range of elements, such as resource use, energy consumption, waste creation, and social equality. This research illuminates the potential of digital twin technology as a useful tool for advancing smart manufacturing towards more ecologically and socially responsible practices.

A case study of sustainable hyper automation in high-tech manufacturing industries, concentrating specifically on linear electromechanical actuators, is presented by Fedosovsky et al. In the context of hyper automation [3], which involves incorporating cutting-edge technology like artificial intelligence, robots, and IoT into manufacturing processes, the authors highlight the difficulties and potential connected with implementing sustainable manufacturing practices. The thesis discusses a number of sustainability-related topics, such as resource efficiency, energy consumption, waste reduction, and environmental impact, and it suggests ways to improve the efficiency of linear electromechanical drives in high-tech manufacturing sectors.

The current scenario, various future aspects, and implementation framework of digital twin technology for sustainable manufacturing supply chains are covered by Miehe et al. [4]. The authors emphasise how digital twin technology has the potential to enable environmentally friendly practices all along the supply chain, from design and production to logistics and end-of-life management. Including resource optimisation, waste reduction, energy management, and circular economy practices, the book provides an overview of current trends and emerging applications of digital twin technology in sustainable manufacturing.

The work offers a framework for integrating digital twin technologies into manufacturing supply chains in order to achieve sustainability goals. This study contributes to our understanding of how digital twin technology might enable more sustainable practices in the manufacturing sector and offers useful insights for scholars and practitioners interested in adopting digital twin technology for sustainable manufacturing supply chains.

In a research article published in 2021 [5], Miehe et al. provided a thorough investigation of the function of digital twins in sustainable manufacturing. The authors draw attention to the revolutionary potential of digital twin technology to track, evaluate, and optimise production processes in real time, hence promoting manufacturing sustainability. The study emphasises the significance of considering sustainability factors, such as environmental, social, and economic considerations, in the planning and development of digital twin systems. In order to effectively use digital twin technology to promote sustainable manufacturing practices, the authors emphasise the significance of coordinated efforts between manufacturers, researchers, and policy makers. In general, the study offers insightful information about the ideas behind and potential use of digital twins.

The use of digital twin technology as a tool to modernise the construction sector is explored in research by Greeshma and Philip [6]. The authors examine how digital twins, which are virtual representations of physical assets and systems, might be utilised to improve several building processes, including design, planning, scheduling, and monitoring. They emphasise how real-time monitoring, data analysis, and simulation may be used with digital twins to increase efficiency, accuracy, and sustainability in the construction industry. The report also analyses the difficulties and advantages of using digital twins in the construction sector and offers insightful information on how they can transform the sector.

4.2.1 INSIGHTS

For several important reasons, it is essential to do research on the application of digital twin technology in sustainable manufacturing practices.

First, in order to address concerns like climate change, resource depletion, and environmental degradation, sustainable production has become an important worldwide necessity. Utilising digital twin technology can optimise resource utilisation, cut waste, and promote circular economy principles, leading to more environmentally friendly manufacturing procedures and goods.

Second, digital twin technology has demonstrated to be an effective instrument for enhancing the productivity, sustainability, and efficiency of manufacturing operations. Digital twins offer real-time monitoring, analysis, and optimisation of manufacturing activities by building virtual replicas of physical systems and processes. This improves resource management, waste reduction, and energy efficiency.

Third, it is crucial for practitioners and decision-makers in the manufacturing sector to comprehend the promise and constraints of digital twin technology in the context of sustainable manufacturing. This chapter offers insightful information for the design and application of digital twins for sustainable manufacturing, empowering businesses to take wise decisions and successfully handle various issue in their system.

In conclusion, this chapter adds to our understanding of the role that digital twin technology plays in advancing sustainable manufacturing methods, and it has ramifications for scholars, practitioners, and decision-makers who are interested in advancing sustainability in the manufacturing sector.

The purpose of this chapter's research is to examine how well digital twin technology may be used in relation to green manufacturing techniques. The chapter specifically looks at how digital twin technology can be utilised to promote circular economy concepts within the manufacturing supply chain and optimise resource utilisation, waste reduction, and energy management. The use of digital twin technology to solve sustainability issues in manufacturing, particularly its capacity to provide real-time monitoring, analysis, and optimisation of manufacturing processes, products, and systems, will be studied in detail. The chapter offers direction for industry practitioners and decision-makers and sheds light on the consequences of digital twin-based solutions for sustainable manufacturing [7]. The results of this chapter advance knowledge of the function of digital twin technology in fostering sustainability in manufacturing, with possible repercussions for fostering future environmentally friendly and socially equitable manufacturing practices.

4.3 METHODOLOGY

Figure 4.1 illustrates the key components involved in digital twin technology. The IoT, cloud computing, AI, and digital twin technology are all quickly developing

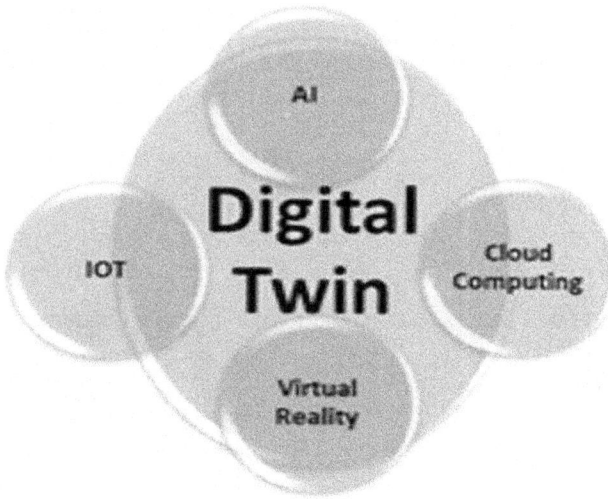

FIGURE 4.1 Key concepts involved in digital twins.

technologies that have the potential to completely transform sustainable production. Each of these technologies can be used in the context of environmentally friendly production in the following ways:

IoT:

IoT stands for the interconnectedness of physical devices, which allows them to communicate with one another and share data. IoT can be used in sustainable manufacturing to track and manage energy and resource utilisation. For instance, sensors can be installed on equipment to monitor energy consumption, and this information can be utilised to streamline manufacturing procedures and reduce waste.

Cloud Computing:

Cloud computing enables businesses to store and process massive volumes of data on distant servers. This can be especially helpful in sustainable manufacturing, where data must be gathered and evaluated in real time from a variety of sources (such IoT sensors). By reducing the demand on on-site servers, cloud computing can also assist businesses in lowering their environmental impact.

AI:

Artificial intelligence describes a machine's capacity to learn from data and make judgements based on that data. AI can be utilised in sustainable manufacturing to streamline production procedures and reduce waste. In order to minimise downtime and the need for replacement parts, AI algorithms can be utilised, for instance, to anticipate machine faults before they happen.

Digital Twin:

A digital twin is a virtual representation of a real-world thing or procedure. Digital twins can be used in sustainable manufacturing to simulate production processes and optimise energy and resource utilisation. For instance, a digital twin of a factory may be used to simulate several production scenarios and pinpoint the method that uses the least amount of energy.

4.3.1 STEPS TO BUILD DIGITAL TWINS IN SUSTAINABLE MANUFACTURING

Creating digital twins for production requires the following fundamental steps.

4.3.1.1 Step 1: List Your Assets

It is of the utmost importance to first identify the assets or processes that need to be digitally copied. These assets could be something as basic as a machine, warehouse, component, or machine part for the industrial sector.

4.3.1.2 Step 2: Identify Material Equivalents

A real-time data flow from physically present systems, processes, or objects that are scanned in the physical world is necessary for digital twins to create a virtual replica. Either a machine code or a QR code can be scanned.

4.3.1.3 Step 3: Build an Instantaneous Flow of Data from IoT Devices and Sensors

In order to generate an accurate depiction, a full understanding of the physical system and the data that must be collected is required. Sensors, IoT devices, and other data collection techniques are some of the possible sources of this information.

4.3.1.4 Step 4: Construct a Statistical Computing Model

To produce a digital model that accurately depicts the physical system, the data must then be processed and analysed. To do this, a 3D model of the physical system and its behaviour is created utilising complex algorithms and software. 3D models of workstations, parts, and equipment are frequently used in the industrial sector for a variety of tasks like creating new products, experimenting with workstation arrangements, and assessing ergonomics. CAD or 3D creation technologies can often be used to build these models.

4.3.1.5 Step 5: Screen the 3D Prototype

The models that are produced are presented using 3D visualisations and AR techniques to replicate real-world situations and visualise the results on any platform, including smartphones, PCs, and AR/VR devices. Frontline employees can examine how the virtual model might interact with the actual world by using augmented reality, which overlays the digital counterpart onto a real-world perspective. This can be extremely helpful for training as well as for prototyping and testing. Digital twin deployment typically starts off on a small scale by concentrating on the performance of a single product component, but it eventually expands and evolves through time.

But there are two different ways that this happens. In order to give a complete picture of the asset, process, or system, a few digital twins are first joined. Second, more sophisticated features are added to the current digital twin to forecast the performance of its physical counterpart in the future.

4.4 USE CASE: DIGITAL TWINS IN VEHICLE MANUFACTURING

A vehicle company might develop a digital twin of its production line, enabling real-time monitoring and analysis of data from IoT devices including sensors, cameras, and other equipment [7]. Figure 4.2 clearly depicts the steps involved in construction of a digital twin for a vehicle manufacturing company. By doing this, the manufacturer would be able to spot possible problems before they arise and optimise the production process, improving productivity, lowering downtime, and boosting sustainability. The producer could, for instance, utilise sensors to track the temperature and pressure of the equipment on the assembly line, and IoT-capable cameras could deliver real-time photos of the manufacturing process. AI systems might analyse this data to find trends and anomalies that could indicate possible issues like equipment failure or flaws in the finished product.

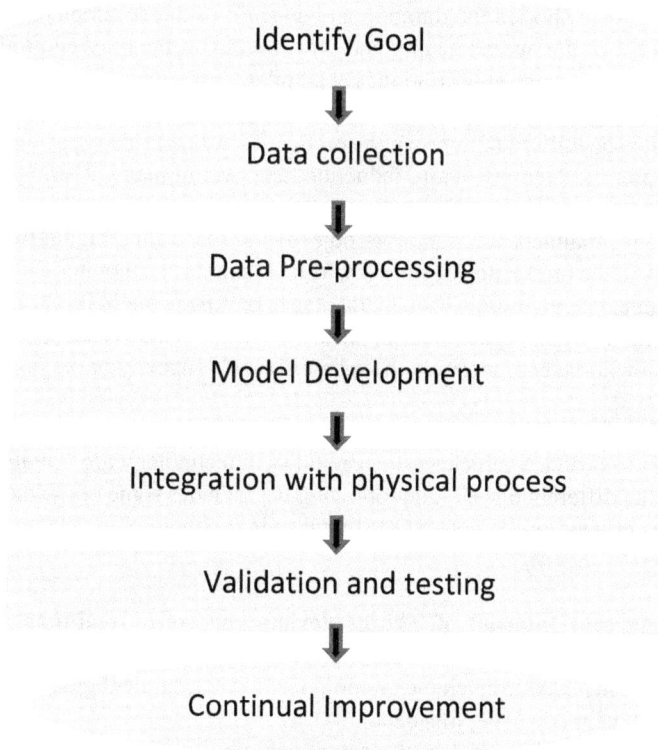

Identify Goal

⬇

Data collection

⬇

Data Pre-processing

⬇

Model Development

⬇

Integration with physical process

⬇

Validation and testing

⬇

Continual Improvement

FIGURE 4.2 Digital twin implementation steps for vehicle manufacturing.

The manufacturer could modify the production process, such as modifying the machinery's temperature and pressure or rearranging the production processes, based on the insights from the digital twin. This might result in increased productivity, lower energy use, and less waste, all of which would help make the manufacturing process more environmentally friendly.

Also, the manufacturer might schedule maintenance with the help of the digital twin through predictive maintenance, minimising downtime and avoiding expensive repairs. The massive amount of data produced by the IoT and digital twins might also be stored and analysed using cloud computing, creating a scalable and adaptable infrastructure for intelligent and sustainable manufacturing [6, 8].

4.4.1 DIGITAL TWIN IMPLEMENTATION PHASES

4.4.1.1 Defining the Goal

There are numerous ways to characterise the objective of utilising digital twin technology in the production of vehicles, including:

- Design and development: With the use of digital twin technology, new cars or component designs can be developed virtually. This enables manufacturers to test and improve their designs before beginning physical manufacturing. This can shorten the duration and expense of the development process and enhance the overall quality and functionality of the finished product.
- Production process optimisation: By simulating the manufacturing process and finding potential bottlenecks or inefficiencies, digital twin technology may also be utilised to improve the production process. This may contribute to greater productivity, waste reduction, and overall manufacturing process efficiency.
- Predictive maintenance is another objective of integrating digital twin technology in the production of vehicles. Manufacturers can monitor and analyse real-time data to foresee when maintenance or repairs might be necessary by building a virtual replica of a physical vehicle or component. This helps to save downtime and improves the reliability and longevity of the product.

4.4.1.2 Data Collection

Data sources to produce vehicles using digital twin technology are versatile and can originate from different places. The following list includes some of the most popular data sources utilised in the context of digital twin technology for the production of vehicles:

- **Sensors and Internet of Things devices** can collect real-time data on temperature, pressure, vibration, and other pertinent factors when they are installed in a physical vehicle or component. An accurate digital twin duplicate can be made using this data.
- **CAD data**: A 3D model of the actual vehicle or component can be made using CAD data. To make sure the digital twin replica is authentic, this information can be utilised as a reference.

- **Manufacturing information:** To enhance the manufacturing process and the accuracy of the digital twin replica, manufacturing information, such as production times, material usage, and machine settings, can be gathered and analysed.
- **Simulation data:** By using simulation data, a digital twin model that faithfully depicts the real vehicle or component can be produced. The performance of the digital twin in various settings can be tested and validated using this data.
- **Data on past maintenance**, repairs, and replacement parts can be utilised to forecast when maintenance or repairs would be necessary, reducing down-time and extending the useful life of the device.

In general, the data sources for digital twin technology in car manufacturing are concentrated on obtaining real-time data and using that data to produce an exact digital twin duplicate. Predictive maintenance is made possible by using this data, which can also be utilised to optimise production and design processes.

4.4.1.3 Data Pre-Processing

Data pre-processing is an essential stage in getting data ready for analysis, especially when considering digital twin technology to produce vehicles. Here are a few typical techniques used to pre-process the data gathered for digital twin technology:

- **Data cleansing** entails eliminating any incorrect or useless information. It could involve getting rid of duplicate records, dealing with missing information, or fixing data that was entered erroneously.
- **Data transformation** is changing the data's format to one that is more practical, such as by scaling or normalising the data to make it more uniform or by converting categorical data to numerical data.
- **Feature selection** entails choosing only the most crucial characteristics or elements that significantly affect how well the digital twin model works.
- **Data integration** is the process of merging data from several sources to produce a dataset that is more complete. This can entail combining manufacturing data, simulation data, and data from sensors in the context of digital twin technology.
- **Data reduction** is the process of lowering the amount of data to make the analysis process more effective. Techniques like principal component analysis (PCA) or clustering can be used to accomplish this.

In general, data preparation is crucial for building a precise and trustworthy digital twin model. The pre-processing procedures aid in ensuring the consistency, accuracy, and relevance of the data used in the model with respect to the research issue.

4.4.1.4 Digital Twin Model Development

Depending on the application and use case, a variety of digital twin simulation model construction techniques can be employed in the manufacturing of vehicles. Here are some popular techniques for creating digital twin simulation models that are appropriate for the automobile industry:

- **Finite-element analysis (FEA)**: FEA is frequently used in the production of vehicles to model and examine the structural behaviour of parts, including chassis, body panels, and suspension systems.
- **Computational fluid dynamics (CFD)**: CFD is important in the manufacture of vehicles to model and examine the flow of fluids through engine parts, exhaust systems, and other parts, including gasoline and air.
- **Multi-body dynamics (MBD)**: MBD simulates and analyses the behaviour of drive trains, powertrains, and vehicle dynamics during the construction of vehicles.
- **System dynamics**: In the manufacture of vehicles, system dynamics is useful for simulating and analysing the behaviour of the complete vehicle system, including interactions with the outside world like weather, traffic, and road conditions.
- **Co-simulation modelling:** Real-time data analysis and feedback are made possible by co-simulating the behaviour of the physical system and the digital twin model at the same time. The performance of the digital twin simulation model in real-world scenarios can be tested and validated via co-simulation modelling in the automobile manufacturing industry.
- In general, the type of data that is available and the specific application determine the simulation model construction process. An accurate and trustworthy digital twin simulation model for car manufacturing is often developed using a multi-disciplinary approach including professionals in engineering, data analysis, and modelling.

4.4.1.5 Integration with Physical Process

Depending on the application and the kind of data that has to be shared, a different interface may be utilised to link the digital twin model to the physical system. In certain circumstances, a software program can be used to link the digital twin and the real system, enabling real-time data transfer over a network. When the physical system is outfitted with sensors and other data gathering devices that can transfer data over a network, this form of interface is frequently utilised.

In other situations, a hardware component might be utilised to link the virtual system to the real one. Devices that can read data from the physical system and provide feedback to manage its behaviour include data acquisition systems and control systems. In some instances, the hardware component may also comprise sensors and other data gathering tools that are physically mounted on the system [8]

Regardless of the interface employed, it is critical that the physical system and digital twin model be linked in a way that enables real-time data transmission. With real-time monitoring, analysis, and control of the physical process made possible, the digital twin can gain insightful knowledge and enable proactive maintenance and optimisation.

4.4.1.6 Validation and Testing

The development and execution of a digital twin need several crucial processes, including validation and testing. Here are a few typical methods for evaluating and confirming a digital twin:

Verifying the accuracy with which the digital twin model represents the real system is known as model verification. This can be accomplished by comparing the output of the model to existing data or by contrasting the behaviour of the model with that of the underlying physical system. Sensitivity testing examines how the digital twin model reacts to modifications in the input variables. This can assist in determining which parameters are most crucial and which can be changed to increase the model's accuracy.

Calibration entails modifying the digital twin model's parameters to increase accuracy. Typically, to do this, the output of the model is compared to data gathered from the actual system, and the model's parameters are adjusted to minimise any discrepancies. The digital twin model is simulated in a variety of settings in order to assess its performance.

To make sure the model functions as anticipated, this can involve testing it under both typical operating settings and unusual or extreme ones. Testing the digital twin in the physical system's actual environment is known as "field testing". This may entail keeping track of how the real system behaves and contrasting it with the forecasts provided by the digital twin model.

4.4.1.7 Continual Improvement

Building digital twin models that are continuously improved is crucial because it allows the model to evolve over time and respond to changes in the physical system. Here are several tactics for developing digital twin models that are continuously improved:

- Data gathering: The first step in continuous improvement is gathering precise and thorough data from the physical system. The digital twin model may be calibrated and validated using this data, and potential improvement areas can be found.
- Feedback circuits: In order to construct digital twin models that are continually improved, feedback loops are essential. Feedback loops allow the model to change and advance over time by tracking the behaviour of the physical system and comparing it to the predictions given by the digital twin model.
- The accuracy of the digital twin model can be continuously increased using machine learning methods. Machine learning techniques can be used to update and improve the digital twin model by examining the physical system's data and looking for patterns and trends.
- Integration with other systems: The accuracy and efficiency of the digital twin model can be increased by integrating it with other systems, such as predictive maintenance or optimisation systems. Systems that share information and insights are better able to take use of one another's strengths.
- Collaboration: For continuous progress in the development of digital twin models, collaboration between many stakeholders, including engineers, data scientists, and domain experts, is crucial. The digital twin model can be honed and enhanced over time by combining various viewpoints and areas of expertise. Figure 4.3 shows the transformation of a physical assert into a digital object in vehicle manufacturing.

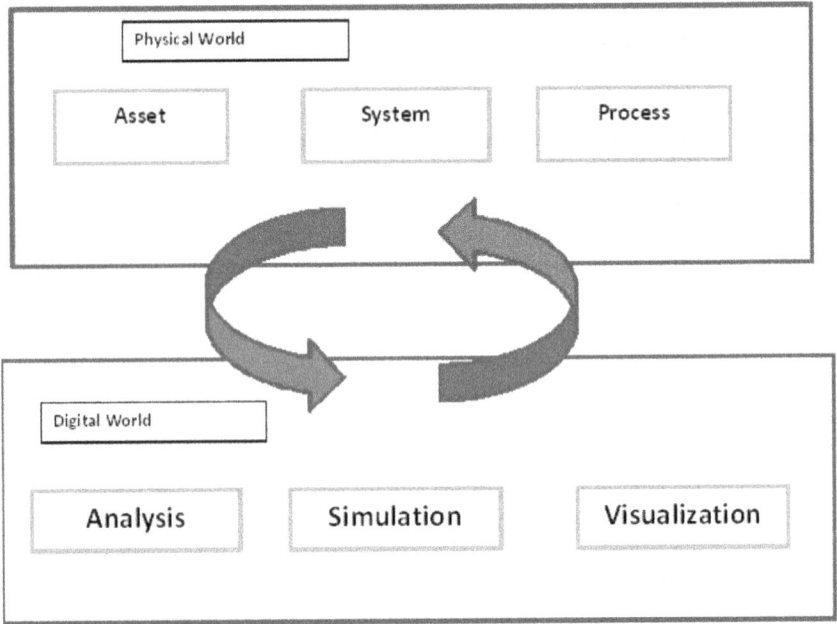

FIGURE 4.3 Sample scenario of a digital twin in vehicle manufacturing.

Digital twin technology is a formidable tool that can enable sustainable production since it provides real-time insights into the manufacturing process. As a result, waste, energy use, and emissions can be optimised and reduced. Digital twins are virtual representations of actual systems or items that can be used to replicate and evaluate them before putting certain situations into reality in the real world.

Digital twins can be utilised in sustainable manufacturing to maximise resource efficiency, cut waste, and enhance product quality. The manufacturing process can be simulated in a digital twin, which allows producers to spot inefficiencies and make changes to improve performance. They can spot opportunities to utilise more sustainable products or cut back on energy consumption, for instance.

Along with monitoring and enhancing product performance throughout its full lifecycle, from conception to disposal, digital twins can also be utilised. This can involve anticipating a product's maintenance requirements to reduce downtime and increase its useful life as well as optimising end-of-life procedures to cut waste and encourage recycling.

Overall, digital twin technology can be a major contributor to sustainable manufacturing by enabling real-time monitoring and optimisation of resource consumption, waste reduction, and emissions reduction throughout the entire production process.

4.5 CONCLUSION

Data from the plant, systems, supply chain, and equipment are essential to smart manufacturing. Manufacturers utilising Industry 4.0 apps employ the real-time data power of digital twins to monitor and analyse constantly changing data in their production process.

Comparatively smart manufacturing [8], or manufacturing assisted by a digital twin, has various benefits over traditional manufacturing, including a shorter time to market for products, improved process and product performance, higher production efficiency, the ability to perform predictive maintenance, and virtual commissioning. Digital twin technology can be applied in production on different levels:

- At the component level, concentrated on a single, extremely important manufacturing process component.
- At the asset level, a single piece of production line equipment can be given a digital twin.
- At the system level, a production line can be monitored and improved using a digital twin.
- At the process level, this examines the sequence of the process, from the design and development of products and processes to manufacturing and production. Throughout the whole life cycle of the finished product, it also applies to its distribution and use by customers or patients, as well as to the creation of new products.

Due to numerous studies, digital twins have a number of advantages, such as predictive maintenance, which is used to keep machinery, production lines, and buildings in good condition; monitoring products in real time as they are used by actual customers or end users to gain a better knowledge of them optimising manufacturing processes; process improvements for product traceability; testing, confirming, and improving hypotheses increasing system integration between disconnected ones; and equipment troubleshooting via remote access, regardless of location.

The capabilities of DTs could be increased by service-oriented design. DTs can have high potential use through services in design, production, and PHM. Together with services and the digital twin, it is specified how the different parts of the digital twin are encapsulated in services and used as services. The research is now in its very early stages. The methodologies of DT modelling still need to be improved and expanded upon.

REFERENCES

BOOKS

[1] Grieves, M. (2019). *Digital Twin: Manufacturing Excellence through Virtual Factory Replication*. CRC Press.

JOURNALS

[1] Li, L. et al. (2020). Sustainability assessment of intelligent manufacturing supported by digital twin. *IEEE Access*, 8, 174988–175008. https://doi.org/10.1109/ACCESS.2020.3026541.
[2] He, B., and Bai, K. J. (2021). Digital twin-based sustainable intelligent manufacturing: A review. *Advances in Manufacturing*, 9, 1–21.

[3] Fedosovsky, M. Ezy., Uvarov, M. M., Aleksanin, S. A., Pyrkin, A. A., Colombo, A. W., and Prattichizzo, D. (2022). Sustainable hyperautomation in high-tech manufacturing industries: A case of linear electromechanical actuators. *IEEE Access,* 10, 98204–98219. https://doi.org/10.1109/ACCESS.2022.3205623.

[4] Miehe, R., Waltersmann, L., Sauer, A., and Bauernhansl, T. (2021). Sustainable production and the role of digital twins–Basic reflections and perspectives. *Journal of Advanced Manufacturing and Processing,* 3, e10078. https://doi.org/10.1002/amp2.10078

[5] Naseri, F., Gil, S., Barbu, C., Cetkin, E., Yarimca, G., Jensen, A. C., Larsen, P. G., and Gomes, C. (2023). Digital twin of electric vehicle battery systems: Comprehensive review of the use cases, requirements, and platforms. *Renewable and Sustainable Energy Reviews,* 179, 113280. ISSN: 1364-0321. https://doi.org/10.1016/j.rser.2023.113280.

[6] Qi, Q., Tao, F., Hu, T., Anwer, N., Liu, A., Wei, Y., Wang, L., and Nee, A. (2021). Enabling technologies and tools for digital twin. *Journal of Manufacturing Systems,* 58, 3–21. https://doi.org/10.1016/j.jmsy.2019.10.001.

[7] Greeshma, A. S., and Philip, P. M. (2022). Digital twin as a revamping tool for construction industry. In *Digital Twin Technology,* M. Vohra (Ed.). Wiley. https://doi.org/10.1002/9781119842316.ch6

[8] Ma, S., Ding, W., Liu, Y., Ren, S., and Yang, H. (2022). Digital twin and big data-driven sustainable smart manufacturing based on information management systems for energy-intensive industries. *Applied Energy,* 326, 119986.

5 Overview of Cyber Security in Intelligent and Sustainable Manufacturing

Aditi Paul[1], Somnath Sinha[2], Parveen Kumar[3], Shirasthi Choudhary[1], Krishna Samdani[1], and Saumya Mishra[1]

1 Banasthali Vidyapith (Deemed to be University), Rajasthan, India

2 CHRIST (Deemed to be University), Bengaluru, Karnataka, India

3 Department of Mechanical Engineering, Rawal Institute of Engineering and Technology Faridabad, Haryana, India

5.1 INTRODUCTION

Smart manufacturing systems (SMSs) as part of Industry 4.0 have tremendous potential to enhance the socio-economic structure of a nation. Smart farming, smart cars, smart healthcare, and smart classrooms are some effective outcomes of the Industry 4.0 revolution. With the advancement of the Industrial Internet of Things (IIoT), operational technology (OT) has merged with information technology (IT) through numerous intelligent and connected devices. These devices collect data from the surrounding environments and send it to a coordinator device, which sends this accumulated data to a server machine for further processing. All of these need network connectivity and remote access to the devices. This connectivity removes the isolation of the manufacturing system from the open network and accelerates production many times over. However, the subtle component of SMSs should be secured from network threats and attacks.

The adverse side of smart processing is the increased risks of cyber attacks [1] on peripheral devices connected through public networks. Thus, security breaches are prominent in such devices and have a higher risk. The use of IIoT and robotics in smart manufacturing brings the attackers closer to the target and increases the risk of distributed denial of service (DDoS) [2] attacks and authentication breaches because of weak or no security. The reason behind the poor security structure of

DOI: 10.1201/9781003405870-5

SMSs cannot be underestimated but needs to be addressed. The challenge of adding a security component to a machine or process is that it obstructs production and affects the overall equipment effectiveness (OEE) [3]. Security needs continuous patching of the security add-on to a system, performing authentication during data collection from various external sensors or connected objects, verifying the identity of the supply chain each time during the transaction, and many other access controls which directly affect the production timeline and process. Also, physical patching of machines needs experts who are hardly hired due to improper policy or priority given by higher management. Thus, machines are made smart with little or no security at all. Even with the Industry 4.0 revolution, statistically, the safety of SMSs is still challenging. This chapter aims to represent the inherent cyber security threats to smart manufacturing systems and proposed mitigation strategies. The most important part of the chapter are the case studies of cyber security incidents on SMSs that have happened so far. Finally, a novel machine learning–based intrusion detection system is proposed to show how researchers can implement a defense framework to safeguard SMSs.

5.2 DIFFERENCE BETWEEN SMS AND IT SYSTEMS

One of the primary reasons for the need for more security practices in SMSs is their rigidity and complexity. Traditional manufacturing systems were isolated systems since they were human driven. Thus, data collection, processing, control, and supply were done manually or with human-driven robots. Therefore, security was only needed for the safety of human lives, physical damage, and eco-friendliness. With intelligent manufacturing being evolved, these remote machines are now attached to the Internet for smooth and speedy workflow, reduced cost, and higher production. Therein lies the difference between information technology and operational technology. It is based on data-driven software and hardware. Thus, putting security into IT systems is a few steps to be executed on the system software or application interfaces. But being the hardware machine, it is a cumbersome task to add security components to the machine-driven systems. This may create catastrophic destruction if attacks happen to critical details, like manipulating sensor readings in a chemical plant or nuclear station. Thus, the necessary infrastructure must balance system security and its production or response in time. Cyber attacks [4] on any domain are based on three crucial aspects identified by the attackers. These are:

Attack vectors: it is a channel or path through which attackers try to gain unauthorized access [5] to the system. For SMSs, attack vectors include malware like worms/Trojans, poor or no authentication, or social engineering.
Vulnerabilities are weakness or loopholes of a system, machines, or devices that attackers exploit to launch attacks. For SMSs, vulnerabilities can be a lack of access control [6], untrained/compromised workers, vulnerable automation logic, weak credentials in the human–machine interface (HMI),

vulnerable software extensions/add-ons, lack of patching or updating of machines, or poor authentication mechanisms.

Attack targets: these are the target machines, interfaces, devices, and sensors that the attackers want to victimize. Attack targets in SMSs include workstations, machines/robots, mobile HMI, IoT devices, SCADA systems, computerized maintenance management systems (CMMSs), and intellectual property such as CAD design or specification files.

5.3   CYBER SECURITY PRINCIPLES FOR SMART MANUFACTURING SYSTEMS

In another aspect, the integration between IT and OT extends IT security to the OT. However, this extension does not directly employ the IoT. Industry 4.0 was envisioned as merging initiatives with the IoT and cloud-based virtualization. But the confidentiality, integrity, and availability (CIA) triad [7], as applied to the IoT, could not be merged into intelligent machines due to the inherent complexity of configuration, processing, control, and practices. Instead, the performance of SMSs is measured in terms of availability, performance, and quality of products. Cyber threats to these parameters are crucial and can be parameterized as violations of availability, integrity, confidentiality [8], and authenticity. We discuss these performance metrics and the related security threats (Table 5.1).

Availability: Availability indicates that the system or machine should not be stopped, hung up, or broken down during critical processes. The immediate effect of loss of availability is a considerable loss of production or goods due to a machine halt for a more extended period (a few minutes to a few days). Cyber attackers may disrupt a server or central control unit like the programmable logic controllers (PLCs) of an SMS by manipulating the programming or commands executed in these circuits. The most popular cyber attack on availability is Stuxnet, a worm injected into an Iranian nuclear enrichment plant in Natanz in 2009 and 2010. This worm checks for the PLCs connected to the computer systems. It alters the PLCs' programming, which in turn causes centrifuges to spin faster for a longer time, causing catastrophic destruction of the equipment [9, 10]. Another incident occurred in 2014 in Germany in which the attack targeted the industrial control system of a German steel mill and disabled multiple components of the system [11].

Loss of availability can also be caused by injecting malware or A Trojan into a system component. One such attack was executed on a manufacturing plant of the reputed car company Honda by injecting the WannaCry virus. This virus targeted the vulnerable computer systems of legacy systems, took control of these, and demanded money for releasing the production shutdown [12].

Integrity: Integrity and performance are critically related in SMSs. The performance of an autonomous machine can be decreased due to changes in the critical input parameters such as yield rate, spindle or feed rate, nozzle speed, or motion of a printer. For example, in a subtractive manufacturing system, by changing the geometry (G)-code or miscellaneous (M)-code, an attacker can change or reverse

TABLE 5.1

Attack Vectors and Threats on CIA Triad in Smart Manufacturing Systems

Cyber Security Principles	Attack Vectors	Attack Types
Availability	Overloading node resources (memory, computation power, battery), using cloud services at the cost of other users, exploiting hypervisor vulnerabilities	DDoS attacks on cloud services, service theft attacks, and interruption of information flow in a time-critical situation.
Integrity	Malware code injection to legitimate virtual machines, compromising weak integrity measures on big data, changing machine instruction through malicious logic	DDoS attacks, data tampering, changing, taking control of critical systems, production shutdown, internal and external attacks.
Confidentiality	Node IDs, personal/health information of wearable devices (intellectual property), manipulating routing information, neighbor state, available radio links, node identity manipulation, weak authentication, using keys by man-in-the-middle attacks, compromised service providers, collection and analysis of information through a side channel, extracting information by malicious agents	Man-in-the-middle attacks, identity theft, data falsification attacks, insider attacks, side channel attacks.
Authentication	Inappropriate access control, poor authentication, unauthorized access, false website design to collect authentication information, phishing	Phishing attacks, social engineering.

the mechanical properties of the machines, which may result in a longer production time than expected. Thus, by disrupting the machine, integrity performance is compromised. Additive manufacturing systems where 3D printing or computer-aided design (CAD) are used can be worse if the attacker changes the machine instruction to slow down [13] the nozzle head or print head speed for creating layers of melted alloys. This manipulation can worsen [14] the quality of the product due to defective or poor metal composition. Supply chain vulnerabilities are another major cause of performance degradation in SMSs. As supply chain in SMSs is connected externally through the cloud and the public Internet, an attacker may inject a virus or Trojan into the weak machines or computer systems through insecure network. This may disrupt the supply chain due to shutdown [15] of production, even for days.

Confidentiality: Attacks on confidentiality in SMSs occur due to the unmanageability of scattered devices and machine interfaces connected through the Internet. The lack of authentication applied [16] to these devices, like sensors, is targeted by cyber attackers. Industrial machines are now connected to the cloud for collecting information and then executing commands for running the machines remotely. Attackers target developers' systems by compromising the device interfaces. An Arduino developer may design a code for a sensor using a library that may already be infected by malicious code. When embedded in the machine interfaces or sensors, this code may change the machine instruction adversely and thus create disruption. Hence, using third-party devices, APIs, and libraries for SMSs interfaces [17] is prone to confidentiality violation. For example, a developer may not be aware of a Trojanized library [18, 19] used for a temperature sensor that causes incorrect temperature readings and turns the alarm off. For a chemical plant, this could be devastating. Some third-party apps may leak authentication information, which can be used later to control a machine interface through the app by an attacker.

5.4 TAXONOMY OF CYBER SECURITY THREATS TO SMSs

Cyber threats to SMSs are no different from IoT attacks. However, the risks [20] associated with these are additional. SMSs use some preoperatory protocols like Modbus, PROFINET, DNP3, and EtherCAT, which are serial protocols. Among these, the widely used protocols Modbus and DNP3 were connected to TCP/IP for data communication, forming the basis of the IIoT. TCP/IP is an insecure protocol, creating vulnerabilities to multiple cyber attacks. Hence, these systems lack availability, data integrity, and authentication, the primary security triad for SMSs. The integration of manufacturing systems with low-powered, low-cost IoT devices has made SMSs work more flexibly with cost effectiveness. To achieve this goal, most of these IIoT end devices are not tightly secured against attacks common to IoT but can be catastrophic for critical infrastructure systems. To show the nature of these attacks and vulnerabilities, a taxonomy (Figure 5.1) of cyber attacks on SMSs is depicted, and their impacts on protocol levels are explained with targeted SMSs that are affected most.

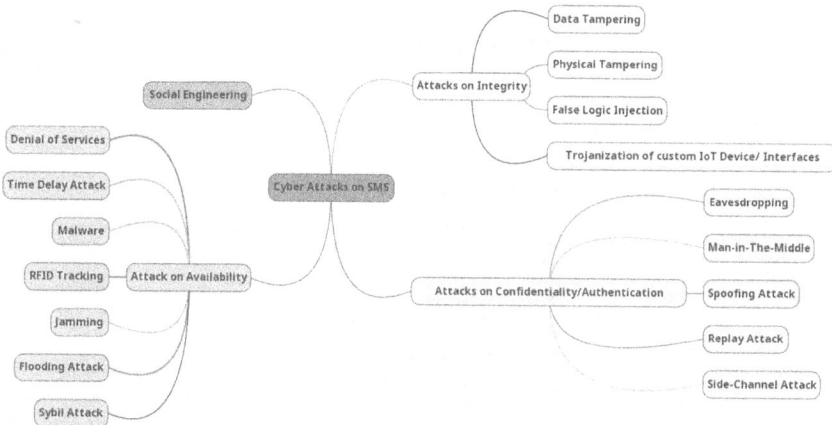

FIGURE 5.1 Taxonomy of cyber security threats on SMSs.

The taxonomy of cyber attacks in intelligent automation, manufacturing, and IIoT has many dimensions [21, 22]. We can categorize these attacks on the CIA triad, attacks on targets/locations [23], attacks on targeted domains [24], and so on. Thus, considering a single dimension may ignore some important attack vectors. However, this chapter considers only the CIA triad irrespective of attack dimensions and includes or matches other dimensions accordingly. Figure 5.2 has almost all the attacks that occur in SMSs categorized according to their effects on the CIA triad. Details on each attack are discussed in the following subsections.

5.4.1 ATTACKS ON AVAILABILITY

Denial of service (DoS)/DDoS: One of the most common attacks on SMSs and IIoT is a denial of service attack that targets the server machines or brings interfaces crashing down. The more severe variant is distributed denial of service, which is executed by the attacker from multiple bots or machines automatically. Thus, identifying their IP addresses is a very complex task. Since numerous components of SMSs are connected to the Internet and exchange information with industrial sensor nodes, launching DDoS attacks is easier. The attacker compromises weak industrial devices or sensors by injecting malicious code into these and launching attacks on the targeted machines by crashing them or flooding them with unnecessary traffic. The Mirai botnet attack [25] is an example that compromised millions of IoT devices and launched the most significant DDoS attacks on targeted systems. Another type of botnet, such as the Brickerbot [26] and the Reaper [27] botnets, exploits automated IoT devices which are not monitored once deployed.

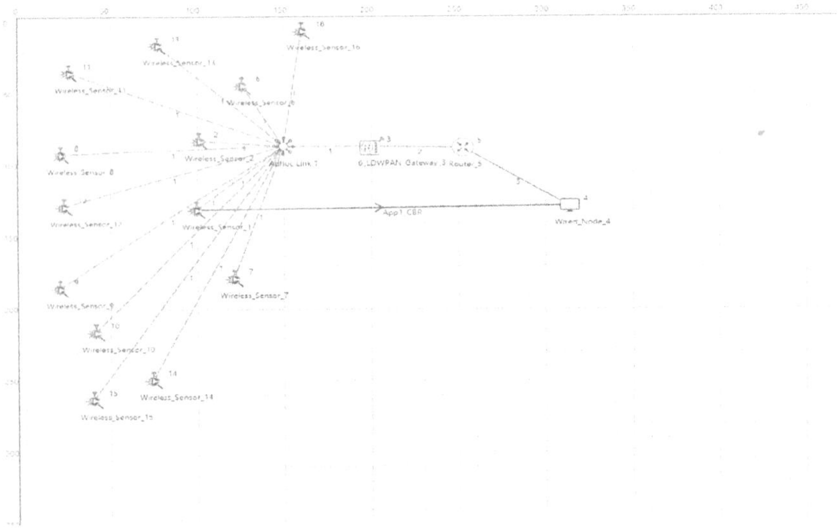

FIGURE 5.2   SYN flood attack with ten legitimate nodes and three attacker nodes (red).

5.4.2 ATTACKS ON INTEGRITY

Integrity attacks are of two types: material and structural attacks. In *material* attacks, the physical properties of the machines are changed, such as strength, roughness, color, and other physical properties [22]. In *structural* attacks [28], the geometric dimensions of the machines are tampered with or changed so that the product becomes weak or faulty or less durable than expected, and the overall life cycle of the product decreases. Some examples are scaling up or down a part of a machine to make it unfit to connect with other components, changing vertex coordinates of features to change the desirable mechanical properties, and creating an internal void during the additive manufacturing process to make the final product weak and less durable.

5.4.3 ATTACKS ON CONFIDENTIALITY

Attacks on confidentiality include spoofing, replay, side-channel attack, and man-in-the-middle attacks. Attackers mainly penetrate the system by exploiting weak authentication like simple or default passwords. For example, many critical machines used in healthcare for measuring patients' body parameters occasionally cannot be authenticated in every use. This creates a time lag which may be essential in emergencies. These systems can be a target for attackers. Some attacks, like *replay*, do not break authentication. Instead, the attacker retransmits a packet by altering its contents. Since the packages come from a legitimate source, they remain unnoticed. This attack can be prevented by checking the *sequence number* of the incoming packet. *Side channel attacks* are executed by analyzing information leakage by the industrial hardware or software, such as the amount of power consumption, traffic flow, electromagnetic emission, or the timing of an operation. *Side channel* attacks on 3D printers [29] using smartphone sensor information are one such attack. When a device or interface is compromised to leak information, it is termed a covert-channel attack [30]. An insecure communication channel in which no authentication is applied is prone to eavesdropping attacks. In this attack, the attacker gains sensitive information by tapping the track, monitoring network traffic flow, or fishing and then uses this information to launch other attacks. This is a type of passive attack that does not harm any equipment or machines directly. In a man-in-the-middle attack, the attacker replaces the keys of communicating parties with its created pairs and continues communication with both parties. Since the attacker now has the same key as the sender, he modifies the message contents and sends it in the sender's name. Since the receiver's key also matches the attacker, the receiver can decrypt the message, which makes him or her believe the message must have come from the sender.

5.5 CASE STUDIES ON DoS ATTACKS IN SMSs

Cyber attacks have targeted smart manufacturing and Industry 4.0 multiple times using various attack vectors. This section represents popular cyber attacks on SMSs, which include SCADA system attacks, smart buildings, electric and thermal power grids, automotive, oil and gas industries, rail and water companies, and many more.

5.5.1 SCADA System Attacks

5.5.1.1 Stuxnet

Stuxnet [31] is a computer worm developed jointly by the United States and Israel. The worm was first created in 2010 and targeted Iran's nuclear program and control systems. Stuxnet was notable for its complexity and sophistication. It spread via USB flash drives and network shares, and it exploited zero-day vulnerabilities in Microsoft Windows to gain access to target systems. Once inside a target system, it could manipulate industrial processes and cause physical damage to equipment. Stuxnet is widely believed to have successfully delayed Iran's nuclear program. However, it also raised concerns about the potential for cyber attacks to cause physical damage and highlighted the need for better cybersecurity measures for critical infrastructure.

5.5.2 Attacks on Power Grid

CRASH OVERRIDE: CRASH OVERRIDE, also known as "Industroyer", is sophisticated malware created to damage and disrupt industrial control systems (ICSs) used in electric power grids. It is believed to have been responsible for the December 2016 power outage in Ukraine, which left approximately 230,000 people without electricity for several hours. CRASH OVERRIDE is modular malware targeting various ICS equipment, including substation switches and circuit breakers. Once it gains access to a target system, it can issue commands to disrupt power flow and cause physical damage to equipment. The malware is believed to have been developed by a Russian state-sponsored hacking group known as Sandworm, which has been linked to other cyber attacks against Ukraine, as well as attacks against other countries. Attacks on power grids pose a significant threat to national security, as they can cause widespread disruption and economic damage. To mitigate these risks, governments and utilities must implement robust cybersecurity measures, such as regular vulnerability assessments, network segmentation, and incident response plans.

5.5.3 Automotive and Oil Tank Attacks

Zotob: The Zotob worm was a computer worm that caused widespread disruption in August 2005, targeting Microsoft Windows operating systems. It exploited a vulnerability in the Windows Plug-and-Play system, allowing it to spread rapidly across networks. The Zotob worm infected several high-profile organizations, including ABC News, CNN, and the *New York Times*. It also caused disruptions in several airports, including the Paris-Charles de Gaulle airport ticketing system. While the Zotob worm did not specifically target the automotive or oil tank industries, it did highlight the vulnerability of critical infrastructure to cyber attacks. As such, it served as a wake-up call for organizations to improve their cybersecurity practices and implement measures to protect against malware and other cyber threats. Oiltanking, a significant independent oil and gas storage company in Germany, acknowledged on January 29, 2022, that it had experienced an attack on its IT systems, which harmed its inland supply and resulted in terminals running at a reduced capacity. According to claims in the German media, the incident has influenced operating technology.

The incident happened on January 29, and it also hit its sister company, the oil trading corporation Mabanaft. Oiltanking, a significant component of the €14-billion Marquard & Bahls group, runs 47 tank terminals with a combined capacity of 18.5 million cubic meters across 21 countries.

Along with other small and medium-sized petrol station companies, Shell fuel stations in Germany are supplied by its 13 fuel storage facilities. According to an early report in Handelsblatt that was later picked up by Der Spiegel (both in German), the attack virtually halted all loading and unloading activities at Oiltanking. It is still unclear exactly how the incident affected Mabanaft. After the Oiltanking cyberattack, the corporation declared force majeure.

Shamoon: Saudi Aramco and RasGas: Shamoon is destructive malware [32] used in several high-profile cyber attacks, including attacks on Saudi Aramco and RasGas, two major energy companies in the Middle East. The first Shamoon attack was carried out against Saudi Aramco in 2012, and a second attack targeted RasGas in 2012. The Shamoon malware is designed to overwrite data on infected systems, rendering them inoperable. It is believed to have been developed by a group with ties to Iran and has been used in other attacks against targets in the Middle East. The attack on Saudi Aramco was particularly significant, as it destroyed tens of thousands of computers and forced the company to shut down its network for several days. The attack significantly impacted the company's operations, and it is estimated that it cost Saudi Aramco hundreds of millions of dollars in lost productivity and recovery costs. The attacks on Saudi Aramco and RasGas demonstrate the significant threat that cyber attacks pose to critical infrastructure and highlight the need for organizations to implement robust cybersecurity measures to protect against such threats.

BlackEnergy3: BlackEnergy3 was malware in a series of cyber attacks against the Ukrainian [33] power grid in December 2015. The attacks resulted in a widespread power outage that affected approximately 225,000 customers and lasted several hours. The BlackEnergy3 malware was used with other tools and techniques, including spear-phishing emails and remote access trojans, to access the targeted systems. Once inside, the attackers could take control of the power grid's SCADA systems, allowing them to issue commands to shut down power distribution equipment. The attack on the Ukrainian power grid was highly sophisticated and demonstrated the ability of cyber attackers to cause physical damage and disrupt critical infrastructure. The incident highlighted the importance of implementing strong cybersecurity measures for critical infrastructure, including network segmentation, access controls, and incident response plans. The Ukrainian government and security experts have attributed the attack to Russian state-sponsored hacking groups, though the Russian government has denied involvement. The incident increased tensions between Ukraine and Russia and heightened concerns about the loss and disaster caused by cyber attacks.

Night Dragon: Night Dragon was a series of coordinated and targeted cyber attacks that were carried out in 2010 against several major energy companies around the world, including ExxonMobil, Marathon Oil, and ConocoPhillips. The attacks focused on stealing sensitive information related to oil and gas field exploration and production and the energy infrastructure. The attackers behind Night Dragon used a combination of tactics, techniques, and procedures (TTPs) to gain access to the targeted systems, including spear-phishing emails, malware, and social engineering. Once inside, the attackers could exfiltrate sensitive data, such as confidential documents and login credentials. The Night Dragon attacks were highly sophisticated and well coordinated, indicating the involvement of a well-funded and organized group. The attacks also demonstrated the significant threat that cyber attacks pose to the energy sector, a critical infrastructure sector with important economic and national security implications. To mitigate the risk of similar attacks, energy companies must implement robust cybersecurity measures, including regular vulnerability assessments, employee training, and incident response plans. They must also collaborate with government agencies and industry partners to share threat intelligence and best practices for cybersecurity.

Duqu, Flame, and Gauss: Hungarian cyber security researchers (2011) exposed three instances of malware: Flame, Duqu, and Gauss. Duqu was an information-stealing virus that could collect confidential information from HTTP. Flame was also a stealing virus like Duqu, but its target was microphones, key loggers, webcams, and others. Gauss virus could capture passwords, cookies, and browser history by intercepting sessions, network connections, BIOS, local, network, and USB drive.

5.5.4   ATTACKS ON SMART BUILDINGS

Smart buildings are structures that incorporate advanced technologies and automation systems to enhance the building's functionality, efficiency, and sustainability. These buildings have integrated sensors, controls, and communication networks that enable them to monitor and optimize energy usage, environmental conditions, and occupant comfort. Smart buildings use data analytics and machine learning algorithms to continuously adjust and optimize their operations. They can also interact with the occupants through various devices and platforms, providing real-time information and control over building systems. Recent years have seen multiple cases of cyberattacks on intelligent buildings. Attackers target the sensors, controls, and communication networks using online and offline attacks that cause significant harm to people. One such attack was executed in 2016 in Austria, where people were locked inside their rooms at a hotel in Austria until a ransom was paid [34]. Another such incident was in Finland, where a DDoS attack was performed on the heating system of a building during the winter [35]. The consequences would have been worse if the target had been critical structures such as hospitals, data centers, or government buildings. In 2018, a hacker disabled the cooling system of a pharmaceutical company [36]. This was an insider threat, and it could have been severe if not detected early. We need to utilize the knowledge and understanding of AI/ML to implement effective cybersecurity strategies to mitigate or reduce the potential risks associated with smart buildings.

5.5.5   ATTACKS ON WATER COMPANIES

New York Dam: In 2013, hackers penetrated a small dam in New York. However, this attack was not devastating, as it was a simple test by the hackers to check their power. The dam was controlled remotely by a utility, Bowman Dam, which was connected to the Internet. During the attack, the system was at maintenance and hence had no control features enabled, which the attackers exploited.

"Kemuri" Water Company: In 2016, the Verizon water company stated that a cyber attack had happened on its SCADA system. Verizon never disclosed its identity for security reasons, and it was known as "Kemuri" Water Company. The attackers got control over the PLCs and access to the district's water valve and flow control applications. The attackers slowed down chemical processing by altering chemicals used for purifying water. This, in turn, affected the water treatment and production process, causing delays in water supply recovery. The "Kemuri" breach could have been more dangerous if the attackers had had more time and information about the SCADA system.

German Steel Mill: In 2014, a German steel mill was attacked, and the attackers infiltrated the network interface of the steel plant and got control over the control systems. The attack caused multiple failures at different parts of the control systems. This attack forcefully controlled the blast furnace, preventing it from shutting down due to temperature increase, and this caused massive damage to the plant.

5.5.6 HEALTHCARE CYBER ATTACKS

Emsisoft, a cybersecurity company, reported 25 ransomware attacks on healthcare institutions in 2022, affecting as many as 290 hospitals nationwide, including "hospitals and multi-hospital health systems". Compared to 2021, when Emsisoft counted 68 attacks on healthcare providers, this figure has significantly decreased. But US hospitals and medical facilities remain a favorite target for threat actors. The industry was the one that ransomware attacks targeted the most during the third quarter of 2022, according to CheckPoint Research. Even some cybercriminals believe that attacking the healthcare industry is unethical because doing so frequently disrupts medical services and patients' health. Yet some ransomware groups are unaffected, given the potential profit in targeting healthcare. Ransomware actors often target lucrative data in hospital computer systems in addition to encrypting IT systems. Threat actors view the healthcare industry as a large terrain for monetization, as electronic healthcare record databases expand by the second.

Furthermore, due to technological advancements, cybercriminals can profit from endpoints by paying a ransom or using sensitive data from the dark web. According to Deryck Mitchelson, field CISO for Check Point Software Technologies in EMEA, "All cyberattacks result in significant financial gains; however, they value the healthcare information so much. On the dark web, this information is incredibly easy to commoditize". According to the threat detection vendor Sopho's "State of Ransomware in Healthcare 2022" report, 66% of the 381 healthcare companies were targeted by ransomware attacks showing an increase of around 50% from 2021. Healthcare computer systems are challenging to secure. Sometimes the appropriate precautions are overlooked to allow quick medical procedures, leaving data accessible to intruders.

As explained by Chester Wisniewski, Sophos's field CTO for applied research, "It is essentially a matter of life and death if you can't give someone their medication promptly in an emergency department or if you can't access a record to find out if they may have allergies". Systems are frequently left far more exposed for the ease of delivering care. Additionally, infosec specialists explained that the conditions in which medical personnel operate cause them to open emails rapidly, rendering them vulnerable to phishing assaults. Mitchelson added, "They tend to react and click rapidly. They frequently move around while working because they walk between tasks. Because of this, phishing emails are one of the most prevalent attack vectors".

5.5.6.1 Rail System Attack

On November 29, 2016, the San Francisco Municipal Transportation Agency (SFMTA) experienced a ransomware attack [37]. The hackers were able to get into the SFMTA system and deploy a virus that scrambled and encrypted the data on the computer device, making it inaccessible and unavailable. The data could only be decrypted using the key owned by the hackers, and they were demanding ransom for the digital key and for releasing their hold on the system. Per the information, SFMTA reported that the ransom attack had no impact on the transportation service system.

Per the reports, the fare machines inside the station were lit up with the message "OUT OF SERVICE" for the whole weekend. Passengers during the attack duration were able to travel for free. Per the information shared with news and media by the *Examiner*, a station agent's computer showed the message "You Hacked, ALL Data Encrypted. Contact For Key (cryptom27@yandex.com) ID:681, Enter". Someone in charge of the email address cryptom27@yandex.com claimed responsibility for the attack and demanded SFMTA hand over more than $70,000 in bitcoins, a digital currency. The email claimed SFMTA systems lacked adequate security and called the attack a proof of concept. The attacker could access 30 gigabytes of information, including employee data and contracts, which they threatened to release in the open media. Their ransom demands were not met.

Belarus Railway Hack: On January 25, 2022, activist hackers in Belarus [38] are said to have taken control of the systems that run the nation's trains and disrupted some rail services. According to the activist hackers known as Cyber Partisans, this action was taken to "slow down the transfer" of Russian forces invading Ukraine. According to Bloomberg, the attack served to give Ukrainians more time to fend off Russian aggression. According to the scheme, hackers commandeered the train system and switched it to "manual control", which "substantially slows down the movement of trains but does not generate emergencies". As a result, there were delays in the movement of trains in the towns of Osipovichi and Orsha. The encryption used by the hackers to breach the networks rendered the routing and switching equipment for the train system useless. It was also hard to buy tickets since many websites connected to Belarus's rail system displayed error signals. Bloomberg News could not independently verify the claims made by the hacker organization. Former employee Sergei Voitehowich of Belarus Railway Company reported that the traffic control systems had been repaired following the intrusion. Though the external train network websites weren't accessible, Voitehowich and other methods weren't working.

5.6 MITIGATION STRATEGIES FOR CYBER ATTACKS ON SMSs

Cyber attacks on SMSs cause catastrophic damage to machines and control systems. Since SMSs are considered remote machines, security is given the lowest priority. Most of these systems do not consider authentication, authorization, access control, and repudiation aspects, which make these systems vulnerable to attacks. Prominent denial of service and distributed denial of service attacks can disrupt a system and prevent it from acting. This may lead to a physical explosion or even loss of life. Many such incidents are caused due to lack of security in such SMSs. However, applying defense mechanisms or detection algorithms on such machines is challenging. This task differs from IT systems as OT needs human intervention until now for such security patching. Also, detection, prevention, or recovery steps need to be followed rigorously in such scenarios. Most of the security solutions of SMS industries follow conventional guidelines or policies which are never updated with new threats and vulnerabilities.

On the other hand, some dynamic detection and prevention techniques like cryptography, intrusion detection systems, and incident response systems are prevalent and need to be correctly implemented and maintained manually. This section focuses on the existing cyber attack mitigation techniques for SMSs and their strengths and weaknesses. These techniques are divided into two categories, *static* and *dynamic*. Static techniques follow the rules and regulations employed by the government and followed by companies, whereas dynamic processes follow the implementation of detection strategies using various methods.

5.6.1 CYBER SECURITY STANDARDS, GUIDELINES, AND REGULATIONS

One of the significant differences between IT and SMS security is that a machine can only perform with total efficiency by applying authentication and access control mechanisms. Thus, most smart manufacturing companies need to pay more attention to the regulations and security policies imposed by the organization. The National Institute of Standards and Technology (NIST) has proposed cybersecurity implementation policies, guidelines, metrics, and tools to establish a balance between security, performance, and reliability for smart manufacturing systems. These guidelines focus on measuring cyber security mechanisms' impact on intelligent machines' performance and reliability. The tools and metrics are designed to propose the appropriate approaches, security, and implementation best practices; predict the effect of security measures on the smart machines; and monitor the OT to ensure desired results. Following are some standards specifically designed for the secure and reliable operation of SMSs.

5.6.1.1 ISA/IEC 62443

The standard provides security best practices for electronically secure industrial automation and control systems (IACS). This is the benchmark standard followed in almost every smart manufacturing sector, such as smart buildings, electric power grids, smart healthcare, oil and gas, and chemical plants. The standard provides detailed guidelines for assessing cyber security performance on SMSs, business risk assessment methodologies, control systems, setting up a typical cyber security framework, and security life cycle for the manufacturers [39, 40].

5.6.1.2 ISO/IEC 27019–2013(2017 Revised)

The standard is specific to the energy utility industry. The standard provides general principles for information security management for process control systems in energy sectors. This includes controlling and monitoring the generation, distribution, and storage of electrical power, gas, and heat. The standard also provides the guidelines for process control of various energy utility systems, devices, and applications to maintain safety and protection of PLCs; monitor the programming logic and parameter values used in the apparatus and the work of the connected sensors and actuators; monitor data visualization, logging, and documentation tasks for process control; control and monitor network connectivity, application interfaces, and remote control technologies; and many more [41].

5.6.1.3   ISO/IEC 27033–1:2015

This is the standard for IT network security of devices, security management processes, services and interfaces, and end-user communication. The standard provides guidelines to identify network security risks and possible countermeasures, the basics of network security architecture, and an overall baseline for developing and designing quality network security structures with risk control capacity [42].

5.6.1.4   ISO/IEC 29180:2012

This standard provides a security framework for sensor networks which connects the manufacturing components to the IoT or cloud through information exchange [43].

5.6.1.5   IEC 61508

This standard for the electronics industry provides guidelines for all types of electrical and electronic components' functional safety [44].

5.6.1.6   IEC 61784

It is a standard for industrial communication networks. It specifies a set of safety and security guidelines applicable to communication in factory manufacturing and process control [45].

5.6.1.7   ISO/IEC 27000

This standard comprises all domains and provides an information security management framework. It gives with several means of securing information in organizations. The standard sets guidelines and best practices to ensure and manage critical assets like financial information, intellectual property (IP), and employee data [46]. The standard is designed to provide information security and risk management for organizations of all domains.

5.6.2   Cryptographic Techniques

Using cryptographic authentication for SMSs is challenging because it hinders the normal flow of machine operations, and thus performance degradation occurs frequently. Operators or staff handling smart machines do the critical management manually, like revoking keys, updating new keys, and so on, which could be a more logical fit for deploying various devices with unique functionalities. Thus, encryption and context-based access control are not widely used in the automation and manufacturing industry. Also, for smart manufacturing, keyspaces should be large enough to be used in multiple machines connected to the IoT. Moreover, interoperability among interfaces and systems and real-time availability of the processes are significant concerns for crucial exchange procedures. The most prominent drawback of encryption techniques is computational overhead, which must be optimized to maintain the performance standard of the machines. Automatic key revocation during an attack also needs to be addressed.

5.6.3   INTRUSION DETECTION SYSTEMS

The sensor nodes connecting industrial machines with the IoT are vulnerable due to their inherent characteristics like insufficient memory, battery, and computing power. To safeguard this part of the IIoT, individual sensor nodes are scanned to identify the occurrence of an attack. This is done by the intrusion detection system (IDS). However, implementing IDSs is also computationally and resource intensive. IDSs have different types: in an *anomaly-based IDS*, any abnormal behavior is identified as an attack; no abnormality indicates no attack. For this, the IDS is trained with the standard and *expected* behavior of the system, and any deviation from the desired behavior during real-time data scanning is termed an anomaly.

On the other hand, a *signature-based IDS* matches the incoming data with a predefined signature database and identifies the attack type. This is also called classification of attacks. The disadvantage of this type is that it can only encounter attacks that match the signatures. Any new attack cannot be identified. Another way IDSs are categorized depends on the deployment strategy. When deployed on a central point like a router or switch, it is called a *centralized IDS*, whereas installing an IDS on multiple hosts makes it a *distributed IDS*.

Implementing IDSs in SMSs is challenging and little explored due to limited data available for training. The efficiency of IDSs in securing smart manufacturing has also yet to be evaluated in terms of accuracy and reliability. Thus, there has been a gap between proposed IDSs and their real-time implementation. However, multiple kinds of literature show the use of IDSs in SCADA systems [47], embedded systems [48], wireless industrial networks [49], power transmission systems [50], fluid control systems, and electric power grids [51].

5.6.4   POST-INCIDENT MANAGEMENT

Incident response (IR) is one of the significant actions to be taken immediately after a cyber attack occurs in an organization. An expert team executes this cyclic process to neutralize the attack. Incident response teams start the recovery process when an attack is reported. This begins with interviewing the employees using the attacked systems, identifying machines' status and configurations, scanning the network resources, and recovering the designs from their normal state. The critical aspect of IR is to ensure an attack does not occur in the future, and thus, they generate a report for further maintenance of the attacked systems. However, IR is an expensive and continuous process. Depending on the organization's business needs and security policy, IR is deployed. Thus, it needs appropriate policies to be followed.

5.7   CASE STUDY: ML-BASED IDS FOR DoS ATTACK DETECTION

The mitigation strategies for cyber attacks on SMSs discussed in the previous section show the efficiency and difficulties in implementation. Among all these strategies, IDSs have a considerable scope.

In this section, a machine learning–based IDS is designed to detect DoS attacks on the transport layer on smart devices. The study shows a general IDS that can be easily applied on SMSs connected to smart devices through the Internet. The assumption is that the machines are connected to smart sensor nodes through the IoT, which collect data and send it to a server machine. During this communication, a three-way TCP handshake happens between the sender and the receiver:

- The sender sends a SYN request message to connect to the server
- The server opens a connection and sends a SYN-ACK to the sender to acknowledge the request.
- The receiver responds with an ACK message to establish the connection.

After sending a SYN-ACK message to the sender, the server waits for the ACK from the receiver to open a connection and thus remains busy with this until an ACK message comes for a specific time interval. It takes the server's energy to hold up the communication. Other nodes must wait until the server becomes free.

An attacker can manipulate this process to launch an attack. Here the attacker sends massive SYN requests to the server but never sends ACK messages. Thus, the server waits for the ACK until timeout, which consumes server energy. When a massive number of request messages come to the server, it gets busy processing these, causing the server energy to quickly drain. In parallel, other legitimate nodes remain in the queue to get server connections, which never happens, and finally, the whole system crashes.

5.7.1   EXPERIMENTAL SETUP

Two stages are required to detect SYN flood attacks using ML-based IDS: generate attack data and design IDSs using ML models for detecting anomalies in the dataset.

For the first stage, an IoT transport layer SYN-flood attack is designed using a NetSim simulator (Figure 5.2), generating trace files (.CSV file). The readers may use many other open source simulation tools for attack generation. These include Cooja, NS2, NS3, and many others. The network consists of ten legitimate nodes and three attacker nodes, and all the nodes are connected to a single ad hoc link. The routers and 6LowPAN Gateway are connected to the network through wireless. Sensor nodes can send data to the wired node/server through application–constant bit rate (CBR). Readers are suggested to explore the NetSim manual for further reading [52]. Nodes 2, 6, and 10 are the attacker nodes. After setting up the network scenario with the attack environment, traces are generated for taking multiple simulation times (50, 70, and 100 s). Similarly, a non-attack climate without attacker nodes is designed, and the simulation is run for 50, 70, and 100 s. From each simulation, one trace file is generated (we consider the trace for the 100-s simulation), and finally, attack and non-attack paths are merged to get a single trace file with 35 features and 195,049 rows.

After the dataset is generated, the next step is to clean the dataset and extract relevant features contributing to finding anomalies. In machine learning, multiple stages must be followed to develop the final dataset with all the relevant parts. The steps are as follows:

Data cleaning: The generated trace file with 35 columns and 195,049 rows is cleaned by omitting NA fields, missing values, and irrelevant fields, which yields 17 features from 35 features.

Converting categorical data to numerical data: In the next step, One Hot encoding is applied to convert categorical data (packet type, control packet type, and packet status in our case) to a numerical value. This yields 21 features.

Extracting correlated features: In the dataset, several components may have high correlations among themselves. When two or more features correlate, one can be predicted from another. If this happens, all these features have the same contribution toward the target variable, and only one should be considered. This helps us reduce the feature set's dimensionality, along with selecting more relevant and important features for the ML models. The current model uses the Pearson correlation technique to determine the highly correlated features (correlation coefficient taken as 0.9), reducing the dimensionality to 13 from 21. Thus, the final dataset consists of 14 features and 195,049 samples.

Adding target variable: The last step is to add the target column in the dataset, which represents the value of the target variable. Since this is an anomaly-based IDS, the target variable is 0 for no-attack/regular and 1 for attack or anomaly.

After the dataset is ready, the second stage is to design the IDS. The proposed IDS is based on ML models, which are used to predict the attack on an IIoT system. Machine learning-based IDSs can detect attacks from a large dataset with minimum effort and time. Also, the available supervised ML models apply to both binary and multiclass classification problems. Thus, the current study aims to design an anomaly-based IDS using supervised ML models.

Selecting ML models: At this step, selected ML models were trained with the dataset from the first stage. The selected ML models are random forest (RF), decision tree (DT), logistic regression (LR), K-nearest neighbor (KNN), gradient boosting (GB), and stack ensemble classifier (SC). Each model except SC is trained with the dataset, and the accuracy percentages are observed. To enhance the accuracy, all these models are ensembled by using SC. Ensembling is a technique used to improve the accuracy of the ML models. Table 5.2 shows that SC has a more significant training accuracy than the other models. Figure 5.4 shows a graphical representation of the exactness of each model in which SC has the highest training accuracy.

Test the model: After ensembling with SC, the next step is to train and test the model. Training is done with the final dataset from step 1, and testing is executed

TABLE 5.2

Accuracies of ML Models during Training

ML Models	SC	RF	LR	KNN	GB	DT
Accuracy	0.94806	0.93612	0.93299	0.93402	0.9476	0.9415

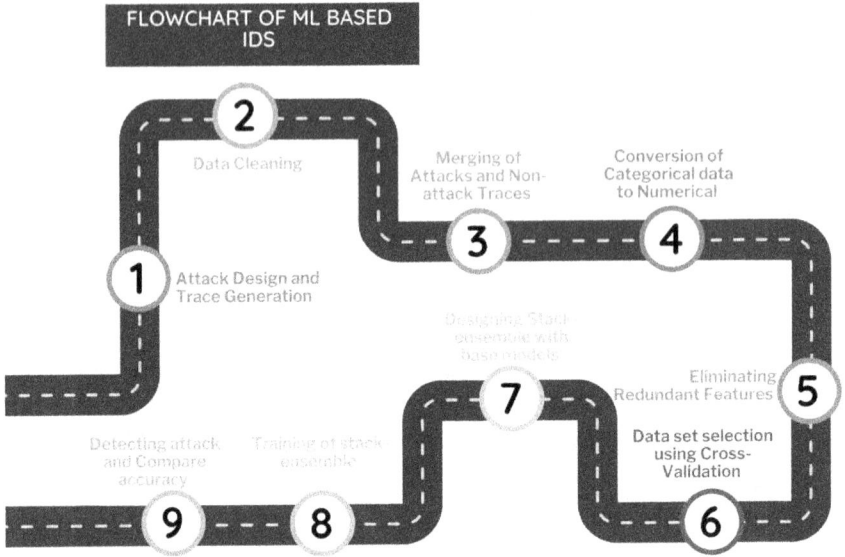

FIGURE 5.3   Flowchart of ML-based IDS for detecting SYN flood attacks on IoT.

on two more datasets (Dataset 1 with 97,000 rows and Dataset 2 with 80,000 rows) generated in the same way as stage 1 but with a varied number of attackers. The results of both datasets are shown in Figures 5.5 and Figure 5.6. Figure 5.3 shows a flowchart of the proposed ML-based IDS and its internal steps.

5.7.2   RESULT ANALYSIS

Figure 5.4 shows a comparative analysis of all the ML models' accuracies concerning the training dataset. The graph shows that SC is much higher than RF, LR, DT, and KNN. However, GB has a little lower accuracy than SC. Thus, enabling weaker ML models gives much better results, as stated in the previous section.

Figures 5.5 and 5.6 show the testing results of the SC and other models concerning test Datasets 1 and 2, respectively. Here also, we observe SC has a much larger accuracy than other models except GB, which has a little lower accuracy than SC.

Figure 5.7 compares the accuracy of testing Datasets 1 and 2. With a more significant number of test datasets, the model's efficiency can be verified with a wider variety of data.

Training accuracy-ML Models

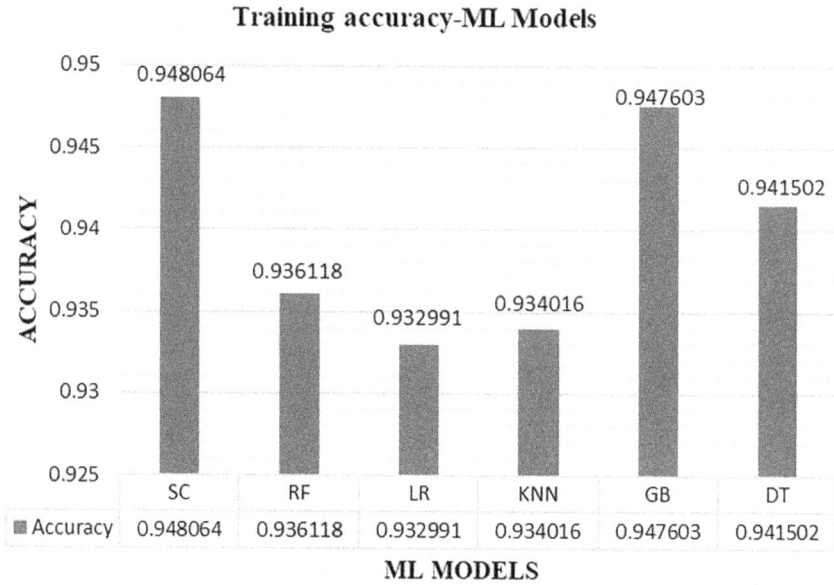

	SC	RF	LR	KNN	GB	DT
■ Accuracy	0.948064	0.936118	0.932991	0.934016	0.947603	0.941502

ML MODELS

FIGURE 5.4 Training accuracies of ML models.

Testing accuracy -Dataset1

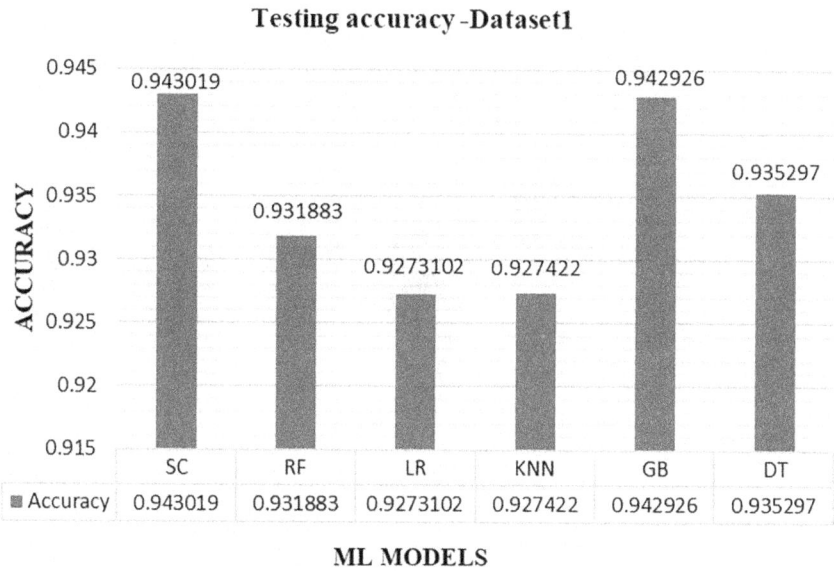

	SC	RF	LR	KNN	GB	DT
■ Accuracy	0.943019	0.931883	0.9273102	0.927422	0.942926	0.935297

ML MODELS

FIGURE 5.5 Testing accuracy concerning Dataset 1.

Testing Accuracy-Dataset2

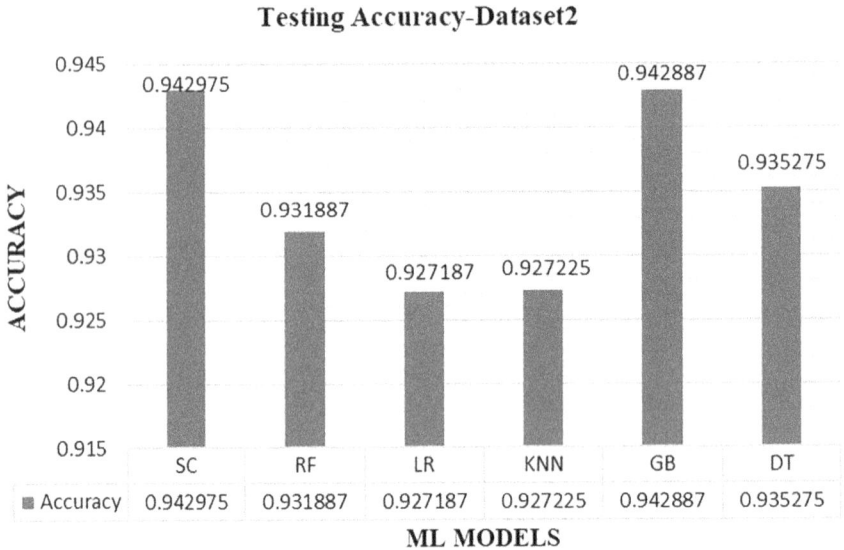

	SC	RF	LR	KNN	GB	DT
▪ Accuracy	0.942975	0.931887	0.927187	0.927225	0.942887	0.935275

ML MODELS

FIGURE 5.6   Testing accuracy concerning Dataset 2.

Accuracy comparison-Dataset1,Dataset2

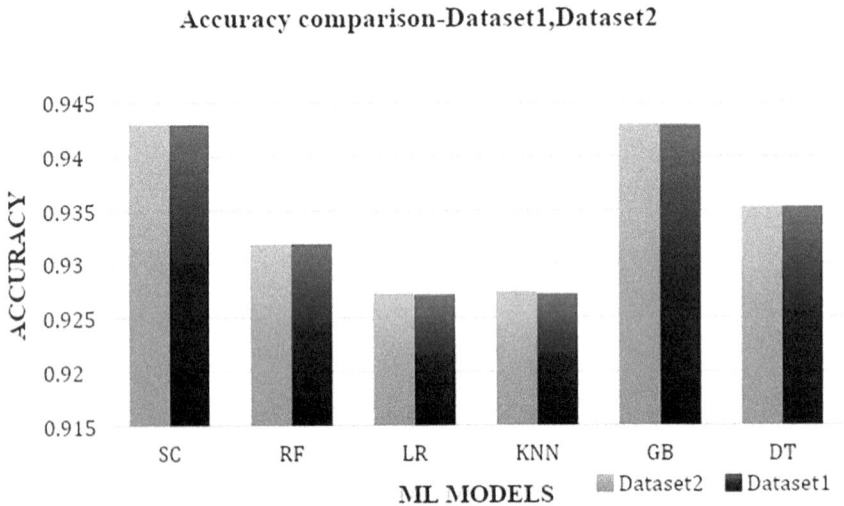

ML MODELS   ▨ Dataset2   ■ Dataset1

FIGURE 5.7   Accuracy comparison of Datasets 1 and 2.

5.8   CONCLUSION AND FUTURE SCOPE

This chapter gives a detailed discussion of cyber threats on smart manufacturing systems. The security issues and types of vulnerabilities of industrial automation, smart machines, and IIoT are represented. The difference between SMS and IoT threats are explained with examples, and how the security principles differ is also described. The most valuable part of this chapter are the case studies on cyber incidents happening in industrial automation, smart manufacturing, and IIoT to date. These case studies are examples of cyber attacks on SMSs and the types of vulnerabilities associated with these systems due to the connection with the Internet/cloud through smart objects. These case studies also show the threat landscape of cyber war worldwide. The chapter's last and most technical representation is the design and implementation of an ML-based IDS to detect DoS attacks like SYN flood attacks in IoT, which can disrupt SMSs. The proposed IDS is designed on ML models and has considerable detection accuracy. Although IDSs for SMSs are in the early stages, the scope of the detection technique is vast. IDSs can be implemented for any smart machine, provided data is available. Thus, some of the limitations of IDSs are also important to discuss for future researchers to help them understand the crux of IDSs.

- ML-based IDSs are efficient in detecting attacks on SMSs, provided datasets are available. In the case of the IoT, datasets are easily achievable through sensor reading or by simulation. However, it is challenging for machines to generate traces through simulation. Also, the traces may need to be sufficiently large for training.
- In the case of real-time systems, accurately identifying the anomaly is essential, which cannot be achieved by IDSs, as IDSs generate false alarms frequently.
- Implementing IDSs in smart machines is more complex than in smart devices. It may slow down the machine's performance, which causes a loss of time and money.

SMS security is crucial to implement due to a need for standards and interoperability. However, detecting anomalies is a general and secondary defense mechanism that can identify attack behavior in a system. But the accuracy and reliability of IDSs play essential roles in the context of reliable and robust defense mechanisms. ML-based IDSs are more promising than existing IDSs in the sense that these models can be trained with large amounts of data with a variety of attacks at once and can be used to detect anomalies and types of attacks with higher accuracy.

The ML-based IDS for detecting DDoS attacks on SMSs is more effective than other approaches like cryptographic methods. Because of the manual intervention required to create or update keys and the lack of storage capacity of the tiny sensor nodes attached to the SMSs, cryptographic techniques become more complex. In contrast, ML-based IDSs are more flexible and can be implemented at network hubs. This minimizes the implementation hazards, along with human intervention. The alarms generated by IDSs can be used by system administrators to either block or isolate victim nodes from data exchange with the critical component of the SMS. Also, a more significant number of attacks can be combined than a single attack to the traces, which can be used to train models for multiclass classification. This will be more powerful and efficient than designing a single attack detection at a time.

Despite the complex behavior and lack of flexibility issues, SMSs can become responsive to cyber security defense mechanisms. The future aspect of SMS security is inclined to the ML-based IDSs that can become a bridge between remote machines and IoT to identify anomalies and attacks efficiently.

REFERENCES

1. Sinha, S., & Paul, A. (2020). Neuro-fuzzy based intrusion detection system for wireless sensor network. *Wireless Personal Communications*, *114*, 835–851. https://doi.org/10.1007/s11277-020-07395-y

2. Carreras Guzman, N. H., Wied, M., Kozine, I., & Lundteigen, M. A. (2020). Conceptualizing the key features of cyber-physical systems in a multi-layered representation for safety and security analysis. *Systems Engineering*, *23*(2), 189–210. https://doi.org/10.1002/sys.21509

3. Liu, J., Yuan, C., Lai, Y., & Qin, H. (2020). Protection of sensitive data in industrial Internet based on three-layer local/fog/cloud storage. *Security and Communication Networks*, *2020*, 1–16. https://doi.org/10.1155/2020/2017930

4. Horak, T., Strelec, P., Huraj, L., Tanuska, P., Vaclavova, A., & Kebisek, M. (2021). The vulnerability of the production line using industrial IoT systems under DDoS attack. *Electronics*, *10*(4), 381. https://doi.org/10.3390/electronics10040381

5. Zheng, P., Wang, H., Sang, Z., Zhong, R. Y., Liu, Y., Liu, C., . . . & Xu, X. (2018). Smart manufacturing systems for Industry 4.0: Conceptual framework, scenarios, and future perspectives. *Frontiers of Mechanical Engineering*, *13*, 137–150. https://doi.org/10.1007/s11465-018-0499-5

6. Zografopoulos, I., Ospina, J., Liu, X., & Konstantinou, C. (2021). Cyber-physical energy systems security: Threat modeling, risk assessment, resources, metrics, and case studies. *IEEE Access*, *9*, 29775–29818. https://doi.org/10.1109/ACCESS.2021.3058403

7. Zarreh, A., Wan, H., Lee, Y., Saygin, C., & Al Janahi, R. (2019). Cybersecurity concerns for total productive maintenance in smart manufacturing systems. *Procedia Manufacturing*, *38*, 532–539. https://doi.org/10.1016/j.promfg.2020.01.067

8. Leander, B., Čaušević, A., Hansson, H., & Lindström, T. (2020, September). Access control for smart manufacturing systems. In *Software Architecture: 14th European Conference, ECSA 2020 Tracks and Workshops, L'Aquila, Italy, September 14–18, 2020, Proceedings* (pp. 463–476). Springer International Publishing.

9. Fruhlinger, J. (2017). What is Stuxnet, who created it and how does it work. *CSO Online*, *22*.

10. Karnouskos, S. (2011, November). Stuxnet worm impact on industrial cyber-physical system security. In *IECON 2011–37th Annual Conference of the IEEE Industrial Electronics Society* (pp. 4490–4494). IEEE. https://doi.org/10.1109/IECON.2011.6120048

11. Lee, R. M., Assante, M. J., & Conway, T. (2014). German steel mill cyber attack. *Industrial Control Systems*, *30*(62), 1–15.

12. Kumar, A., Singh, H., Kumar, P., & AlMangour, B. (Eds.). (2023). *Handbook of Smart Manufacturing: Forecasting the Future of Industry 4.0*. CRC Press. https://doi.org/10.1201/9781003333760

13. Wu, D., Ren, A., Zhang, W., Fan, F., Liu, P., Fu, X., & Terpenny, J. (2018). Cybersecurity for digital manufacturing. *Journal of Manufacturing Systems*, *48*, 3–12. https://doi.org/10.1016/j.jmsy.2018.03.006

14. Imani, F., Gaikwad, A., Montazeri, M., Rao, P., Yang, H., & Reutzel, E. (2018, June). Layerwise in-process quality monitoring in laser powder bed fusion. In *International Manufacturing Science and Engineering Conference* (Vol. 51357, p. V001T01A038). American Society of Mechanical Engineers. https://doi.org/10.1115/MSEC2018-6477

15. Virus outbreak at iPhone chip plant could delay shipments. (n.d.). *Computer Weekly*. ComputerWeekly.Com. Retrieved 18 March 2023, from www.computerweekly.com/news/252446344/Virus-outbreak-at-iPhone-chip-plant-could-delay-shipments.

16. Maggi, F., Quarta, D., Pogliani, M., Polino, M., Zanchettin, A. M., & Zanero, S. (2017). Rogue robots: Testing the limits of an industrial robot's security. *Trend Micro, Politecnico di Milano, Technical Report*, 1–21.

17. *Malicious Code in the PureScriptnpm Installer—Harry Garrood.* (n.d.). Retrieved 18 March 2023, from https://harry.garrood.me/blog/malicious-code-in-purescript-npm-installer/
18. Canonical GitHub account hacked, ubuntu source code safe. (n.d.). *ZDNET.* Retrieved 18 March 2023, from https://www.zdnet.com/article/canonical-github-account-hacked-ubuntu-source-code-safe/
19. PyPI Python repository hit by typosquatting sneak attack. (2017, September 19). *Naked Security.* https://nakedsecurity.sophos.com/2017/09/19/pypi-python-repository-hit-by-typosquatting-sneak-attack/
20. Tuptuk, N., & Hailes, S. (2018). Security of smart manufacturing systems. *Journal of Manufacturing Systems, 47,* 93–106. https://doi.org/10.1016/j.jmsy.2018.04.007
21. Wu, M., & Moon, Y. B. (2017). Taxonomy of cross-domain attacks on cyber manufacturing systems. *Procedia Computer Science, 114,* 367–374. https://doi.org/10.1016/j.procs.2017.09.050
22. Pan, Y., White, J., Schmidt, D., Elhabashy, A., Sturm, L., Camelio, J., & Williams, C. (2017). Taxonomies for reasoning about cyber-physical attacks in IoT-based manufacturing systems, *4*(3), 45–54. http://doi.org/10.9781/ijimai.2017.437
23. Wu, M., & Moon, Y. B. (2018, November). Taxonomy for secure cybermanufacturing systems. In *ASME International Mechanical Engineering Congress and Exposition* (Vol. 52019, p. V002T02A067). American Society of Mechanical Engineers. https://doi.org/10.1115/IMECE2018-86091
24. Mahesh, P., Tiwari, A., Jin, C., Kumar, P. R., Reddy, A. N., Bukkapatanam, S. T., . . . & Karri, R. (2020). A survey of cybersecurity of digital manufacturing. *Proceedings of the IEEE, 109*(4), 495–516. https://doi.org/10.1109/JPROC.2020.3032074
25. Antonakakis, M., April, T., Bailey, M., Bernhard, M., Bursztein, E., Cochran, J., . . . & Zhou, Y. (2017). Understanding the Mirai botnet. In *26th {USENIX} Security Symposium ({USENIX} Security 17),* (pp. 1093–1110). USENIX Association.
26. Labs, C. (n.d.). *Brickerbot: Mirai-like Malware Threatens to Forever Brick Insecure IoT Devices.* Cyware Labs. Retrieved 22 April 2023, from http://cyware.com/news/brickerbot-mirai-like-malware-threatens-to-forever-brick-insecure-iot-devices-64c02370
27. Eugenio, D. (2017, October 19). *A New IoT Botnet Storm Is Coming.* Check Point Research. https://research.checkpoint.com/2017/new-iot-botnet-storm-coming/
28. Sturm, L. D., Williams, C. B., Camelio, J. A., White, J., & Parker, R. (2017). Cyber-physical vulnerabilities in additive manufacturing systems: A case study attack on the. STL file with human subjects. *Journal of Manufacturing Systems, 44,* 154–164. https://doi.org/10.1016/j.jmsy.2017.05.007
29. Tuptuk, N., & Hailes, S. (2015, March). Covert channel attacks in pervasive computing. In *2015 IEEE International Conference on Pervasive Computing and Communications (PerCom)* (pp. 236–242). IEEE. https://doi.org/10.1109/PERCOM.2015.7146534
30. Song, C., Lin, F., Ba, Z., Ren, K., Zhou, C., & Xu, W. (2016, October). My smartphone knows what you print: Exploring smartphone-based side-channel attacks against 3d printers. In *Proceedings of the 2016 ACM SIGSAC Conference on Computer and Communications Security* (pp. 895–907). https://doi.org/10.1145/2976749.2978300
31. Knapp, E. D., & Langill, J. T. (2015). Chapter 7—Hacking industrial control systems. In E. D. Knapp & J. T. Langill (Eds.), *Industrial Network Security* (Second Edition, pp. 171–207). Syngress. ISBN: 9780124201149. https://doi.org/10.1016/B978-0-12-420114-9.00007-1.
32. *Major German Oil Supplier Confirms Cyber-attack—"Oiltanking" Says Incident Has Crippled Inland Supply.* (n.d.). https://thestack.technology/oiltanking-cyber-attack/
33. *Blackenergy Used as a Cyber Weapon against Ukrainian Critical Infrastructure.* (n.d.). Infosec Resources. https://resources.infosecinstitute.com/topic/blackenergy-used-as-a-cyber-weapon-against-ukrainian-critical-infrastructure/

34. Nast, C. (n.d.). Could hackers really take over a hotel? WIRED explains. *Wired UK*. www.wired.co.uk/article/austria-hotel-ransomware-true-doors-lock-hackers

35. Ashok, I. (2016). Hackers leave Finnish residents cold after DDoS attack knocks out heating systems. *International Business Times*.

36. dos Santos, D. R., Dagrada, M., & Costante, E. (2021). Leveraging operational technology and the Internet of things to attack smart buildings. *Journal of Computer Virology and Hacking Techniques*, *17*(1), 1–20. https://doi.org/10.1007/s11416-020-00358-8

37. Peterson, A. (2021, December 5). San Francisco's light-rail system was held hostage by hackers. *Washington Post*. www.washingtonpost.com/news/the-switch/wp/2016/11/28/san-franciscos-light-rail-system-was-held-hostage-by-hackers/

38. Nair, S. (2022, March 1). Belarus hackers attack train systems to disrupt Russian troops. *Railway Technology*. www.railway-technology.com/news/belarus-hackers-attack-train-systems/

39. ISA99: Developing the ISA/IEC 62443 series of standards on industrial automation and control systems (IACS). 2017http://isa99.isa.org/ISA9920Wiki/Home.aspx/.

40. ISA99: Developing the vital ISA/IEC 62443 series of standards on industrial automation and control systems (IACS) security. 2001http://isa99.isa.org/

41. ISO/IEC TR 27019:2013: Information technology—security techniques—information security management guidelines based on ISO/IEC 27002 for process control systems specific to the energy utility industry. 2013www.iso.org/standard/43759.html.

42. ISO/IEC 27033–1:2015 Preview Information technology-Security techniques—Network security—Part 1: Overview and concepts. 2015www.iso.org/standard/63461.html.

43. ISO/IEC 29180:2012 Information technology—Telecommunications and information exchange between systems—Security framework for ubiquitous sensor networks. 2012www.iso.org/standard/45259.html.

44. Functional safety and IEC 61508. 2010www.iec.ch/functionalsafety/

45. IEC 61784–1:2014: Industrial communication networks—Profiles—Part 1: Fieldbus profiles. 2014https://webstore.iec.ch/publication/5878

46. ISO/IEC 27000: Family—Information security management systems. 2014https://webstore.iec.ch/publication/5878.

47. Reeves, J., Ramaswamy, A., Locasto, M., Bratus, S., & Smith, S. (2012). Intrusion detection for resource-constrained embedded control systems in the power grid. *International Journal of Critical Infrastructure Protection*, *5*(2), 74–83. https://doi.org/10.1016/j.ijcip.2012.02.002

48. Shin, S., Kwon, T., Jo, G. Y., Park, Y., & Rhy, H. (2010). An experimental study of hierarchical intrusion detection for wireless industrial sensor networks. *IEEE Transactions on Industrial Informatics*, *6*(4), 744–757. https://doi.org/10.1109/TII.2010.2051556

49. Pan, S., Morris, T., & Adhikari, U. (2015). Developing a hybrid intrusion detection system using data mining for power systems. *IEEE Transactions on Smart Grid*, *6*(6), 3104–3113. https://doi.org/10.1109/TSG.2015.2409775

50. Linda, O., Vollmer, T., & Manic, M. (2009, June). Neural network based intrusion detection system for critical infrastructures. In *2009 International Joint Conference on Neural Networks* (pp. 1827–1834). IEEE. https://doi.org/10.1109/IJCNN.2009.5178592

51. Panthi, M., & Das, T. K. (2022). Intelligent intrusion detection scheme for smart power-grid using optimized ensemble learning on selected features. *International Journal of Critical Infrastructure Protection*, *39*, 100567.

52. https://www.tetcos.com/documentation/NetSim/v13/NetSim-Experiments-Manual/index.htm

6 Service Quality Improvement Using Fuzzy Inference Systems and Genetic Algorithm

Minoo Zarghami[1] and Taha-Hossein Hejazi[2]

1 Department of Industrial Management,
Caspian Higher Education Institute, Iran

2 Department of Industrial Engineering,
Amirkabir University of Technology (Tehran
Polytechnic) Garmsar Campus, Iran

6.1 HIGHLIGHTS

This chapter provides a mathematical model that considers real-world restricting factors for a cell phone service provider company, such as budget, workforce, and equipment limitations. These factors are analyzed collectively by assuming them as limitations on the total number of improvement plans to be developed during a specific enhancement process and the possible range of advancement for each individual variable. This chapter demonstrates which variables are critical in defining the excellence of cellphone customer service quality among several important variables.

The chapter also presents seven improvement strategies for the company's success after reaching certain optimized values. It distinguishes the most important variables from the least important ones to promote the efficiency of the plans.

This chapter aims to meet the need for managers to be well informed about their company's level of success and the appropriate amount of consideration and improvement required for every defining factor to advance their careers.

The research approach considers the shortcomings that act as barriers to business improvement, and the outcome directs managers to focus on goals and the importance of causal factors.

6.2 INTRODUCTION

The evaluation of service quality is based on what the client considers important during each interaction. Quality is determined by both what the client expects and what they actually experience. When the service meets the client's expectations, the quality is considered satisfactory, but if it surpasses their expectations, clients will be happy and consider the service quality exceptional. Conversely, if the service falls

DOI: 10.1201/9781003405870-6

short of expectations, the client will perceive the quality of the service as poor. It is difficult to measure service quality, as satisfaction depends on intangible elements rather than objective measurements of a tangible product. Thus, the value and importance of each factor that determines quality can be different based on the type of service [1].

The use of SERVQUAL aids in delivering reliable and consistent service quality that satisfies customer expectations. With the assessment of customer contentment and service quality, organizations can work towards enhancing the services they already provide, as well as creating new ones. Consequently, this leads to elevated customer satisfaction and improved outcomes. Furthermore, SERVQUAL assists organizations in identifying areas for improvement promptly and allows them to concentrate on improving service quality in these specific areas [2]. Here we will look at previous studies done in the quality of services field.

6.2.1 Literature Review

Several studies have contributed valuable findings that have helped develop useful techniques in evaluating service quality across various industries. For instance, [3] utilized VIekriterijumsko KOmpromisno Rangiranje (VIKOR) and PFNs arithmetic operators to assess citizens' satisfaction levels with municipality services. [4] applied the fuzzy TOPSIS method to evaluate the effectiveness of e-commerce supply chains. They identified crucial key performance metrics (KPMs) and assessed the metrics to evaluate the performance of these supply chains. In the context of VinFast, the leading Vietnamese automotive brand, [5] presented a comprehensive approach to assess the quality of automotive services. To address incomplete and ambiguous decision information, they initially applied the grey theory system (GTS) to approximate the subjective viewpoints of decision makers (DMs). Subsequently, they proposed a Grey-DEMATEL approach, integrating a decision-making trial and evaluation laboratory, to identify causal factors and quantify the relationship between the proposed criteria and capture criteria weights based on comprehensive analysis of service quality literature, expert opinions, and the current context. [6] introduced a model for decision-making based on fuzzy inference for predicting the impact of Supply Chain Operations Reference (SCOR) model's performance indicators on customer perceived value (CPV). The SCOR level 1 metrics adopt a multidimensional approach to assessing CPV. A summary of these techniques is presented in Table 6.1.

The aim of this research is to address these inquiries:

How can we evaluate the current situation of the office?
How can we extract the essential parameters?
How is the office's general performance?
What kinds of tools help us recognize and overcome deficiencies?
How can we consider the budget, workforce, and equipment limitations?

In this chapter, we aim to extract critical factors from experts to determine the most significant factor required for efficient cell phone service assessment. Additionally, our goal is to equip leaders with methods to assess the present condition of the firm

TABLE 6.1
Relevant Techniques Used by Previous Research

Literature	Topsis	Fuzzy Logic	Genetic Algorithm	VIKOR	Uncertainty	Grey-DEMATEL	Fuzzy Inference System
(Yildirim & Yıldırım, 2022)				✓			
(Praneeth et al., 2023)	✓	✓					
(Nguyen, 2022)						✓	
(Zanon et al., 2020)							✓
Current Chapter		✓	✓		✓		✓

through the simulation of non-quantitative data. The use of mathematical programming and optimization of values while considering real-world restrictions are two distinguishing features of this research.

Although we focused on financial, time, and workforce constraints, we were able to achieve beneficial improvement plans to implement in real industries using the optimization method of the genetic algorithm. A genetic algorithm was used among the listed techniques for evaluating service quality, and the most effective ones are detailed in the following section.

6.2.2 DATA

We conducted research to evaluate the influence of exceptional service on customer behavior by measuring the perceived value of service quality. In addition, we investigated the relationship between service quality and demand, as well as other factors that influence business decision-making.

In April 2017, we distributed study-based surveys to customers containing questions based on the SERVQUAL model. The surveys were handed out in person by three employees responsible for delivering and processing the surveys. Each survey contained 12 questions, and we received a total of 30 responses. The surveys were conducted in the organization's official setting after customers had received their services. However, due to some data limitations that we encountered during data collection, we had to make some adjustments to ensure the analysis was clear. These limitations are detailed in Table 6.2.

6.3 THEORETICAL FOUNDATIONS OF RESEARCH

In this section, we will present the key concepts that we focused on to accomplish the objective of this research. In the following subsections, we will demonstrate how these concepts are interrelated.

TABLE 6.2
Data, Type, Limitations

Data	Type	Limitation
Up-to-date equipment	5-point Likert scale	Individuals' expectations of technology
Tidy staff	5-point Likert scale	–
Response speed	5-point Likert scale	The time of day a customer uses services affects this variable
On-time services	5-point Likert scale	Breach of promise by other departments affiliated with this organization
Staff's responsibility	5-point Likert scale	–
Reliability	5-point Likert scale	–
Staff's behavior	5-point Likert scale	Personal taste affects this variable
Cost	5-point Likert scale	Market status affects service cost

6.3.1   SERVQUAL

The SERVQUAL model is widely recognized as an important tool for both academics and professionals for assessing and controlling service quality in various service sectors. Its primary objective is to measure customer satisfaction with an organization's service quality performance. The model is composed of the five dimensions of service quality: reliability, assurance, tangibility, empathy, and responsiveness, as well as attributes that identify the gaps between customer expectations and their perceptions of service performance. The model's comprehensive framework makes it a valuable tool for service providers to analyze and enhance their service quality, resulting in elevated levels of customer satisfaction and loyalty [7].

6.3.2   Fuzzy Logic

The procedure of making inferences based on natural language or common sense is referred to as fuzzy logic. The premise of this approach is a fuzzy set, which has an ambiguous and vague border and operates on vague definitions. An element can belong to a fuzzy set with a fractional level of membership, recognizing the potential for partial inclusion, such as the weather being "hot" to a degree of 0.8 or service being "poor" to a degree of 0.5, as illustrated in Figure 6.1. Intersection (AND), union (OR), and complement (NOT) are the most popular fuzzy logic operators [8]. Figure 6.1 illustrates the difference between fuzzy and clear (conventional) quality and emphasizes the need for fuzzy logic.

6.3.3   Fuzzy Inference Systems

When a machine cannot interpret certain mathematical data, fuzzy models come in handy. Non-fuzzy or traditional approaches necessitate a model that is organized and designed in a systematic manner with well-defined parameters. However, there

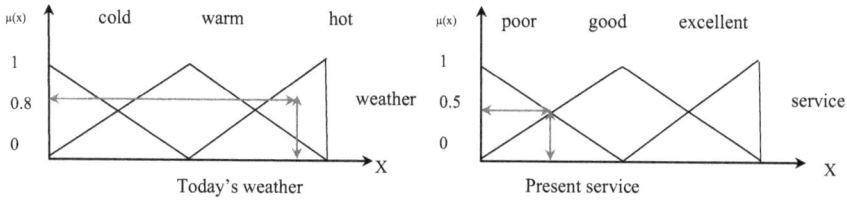

FIGURE 6.1 Membership function [8].

may be unknown variables, unanticipated dynamics, and other identified phenomena that cannot be predicted with the use of mathematical formulas. The most significant benefit of fuzzy modeling theory is its capacity to tackle a wide range of empirical issues that are not easy to explain sufficiently by the classic methods and procedures. The power of many experimental analyses has been the fuzzy modeling of nonlinear theories [9].

Mamdani-type [10] and Sugeno-type [11] fuzzy inference systems have been identified. According to their outputs, these two forms are distinct. Mamdani's efforts were recorded in a study [12] that established a fuzzy method designed for intricate systems. They differ in Sugeno's membership functions of outputs, which either are constant or linear. The Mamdani method produces an output that is a fuzzy set, which facilitates the evaluation of an order's satisfaction level using a continuum of values. Consequently, a Mamdani-type inference scheme is employed to elucidate the fuzzy rules [13].

The term "inference" denotes a mechanism that creates new data from existing one, and it is usually defined in an IF-THEN rule-based format. It explains incomprehensible equations in the way such that if we know a truth (premise, hypothesis, antecedent), we can obtain a different truth known as a conclusion; that is, "If x is a, the resulting outcome would be y as b".

Fuzzification of inputs is one of the steps involved in the fuzzy inference process. When a new value is added, it must be directed to one or multiple linguistic fuzzy sets containing membership values (Figure 6.2). Calculation of rule strength is the degree to which each part of the assumptions is fulfilled. For each rule, it will be determined when the inputs have been fuzzified. The rule strength of the accompanying rule is used to define the degree of a rule. When there are assumptions, the rule strength is determined by applying the standard min operator.

$$\mu_{R_i} = {}^\wedge\left\{\mu_A \sim R_i(x), \mu_B \sim R_i(y), \ldots\right\}, \text{ where } \mu_A \sim R_i(x), \mu_B \sim R_i(y), \ldots \qquad \text{Eq. 6.1}$$

The inputs x, y, and so on have membership values with respect to the premises $\hat{A}$, $\hat{B}$ of rule R_i, as shown. Therefore, the output is a single truth value for each rule, and this is the rule strength of the connected rule ranging from 0 to 1.

Fuzzy output: After computing the rule strength for each rule, the fuzzy output indicated by the rule is the region bounded by the line corresponding to the rule strength calculated by the standard aggregation operator.

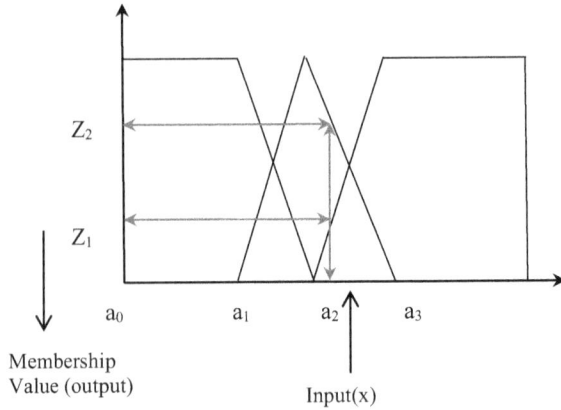

FIGURE 6.2 Fuzzification of inputs [13].

$$\vee\{\mu_A \sim (x), \mu_B \sim (y),...\}$$ Eq. 6.2

Defuzzification: The process of defuzzification converts fuzzy data into a precise value [13]. Mondal and Maiti present the centroid formula, which shows the center of the area under the curve.

$$\text{Output} = \frac{\int x\mu(x)}{\int \mu(x)}$$ Eq. 6.3

6.3.3.1 Definitions

The fuzzy expression $[0,1]^n$ to $[0,1]$, that is, f: $[0,1]^n \rightarrow [0,1]$ is a mapped n-dimensional fuzzy expression function.

Fuzzy logic formulas:

1. Real, binary values, and variable x_i ($\in$ [0.1], $i = 1, 2, ..., n$) are fuzzy.
2. If h is a fuzzy expression, then $\sim h$ is a fuzzy one, too.
3. If f and g were fuzzy, then $f \wedge g$ and $f \vee g$ would be fuzzy expressions.

where $\sim$ (negation), $\wedge$ (conjunction), and $\vee$ (disjunction) are used the same ways as in classical logic; hence, referring to Lukasiewicz, for $a, b \in [0,1]$ [13].

$$\text{Negation} \sim x = 1 - x$$

$$\text{Conjunction } x \wedge y = \text{Min } (x,y)$$

$$\text{Disjunction } x \vee y = \text{Max } (x,y)$$

Up to this point, we had hoped to discover methods for simulating the organization's output and defining non-numeric values. In the next section, we will discuss the implementation of the optimization tool we used after completing the mathematical modeling.

6.3.4 GENETIC ALGORITHMS

A genetic algorithm (GA) is a commonly utilized optimization method of complex and large-scale problems, aiming to find solutions that closely approximate the global optimum. Hence, a GA is an appropriate choice for training artificial neural networks. A classical genetic algorithm is a search algorithm that operates on a population and draws inspiration from natural selection. It is based on the concept of survival of the fittest. Through successive iterations, the algorithm generates new populations using operators on the existing individuals. The main components of GA include chromosome representation, selection, crossover, mutation, and fitness function evaluation.

The genetic algorithm performs the function of training the network. In this context, the bias of the network is represented by a chromosome, which can be seen as a vector of real values. The weights are encoded using an encoder component. Additionally, there exists a decoder that fulfills the opposite task of decoding chromosomes into network bias and weights. Since the GA requires a fitness function, the mean absolute error (MAE) is computed, incorporating the training data errors, as depicted in Equation 7.3. The fitness function is computed in Equation 6.4 [14].

Nonlinear programming problems with several minimum points may have multiple feasible setbacks that can lead to viable solutions. Although dependent algorithms can address such issues, they are often inefficient when dealing with multiple minimum points. Genetic algorithms are one of the most effective and thorough methods for extracting appropriate solutions by mimicking biological evolution. In a GA, a possible solution to a problem is represented as a genetic construction of a chromosome, and it can be interpreted as a person who undergoes genetic evolution. During genetic evolution, outstanding-quality offspring are generated from a population of chromosomes through the selection procedure of crossover and mutation. Strong individuals become superior solutions to the dilemma in a competitive world and are passed down through successive generations.

Solutions are consistently selected using an appropriate measure of the fitness of the objective function value. Genetic algorithms can be tailored to advance solutions for various models by using mapping techniques. GAs have gained popularity, surpassing the industrial engineering area approach, with the combinatorial optimization community taking note. While GAs perform well in a global search, they take a long time to advance to a local optimum. Local progress procedures may find the local optimal solution in a specific area of the space, but in a global search, they are essentially useless.

GAs discover the response through various operations such as pick, cross, mutation, and so on. The parents are chosen randomly from the current population to produce children via genetic operations. To increase the likelihood of discovering local GA optimization, the cross operator converts the genes of two parents and breeds offspring so that the children inherit characteristics from each parent. The mutation operation performs on incipient children to avoid the convergence of all solutions to their respective local optima.

To bring about the evolution of each chromosome in the GA, the crossover operation and mutation operation employed in the current GA depend on reliable operators. The proposed algorithm is specifically designed to solve nonlinear optimization problems with fuzzy relation constraints, distinguishing itself from a generic genetic algorithm. [15].

6.3.4.1 The Procedure of GA
The genetic algorithm flow chart is shown in Figure 6.3.

6.4 PROPOSED METHODOLOGY

This section illustrates the various stages of the research. First, data were collected from customers of a prestigious Iranian mobile network operator, which were then used to simulate the organization's performance. The details of the procedure will also be discussed in this section. The entire research process is depicted in Figure 6.4.

The first box represents a successful model that was used to collect data from customers. The surveys consisted of several metrics that could be used to evaluate an organization's service quality in various ways. The next subsection discusses the fuzzy inference system's inputs based on the survey results. These inputs are divided into two groups: one presents the inputs of an optimization function that will be used in mathematical modeling and genetic algorithm, while the other provides a prediction of the output. Finally, we obtained improvement plans and optimized values that led to the highest level of service quality. The following are some of the most notable advantages of this work:

- Estimation of non-quantifiable values that can be quantified.
- Simulation and assessment of the current state of the office.
- Optimization of crucial factors influencing service quality.
- Considering the limitations that had been imposed on the office.
- Providing development plans to advance one's career.

6.5 CASE STUDY

Given the intense competition among industries and professions, it is crucial to use various evaluations to stay ahead of competitors. Therefore, it is essential to conduct a thorough data analysis to identify the current state of the business. In the initial stage of the assessment, we identified the factors that have a significant impact on service quality. We consulted with experts and carefully selected the critical variables.

6.5.1 Fuzzy Variables and Analysis

Fuzzy logic is a branch of artificial intelligence (AI) that aims to account for uncertainty in machine-generated models by incorporating concepts of truth and falsity from common logic. To create a fuzzy logic model, it is essential to follow three steps: (1) fuzzification, which entails inputting linguistic variables and membership functions; (2) rule evaluation, where fuzzy logic rules are employed to determine the output variable value based on the input variables; and (3) defuzzification, where the output is transformed into a clear outcome by a fuzzy inference system (FIS). Mamdani's fuzzy inference system and the Takagi–Sugeno fuzzy model (Sugeno) are

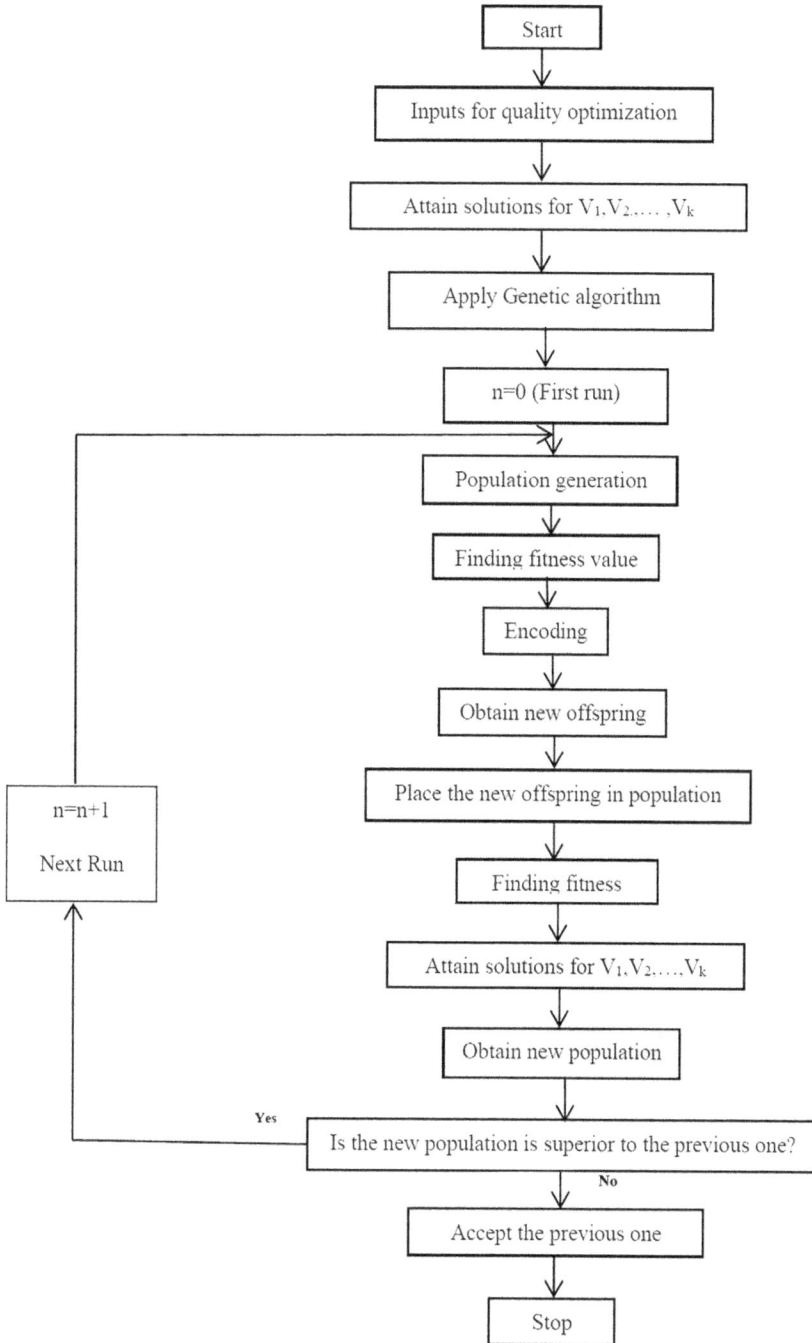

FIGURE 6.3 Genetic algorithm flow chart [15].

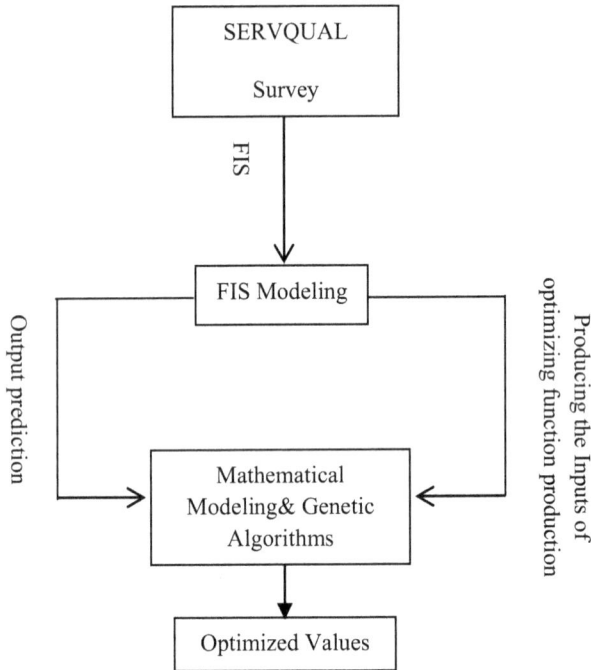

FIGURE 6.4 Research's total cycle.

the two standard types of FIS, with the latter being more computationally efficient. Previous investigations have looked into fuzzy models that utilize data gathered from in-vehicle sensors, including studies that analyzed the correlation between driving behavior and fuel consumption by assessing diverse drivers' performance [16].

MATLAB is a powerful software tool that is widely used for scientific and engineering computations. One of the key features of MATLAB is its ability to provide a fuzzy inference framework and modeling, making it a popular choice for solving fuzzy programming problems. In this chapter, the SERVQUAL survey was used to gather data on eight important variables related to quality services. These variables were identified by experts and applied to real-world situations to obtain accurate perceptions of service quality. MATLAB was used to analyze the data and create an optimization function that could be used to improve service quality. The following are the variables:

V_1 = up-to-date equipment
V_2 = tidy staff
V_3 = response speed
V_4 = on-time services
V_5 = staff's responsibility
V_6 = reliability

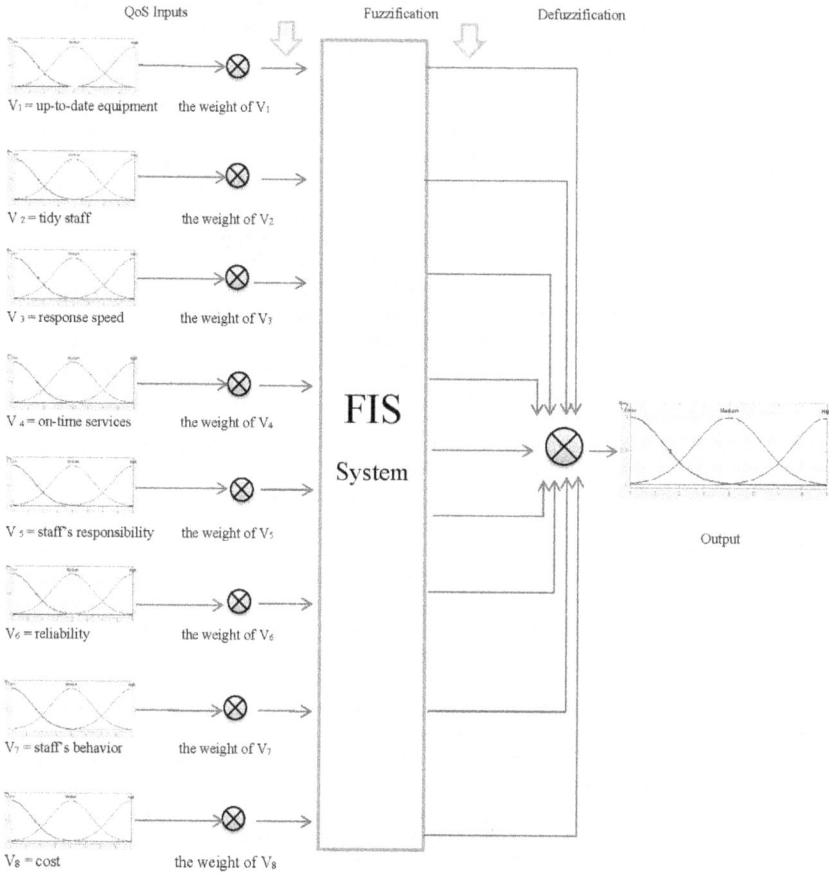

FIGURE 6.5   Fuzzy inference system phases.

V_7 = staff's behavior
V_8 = cost

The 30 surveys were scrutinized, and relevant data were entered into the software as describing rules. The process of the FIS is depicted in Figure 6.5.

The simulated program can predict performance based on additional inputs and given rules. Both weights are equal in the FIS (Figure 6.5). Figure 6.6 shows a surface view of an instance of its efficiency. It demonstrates the connection between V_1, V_8, and the output.

After obtaining a performance score of 6.71 from the survey results, which is in the medium-high range of quality, we attempted to optimize it. This section focuses on programming an optimization function and detecting the improved values of variables to achieve a significant result that contributes to improved service quality.

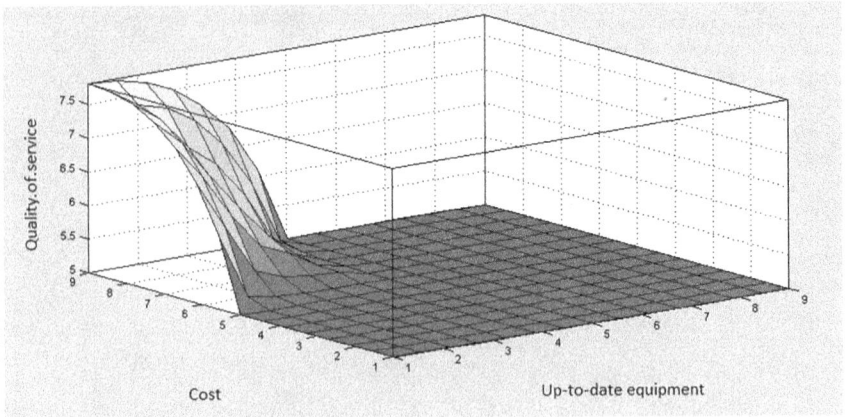

FIGURE 6.6   Surf view of the randomly selected input with output.

6.5.2   MATHEMATICAL MODELING

$$\text{Max Fis}(x)$$

$$\text{St:}$$

$$x_i \leq \mu + \delta_i Z_i \rightarrow \forall_i = 1,2,...,n$$

$$\mu_i \leq x_i \leq 9 \rightarrow \forall_i = 1,2,...,n$$

$$\sum_{i=1}^{n} Z_i \leq \cdots, Z_i \cdot \text{Integer} \cdots, \cdots \forall_i = 1,2,\cdots,n$$

The fuzzy inference inputs derived from SERVQUAL are represented by the letter X. Fis(x) shows the result produced by a fuzzy inference system given a specific input. The average of X_i, found in the surveys, is represented by μ_i. The number of improvement plans for X_i is indicated by Z_i, and the sum of the improved X_i for any improvement plans is represented by δ_i (which is equal to 2). Last, K represents the maximum number of appropriate improvement plans that can be implemented within a planning process (which is equal to 7).

Our goal faced some restrictions. The first specifies that the value of the ith variable should be less than the sum of the observed average of survey values and the number of improvement plans times the amount of improvement for each plan. It is crucial to point out that X is the decision variable. The second constraint we face is that the values of variables should be within the range of their actual values and survey results. The ideal value, on the other hand, is 9. The third constraint shows that the sum of improvement plans should be less than or equal to K.

The parameter δ_i is related to the improvement plan, and it is important to note that each improvement plan aims to increase the value of the decision variable by two units. This increase is intended to reduce the gap between the real performance and the optimized performance.

The following graph was generated after performing data analysis. In the initial stages of the process, the quality of service was valued between 0 and −1, but these values are invalid. In subsequent generations, the system was able to find a valid range of responses, reaching a high of 7, which indicates an acceptable value for this parameter. After exploring 229 generations of responses, a value of 7.5584 was ultimately reached, as shown in Figure 6.7. The actual and optimum values will be compared in Table 6.3.

6.6   STUDY RESULTS

In this chapter, we have determined the optimal values for various variables that significantly contribute to improving service quality. Through the software's data exploration, no improvement plan was found for the first variable, "up-to-date equipment", while one improvement plan was considered for each of the remaining variables.

6.6.1   ACHIEVEMENT

These results indicate an optimized performance of 7.5584, which represents satisfactory service quality, as stated in the previous section. If the company implements and follows through with the improvement plans, customer loyalty and resulting profitability would will increase significantly. This chapter provides valuable insights for managers who aim to advance their careers and achieve financial success. The

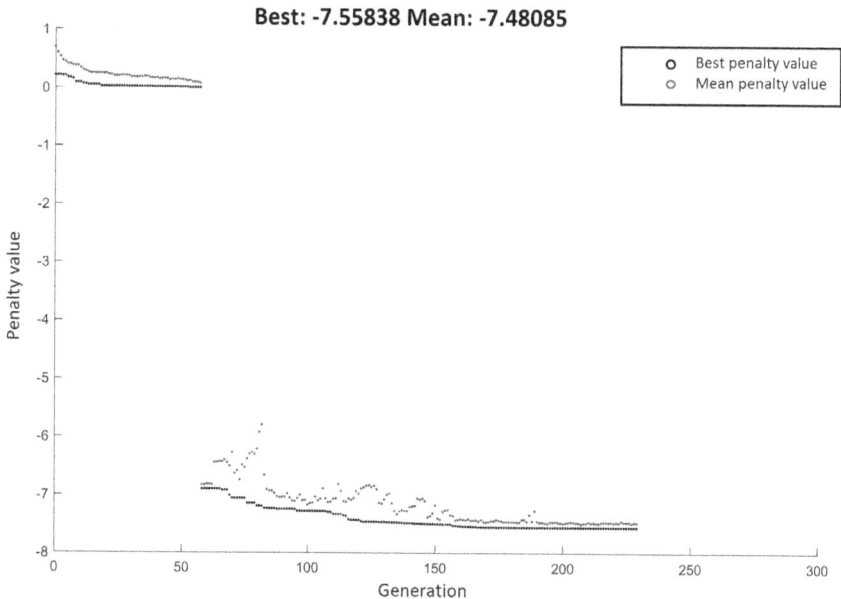

FIGURE 6.7   Generation-output values of optimization process.

TABLE 6.3
Real vs. Optimized Values

Number of Variables

Variable	V_1	V_2	V_3	V_4	V_5	V_6	V_7	V_8
	Up-to-date equipment	Tidy staff	Response speed	On-time services	Staff's responsibility	Reliability	Staff's behavior	Cost
Perceived Value	5.06	7.12	5.92	6.18	6.53	5.19	7.46	6.84
Optimized Value	5.8	8.13	7.8	7.87	7.83	6.19	8.73	7.15
K (number of improvement plans)	0	1	1	1	1	1	1	1
Difference	0.72	1.01	1.88	1.69	1.3	1	1.27	0.31
Importance Rate (Max = 1, Min = 8)	8	5	1	2	3	6	4	7

assessment of the organization's current state, which serves as the foundation for optimization, is the first and most critical step. The proposed strategy offers several advantages, including its realistic applicability and ability to provide solutions in the face of limitations or a lack of quantitative data. Additionally, its application is not restricted to a particular sector, making it useful in various professions.

Future researchers should explore additional limitations that managers may face and implement more development plans over an extended period.

6.6.2 Managerial Implications

In an era where companies are all competing to be the most productive, it is not a futile effort to focus on both external and internal variables simultaneously. In general, this chapter recommends that managers concentrate on assessing their current official situation and diagnosing their flaws. The mathematical model also provides an understanding of quality optimization that can help them advance their careers.

The results highlight the importance of response speed (V_3) in determining the efficiency of an organization's services. As shown in the last row of Table 6.3, speed of response has been ranked at the top of the list of decision factors. This suggests that if managers aim to improve their service quality, the first step they should take is to improve their response time. However, it should not be overlooked that the current state of an organization also plays a crucial role in establishing the sequence of these factors. The subsequent steps to improve service quality will involve addressing "on-time services" (V_4) and "staff responsibility" (V_5).

REFERENCES

[1] Stefano, N., Casarotto Filho, N., Barichello, R. & Sohn, A. (2015). A fuzzy SERVQUAL based method for evaluated of service quality in the hotel industry. *Procedia CIRP*, 30, 433–438, https://doi.org/10.1016/j.procir.2015.02.140

[2] Manunggal, B. & Afriadi, B. (2023). SERVQUAL in higher education institutions. *International Journal of Business, Law, and Education*, 4(1), 107–114, https://doi.org/10.56442/ijble.v4i1.132

[3] Yildirim, B.F. & Yıldırım, S.K. (2022). Evaluating the satisfaction level of citizens in municipality services by using picture fuzzy VIKOR method: 2014–2019 period analysis. *Decision Making: Applications in Management and Engineering*, 5(1), 50–66, https://doi.org/10.31181/dmame181221001y

[4] Praneeth, B.B., Nadeem, S.P., Vimal, K.E.K. & Kandasamy, J. (2023). Performance measurement of e-commerce supply chains using BWM and fuzzy TOPSIS. International *Journal of Quality & Reliability Management*, ahead-of-print(ahead-of-print), https://doi.org/10.1108/IJQRM-03-2022-0105

[5] Nguyen, P. (2022). Automotive service quality investigation using a grey-DEMATEL model. *Computers, Materials & Continua*, 73, 4779–4800, http://dx.doi.org/10.32604/cmc.2022.030745

[6] Zanon, L.G., Arantes, R.F.M., Calache, L.D.D.R. & Carpinetti, L.C.R. (2020). A decision making model based on fuzzy inference to predict the impact of SCOR® indicators on customer perceived value. *International Journal of Production Economics*, 223, 107520, https://doi.org/10.1016/j.ijpe.2019.107520

[7] Lee, C.-H., Zhao, X. & Lee, Y.-C. (2019). Service quality driven approach for innovative retail service system design and evaluation: A case study. *Computers & Industrial Engineering*, 135, 275–285, https://doi.org/10.1016/j.cie.2019.06.001

[8] Rowlands, H. & Wang, L.R. (2000). An approach of fuzzy logic evaluation and control in SPC. *Quality and Reliability Engineering International*, 16(2), 91–98, https://doi.org/10.1002/(SICI)1099-1638(200003/04)16:2%3C91::AID-QRE307%3E3.0.CO;2-9

[9] Eftekhari, M. & Katebi, S. (2008). Extracting compact fuzzy rules for nonlinear system modeling using subtractive clustering, GA and unscented filter. *Applied Mathematical Modelling*, 32(12), 2634–2651, https://doi.org/10.1016/j.apm.2007.09.023

[10] Mamdani, E.H. & Assilian, S. (1975). An experiment in linguistic synthesis with a fuzzy logic controller. *International Journal of Man-machine Studies*, 7(1), 1–13, https://doi.org/10.1016/S0020-7373(75)80002-2

[11] Banga, V., Singh, Y. & Kumar, R. (2007). Simulation of robotic arm using genetic algorithm & AHP. *World Academy of Science, Engineering and Technology*, 25(1), 95–101.

[12] Bellman, R.E. & Zadeh, L.A. (1970). Decision-making in a fuzzy environment. *Management Science*, 17(4), B-141-B-164, https://doi.org/10.1287/mnsc.17.4.B141

[13] Chakraborty, N., Mondal, S. & Maiti, M. (2013). A deteriorating multi-item inventory model with price discount and variable demands via fuzzy logic under resource constraints. *Computers & Industrial Engineering*, 66(4), 976–987, https://doi.org/10.1016/j.cie.2013.08.018

[14] Malik, S., Tahir, M., Sardaraz, M. & Alourani, A. (2022). A resource utilization prediction model for cloud data centers using evolutionary algorithms and machine learning techniques. *Applied Sciences*, 12(4), 2160, https://doi.org/10.3390/app12042160

[15] Khorram, E. & Hassanzadeh, R. (2008). Solving nonlinear optimization problems subjected to fuzzy relation equation constraints with max–average composition using a modified genetic algorithm. *Computers & Industrial Engineering*, 55(1), 1–14, https://doi.org/10.1016/j.cie.2007.11.011

[16] Almadi, A.I., Al Mamlook, R.E., Almarhabi, Y., Ullah, I., Jamal, A. & Bandara, N. (2022). A fuzzy-logic approach based on driver decision-making behavior modeling and simulation. *Sustainability*, 14(14), 8874, https://doi.org/10.3390/su14148874

7 Identifying and Prioritizing Quality 4.0 Practices in Sustainable Manufacturing Using Rough Number-Based AHP-MABAC

Rana Basu[1], Manik Chandra Das[2], Ajay Kumar[3], and Bijan Sarkar[4]

1 School of Management Sciences, Indian Institute of Engineering Science and Technology, Shibpur, Howrah, India

2 Department of Industrial Engineering and Management, Maulana Abul Kalam Azad University of Technology, West Bengal, India

3 Department of Mechanical Engineering, School of Engineering and Technology, JECRC University, Jaipur, Rajasthan, India

4 Department of Production Engineering, Jadavpur University, Kolkata, India

7.1 INTRODUCTION

Quality management (QM) plays a pivotal role in manufacturing processes as it delivers reliable products/services according to customer requirements [1]. With the digital transformation, focus has been shifted to the fourth industrial revolution (Industry 4.0), in which emphasis is given more on emerging technologies such as cyber-physical systems (CPS), artificial intelligence (AI), robotics, and additive manufacturing. Since technology alone cannot provide expected benefits and improvements, it is imperative to have a management system and processes to realize the potential of these digital technologies. In any manufacturing operation, the quality of the manufactured product is a critical component for robustness and consistency of the manufacturing process. The Industry 4.0 (I4.0) ecosystem has the potential to improve the quality of the product; hence, it is important to conduct further research

DOI: 10.1201/9781003405870-7

to link I4.0 with quality management and engineering [2]. Today, advancement of digital technologies from the I4.0 context has transformed the way of doing business, which not only requires a change in organizational culture but also the ability to adapt to new business practices, techniques, and tools. Adopting Industry 4.0 could lead organizations toward success in product innovation, productivity improvement, and product development, which would eventually impact sustainable manufacturing. This is where Quality 4.0 (Q4.0), which is steadily gaining popularity across the globe, comes into play. One of the areas that have significant influence on digitalization is QM. To create improved quality products and services, it is necessary that traditional functions of QM incorporate various concepts of Industry 4.0. Thus, Q4.0 can be referred to as the "digitalization of quality by design, quality of conformance and quality of performance using modern technologies" [3]. The core concept of Q4.0 is to align the QM practices with Industry 4.0 so that organizations can reap the benefits of digitalization to achieve world-class competencies in sustainable manufacturing. In other words, Q4.0 can be referred to as a modern form of quality management that is characterized by digital technology practices, sophisticated methods, and smarter processes. It encapsulates several innovations to develop sustainable manufacturing operations. In Industry 4.0, Q4.0 practices have the potential for sustainable value creation from triple bottom line (TBL) perspectives by improving efficiency of resource in manufacturing systems [4].

Today, India is accelerating forward to become the global manufacturing hub where digital tools and techniques such as ML (machine learning), 3D printing, additive manufacturing, cloud computing, data analytics, AI, and IoT (internet of things) are considered fundamental to industrial growth. In India, large-scale organizations are more competitive because of their advancement in manufacturing technology, driven by automation and robotics, which facilitates many large-scale organizations to partake in global value chains. By integrating with Industry 4.0, even forerunner firms are now migrating to vertically combine their manufacturing systems and processes. Manufacturing companies are facing huge pressure to improve the sustainability in their operations; however, there is a lack of clear understanding as to how these organizations should improve their sustainability from the perspective of people, process, and technology. To visualize manufacturing sustainability in the context of Industry 4.0, it is imperative to indoctrinate quality consciousness in both developed and developing economies, across industry verticals, and in all the functional areas of supply chain [5–7 as cited in 8]. Quality cannot depend on technology alone; aspects from people, processes, tools and techniques, methodologies, and analytical thinking/problem-solving approach are also considered a critical part of present quality revolution.

Based on practice and prior studies, it is evident that traditional approaches of quality management help in error detection with delay in decision making for taking corrective actions [9, 10]. New complexities in production processes are also prevalent with the growing models of mass production and customization with issues related to sustainability that demand new approaches of QM practices. To solve these issues, Industry 4.0 allows integrated model of combining QM with technological processes for managing various aspects of quality [9]. In this juncture, it is vital to comprehend how to accomplish and improve the quality processes in Industry 4.0 environment. More specifically, it is important to identify the critical Quality 4.0 practices and its prioritization for ensuring sustainable manufacturing in the Indian context. Research studies of Industry 4.0 is available with both conceptual and empirical works; however, empirical works pertaining to Quality 4.0 are still in their nascent stage. Thus, it is imperative to carry out further empirical research for the identification of Quality 4.0 practices in Industry 4.0 milieu and its prioritization in the context of Indian sustainable manufacturing perspective. This chapter proposes a framework of Quality 4.0, using a multi-step filtering approach as described in Section 8.4, to explore and identify relevant Quality 4.0 practices and its prioritization using a novel multiple criteria decision making (MCDM) method integrating analytic hierarchy process (AHP) and multi-attribute border approximation area comparison (MABAC) method under a rough environment (RAHP-RMABAC).

7.1.1 Purpose/Objective of the Study

The proposed framework addresses the following objectives:

- To explore and identify the critical Quality 4.0 practices for sustainable manufacturing in the context of Indian manufacturing organizations.
- To evaluate and prioritize the critical Quality 4.0 practices using the multiple criteria decision making (MCDM) method integrating analytic hierarchy process (AHP) and multi-attribute border approximation area comparison (MABAC) method under a rough environment (RAHP-RMABAC).

7.2 THEORETICAL BACKGROUND

7.2.1 Quality 4.0: Definitions and Characteristics

The concept and deployment of Quality 4.0 (Q4.0) is still in its nascent stage [11, 12]. Practical and theoretical understanding is necessary to ensure its implementation. Q4.0 is all about managing quality with reference to Industry 4.0. Therefore, Q4.0 can be construed as a combination of advanced technologies pertaining to industry 4.0 and conventional QM systems, tools, and methods to accomplish operational excellence and innovation [13, 14]. According to the American Society for Quality (ASQ) "Q4.0 brings together Industry 4.0's advanced digital technologies with quality excellence to drive substantial performance and effectiveness improvements" [14]. Antony et al. [15 cited in 12] defines Q4.0 as

> the use of advanced technologies such as IoT, CPS, cloud computing to design, operate and maintain adaptive, predictive, self-corrective, automated quality systems along with improved human interaction through quality planning, quality assurance, quality control and quality improvement to achieve new optimums in performance and operational excellence.

Dias et al. [16] consider Q4.0 as the delivery of superior level of quality by enhancing the competencies of people, quality tools, and techniques. Radziwill [17] has specifically denoted Q4.0 as the "digital empowerment of all the stakeholders of the process for dynamic quality enhancement and sustainment." It is categorically mentioned that the evolution of Q4.0 begins based on traditional quality management approaches and methods as a blending of "people, process and technology" [14].

Q4.0 is considered the fusion of generic QM practices with novel procedures and technologies such as "cloud, machine learning, connectivity devices, big data, artificial intelligence and internet of things" [18]. Jacob [19] defined Q4.0 as leveraging traditional approaches of quality driven by latest technologies to gain competitive advantage at both functional and operational levels. The author also acknowledged that manufacturing organizations that embraced Q4.0 have reached the superior level of quality along with increased market share and innovation and have overcome challenges in supply chain networks with enriched recognition in branding. It is highly recommended that organizations adopt Q4.0; otherwise, business would be under risk [20]. Allcock [21] highlighted the difference between traditional approach of quality and Q4.0, and stated to be a paradigm shift from manual to automated activity with application of sensors, advanced software-based application for controlling process for auto-tuning. Further, Q4.0 is designated as real-time integration and harmonization of data in manufacturing processes with the combination of quality management [22]. Once adopted, it has the capability to connect with real-time systems for monitoring, analyzing, and controlling the value chain for corrective and preventive actions [23]. Based on an exhaustive compilation of literature, a comprehensive, balanced, and hybrid definition has been proposed for Q4.0, which states that it as "an extended approach to quality management, where recent technologies are being integrated with traditional quality practices in order to expand the quality management scope and to improve quality activities' performance and efficiency" [23].

Quality 4.0 pulls benefits by means of cyber-physical systems (CPS) that connect people, machines, and data in novel way. By using advanced technologies, it enables transformation of culture, collaboration, leadership, and compliance. Q4.0 certainly embraces digitalization of QM. Its principles are based on people, process, and technology and are driven by data from sensors and IoT. It creates quality processes and outcomes that are more evident, allied, and pertinent. Apart from a technology perspective, Q4.0 is more about the organization's process that helps to maximize value. The advanced digital tools considered the hallmarks of Q4.0 include artificial intelligence, machine learning, cloud technologies, connected and edge devices, and big data.

LNS Research has identified the following 11 vital components or axes of Q4.0:

1. **Data:** One of the critical barriers that resist organizations toward achieving their quality objectives is bad metrics. Q4.0 is influenced by data-driven quality model that yields real-time quality intelligence and predictive insights to become successful in context of Industry 4.0.

2. **Analytics:** Analytics unlocks the insights captured within data. Accurate and reliable metrics help to decipher patterns and trends for meaningful information by applying BDA (big data analytics), ML/AI. It is important for the organizations to understand the analytics framework that includes descriptive, diagnostic, predictive, and prescriptive analytics. It is recommended that companies struggling to adopt Q4.0 must build their analytics strategy later or concurrent with data strategy.

3. **Connectivity:** Connectivity comprehends the link between "business information technology" (IT) and "operational technology" (OT). Business IT includes enterprise information systems, company-wide quality management systems, and product lifecycle management. On the other hand, OT refers to the technology that is used in laboratory, manufacturing, and service. Connectivity is realized through the creation of reasonable linked sensors that deliver timely response from integrated people, processes, and product and edge devices.

4. **Collaboration:** Collaboration is critical for managing quality since quality by its nature is cross-functional and global. From a Q4.0 perspective, organizations perform conventional quality business practices using digital messaging, mechanized workflows, and portals. With the advancement of modern technologies, collaboration has changed dramatically in every vertical value chain. Social media and blockchain applications are the emerging collaborative tools that are applied for seamless information flow and quality. Deep, versatile supply chain possesses various quality issues that are encountered by the collaboration tools. It is a powerful impetus to innovation and quality improvement and has been intensely transformed by data, analytics, and connectivity.

5. **Culture:** Quality 4.0 creates a quality culture that is more achievable through better visibility, connectivity, collaboration, and comprehension. A company's quality culture is represented by four key elements, namely, process participation, responsibility, credibility, and empowerment. Organizations that strive to attain the culture of quality must emphasize adopting advanced technological tools and techniques to instill among all employees the sense of ownership.

6. **Leadership:** Pertaining to functional aspects of quality, leadership has both an image and a reliability gap with the rest of organization as quality is often seen as departmental affairs without clear alignment with corporate success. On the other hand, industry leaders are embracing Industry 4.0 technologies for quality improvement. In the context of Q4.0, quality teams must emphasize and realign quality initiatives with quality objectives by forming a cross-functional team. It is the quality leadership that would facilitate organizations' performance with better impact on manufacturing, service, finance, and R&D. Q4.0 leadership is characterized by active, visible, and effective leadership by managers at the top.

7. **App Development:** Apps are applications by which organizations fulfill processes, visualize analytics, collect and synthesize data, and establish collaboration. Mobility and interactive apps fuel digital quality management that includes augmented reality (AR), virtual reality (VR), and wearable. Apart from simple web-based user interface, apps also play a significant role in the delivery of intelligence, participation, and adoption. In the context of Q4.0, apps are the powerful enablers of collaboration, competency, and efficiency.

8. **Management Systems:** From a Q4.0 perspective, e-QMS (electronic quality management system) plays a vital role and is the hub of quality management

processes. E-QMS provides scalable solutions to meet the following objectives: workflow automation; connectivity of quality processes; improvement in data veracity; providing integrated analytics for data transparency; and ensuring and fostering compliance and collaboration. Adoption of e-QMS approach helps manufacturers to complement and mechanize processes with software and other systems by leveraging business intelligence/analytics to continuously improve the system.

9. **Compliance:** Compliance is an important attribute of the quality teams across the industry verticals as quality ensures that processes, products, and services conform to specifications. Q4.0 brings many opportunities to automate compliance. In the context of Q4.0, engineering management and analytics can be used to alert organizations to potential compliance breaches or to prevent breaches. Integrated information management model and collaboration technologies such as blockchain by its data-driven approach automate audit ability. Advancement of Q4.0 plays a critical role in compliance management and reducing compliance burden and risk.

10. **Scalability:** Scalability is the capability to upkeep bulk of data volume, users, devices, and analytics on a comprehensive scale.
Without a global scale, traditional quality and Q4.0 become less effective in harmonizing best practices.

   Therefore, "cloud computing" is a significant enabler to scalability. With deployment of cloud technology, manufacturers can procure "software as a service (SaaS)", "infrastructure as a service (IaaS)", and "platform as a service (PaaS)". Data scalability is essential with connected devices and is critical when mechanizing solutions. When deploying Q4.0, the current scalability of in-house systems must be assessed to consider global reach and data complexity.

11. **Competence:** Quality leaders complete the task effectively and efficiently. They embark on improvement, leveraging Q4.0 approaches by using social media applications and machine learning/artificial intelligence. To improve the talent and skills of employees, organizations leverage the mashup apps and virtual/augmented reality (VR/AR), appraisal through connected worker strategies, learning based on systems management perspective, and use of virtual reality (VR) to improve training delivery. Competence, in the context of Q4.0, is one of the critical approaches to improvement.

Quality 4.0 is characterized by lesser waste, quicker setup, reduced lead time, faster decisions, fewer product recalls, and improved strategic agility.

7.2.2 Quality 4.0: Technologies

Organizations are gradually instituting their technology base from production-integrated quality aspect for analyzing and monitoring all the relevant data in real time from a process control perspective [24]. Six Q4.0 technologies have been proposed by Radziwill [25], namely, big data, blockchain, AI, deep learning, machine learning, and data science. Jacob [19] classified Q4.0 technologies into four categories:

connectivity, collaboration, big data, and data presentation. Nowadays, Q4.0 is redefining its approach by presenting adaptability to disruptive technologies [26] with end-to-end connectivity in supply chain network [27, 28] and providing fact-based decisions through feedback mechanisms [29]. Since Q4.0 is a reference to Industry 4.0, supporting technologies such as big data handling, cyber-physical systems, advanced robotics, industrial connectivity, last-generation sensors, IoT, wireless production, big data analytics, cloud computing, virtualization method, and service-oriented technology forms a part of Q4.0 [30]. Javaid et al. [31] identified AI, mobile technology, collaborative technology, smart sensor networks, big data, cloud technology, AR/VR as among the new developments for Q4.0. Q4.0 designates the route to developing technologies that can be embraced by the suppliers. These emerging technologies involve deep learning, statistical analysis, IoT, big data, and cloud analytics. In the context of manufacturing development, concepts such as 3D/4D printing, IoT, virtual learning, and training support the successful implementation of Q4.0 in the manufacturing domain. Big data and AI are the critical technologies of Q4.0 for quality control.

7.2.3 QUALITY 4.0 IN MANUFACTURING

In the era of mass customization, a paradigm shift has been observed in customer order decoupling point (CODP), which is another aspect of Q4.0. The approach toward implementing Q4.0 provides ample opportunities to enhance quality and productivity in manufacturing organizations. Successful implementation of such an approach resolves long-term challenges. It is reported that many manufacturing firms have solved their long-term challenges by implementing Q4.0 through improved data integrity and real-time data-driven perspectives. Javaid et al. [31] cited some of the applications of Quality 4.0 in manufacturing:

1. **Automation of inspection processes:** One of the critical technological innovations from the perspective of Q4.0 in the manufacturing domain is the routine automation of inspection processes. Execution operations of automation of inspection processes are manifested by digital sensors and IoT platform. By automating the inspection process, better-quality products are manufactured. Data-driven quality management and integrated supply chain network, from an Industry 4.0 standpoint, significantly improve quality control that helps to reduce costs of quality, drive agility, and employee skills [32–36].

2. **Quality control:** Adoption of Q4.0 by combining advanced technologies drives manufacturing organizations toward digital quality control that results in better decision making and improved efficiency. Managing quality control operations is done through real-time data analytics and predictive maintenance by analyzing data. Quality control is performed by monitoring and control of data being extracted from sensors. The vast amount of data that gets generated is further assessed to estimate the manufacturing organizations' quality issues followed by corrective and preventive actions for minimizing errors [37–39].

3. **Improved product performance:** To enhance the performance of the product, quality monitoring plays an important role that applies prescribed methods, upholding operator and apparatus. Reducing complications and errors related to process can be easily handled by Q4.0 applications. Its application in manufacturing process makes the operations more manageable, enhancing the quality by emphasizing safety and compliance [40–43].

4. **Business culture and partnership:** Quality 4.0 influences the organizations' culture and partnership by leveraging digital technologies. Systems pertaining to Q4.0 help business to link suppliers and their customers directly through multi-level channels. Digital QMS contributes to higher services and reduced risks based on real-time coordination, simplified reporting, and customer experience [44, 45].

5. **Continuous monitoring:** Quality 4.0 has the capability to monitor continuously by regulating various systems and process parameters that affect the quality of the product. To enable higher performance and time-sensitive mechanization, secure communications have been found to be applicable [46, 47].

6. **Appropriate operations of finished goods:** In the context of the manufacturing processes, Q4.0 depends on inspection and appropriate operations of finished goods. By deploying Q4.0, nonconforming items can be easily detected within the manufacturing process through appropriate actions. Machine learning procedures are found to be useful since they provide computational solutions to streamline process, incongruity assessment, reliability, and security issues, which traditional methods are not able to resolve [48–51].

7. **Quality assurance:** By leveraging Quality 4.0 applications, existing standards of quality and reliability are maintained in all stages of the manufacturing processes. Thus, deployment of Q4.0 is deemed appropriate to augment a company's structural events, productivity, and efficacy [52, 53].

It is reported that organizations that want to perform well to attain competitive advantage through the opportunities generated by Industry 4.0 need to evaluate their readiness levels in the diverse capacities of their business and operations, that is, benchmarking. A framework was developed and validated to evaluate the maturity level of an organization working in "advanced manufacturing and engineering". The framework is named "Industry Competence and Maturity for Advanced Manufacturing (ICMAM)", based on the "Software Engineering Institute's Capability Maturity Model Integration (CMMI)" approach. The ICMAM model deals with various applications in areas such as

strategy and long range planning, human resource, customer and market focus, manufacturing and engineering process, business process, maintenance, logistics process, process in supply chain and product life-cycle, information and knowledge management, cyber assurance process, infrastructure and equipment, actual improvement outcomes and results.

[54, cited in 55]

7.2.4   QUALITY 4.0: IMPLEMENTATION ISSUES

Pertaining to Industry 4.0, Q4.0 is the blending of quality improvement models, methods/procedures, and technological tools that foster critical competencies and factors for organizational success [56]. Q4.0 practices include the integration of people, process, and technology, along with quality management with reference to Industry 4.0. As reported in literature, enablers of Q4.0 are data management and intelligence, models and algorithms for data evaluation, manufacturing process and facts related to product [57]. Antony et al. [12] identified Q4.0 readiness factors, skill factors, and critical success factors, which are represented in Table 7.1.

Based on qualitative findings, three factors have been identified as critical for Q4.0 implementation: top management support, management engagement and commitment, and change management [12]. Issues related to human involvement and expertise in Q4.0 have been raised in the literature [58]. By taking an example of Ralco industries, the literature discussed how the organization has developed an intelligent mistake proofing system by vertical integration between standard operating procedures tracked at each stage with their cloud-based Enterprise Resource Planning (ERP) system. Sahu et al. [59] show the usage of augmented reality in assembly and maintenance operations through AI that resulted in cost-effective and improved quality output. In the context of Q4.0, radio frequency identification device (RFID) and smart sensors have become imperative to understand the status of products to ensure

TABLE 7.1
Quality 4.0—Readiness, Skill, and Critical Success Factors

Quality 4.0 readiness factors	Quality 4.0 skill factors	Quality 4.0 critical success factors
Workforce management	Focus on data science and data mining	Data availability
Automated inspection process	Communication, time management	Organizational strategy
Data quality	Assessment and analysis	Leadership commitment and support
Big data handling	Leadership	Adequate training
Cultural transformation	Customer focus	Receptive organizational culture
Clear vision and strategy	Digital skills	Investment in technology
Customer and supplier focus	Use of smart and reliable data	Proper training and knowledge
Big data connectivity	Problem-solving approach	Cybersecurity issues
Collaboration and data presentation	Teamwork	Management support
Change management	Analytical skills	Appropriate skill sets
Horizontal, vertical, and end-to-end system integration	Leadership and digital literacy	Resistance to change

Adapted from Antony et al. (2022)[12]

that they are functional and conform to specifications [60, 61]. Many authors have reported that cyber-physical systems may be used to reduce human errors explicitly in more labor-intensive stages [62–67]. A theoretical Q4.0 model has been proposed with 11 themes across three categories—people, process, and technology. The 11 themes under each category are as follows:

TABLE 7.2
Model for Quality 4.0

People	Process	Technology
Top management	Q4.0 based on I4.0	Artificial intelligence and predictive software
Digital skills for quality staff	Q4.0 based on ISO 9001	Machine-to-machine communication
	Process mapping	Smart technologies for identification and traceability
	Automatic data collection	
	Data integration with ERP	
	Automated document control	

Adapted from Chiarini and Kumar (2022) [28]

In their analysis, Srinivasan et al. [68] highlighted the issues about the roles of employees, problem-solving approach, and cognitive skills as crucial for Q4.0 implementation. The findings also indicated the role of top management for creating quality culture based on trust, empathy, and emotional intelligence. To implement Q4.0, the findings specifically mentioned the need to upskill the employees not only in digital talents but also in social and communication skills [68, 69]. In the context of Q4.0, it is necessary that quality managers understand the type of data they need and the methods/techniques to be used to collect data from various digital information systems (ERP/MES/SCADA/CRM/PLM software), how to leverage knowledge of AI and machine learning for doing proper analytics, and how to identify potential signals of failure that might happen in the future. The research found that the new trends, namely, big data and analytics, can be used to deploy web-based cloud solutions for the improvement of quality management [70]. AR technologies are found to be efficient options for quality control at different stages of production [71]. Based on a review of the literature, eight key ingredients of Q4.0 were identified, which include:

handling big data; improving prescriptive analytics; using Quality 4.0 for effective vertical, horizontal and end to end integration; using Quality 4.0 for strategic advantage, leadership in Quality 4.0; training in Quality; organizational culture for Quality 4.0 and lastly top management support for Quality 4.0.

[3]

A proper methodical approach is recommended for deploying these ingredients for successful Q4.0 implementation. Security issues of the organizations' employees also need to be emphasized while implementing.

7.3 SUSTAINABLE MANUFACTURING

Emerging technologies have significant impact on sustainable value creation from triple bottom line (economic, social, and environmental) perspective [72]. Sustainable manufacturing is defined as the

> integration of processes and systems capable to produce high quality products and services using less and more sustainable resources (energy and materials), being safer for employees, customers and communities surrounding, and being able to mitigate environmental and social impacts throughout its whole life cycle.

[73]

Benefits of sustainable manufacturing include reduction of cost through improvisation of resources and regulatory compliance, enhanced brand repute and better accessibility to new market, reduced labor turnover, and opportunity creation toward access to capital and finance through long-term business strategy [73–75].

7.3.1 LINK BETWEEN QUALITY 4.0 AND SUSTAINABLE MANUFACTURING

Digitalization of quality management and sustainability cover the entire gamut of production chain. Both approaches follow common practices such as remanufacturing and applied life cycle management, reverse logistics, lean and green management, sustainable design, and eliminating toxic parts in product and production processes [76, 77]. Sustainability benefits of Q4.0 are expected in productivity improvement, flexibility, managing large chunks of data for predictive maintenance and fast production system configuration for resource efficiency, waste reduction, energy consumption, servitization, collaboration in the form of closed-loop production system that connects IT systems, products and people in network, and improvement in quality of working environment (creating employment opportunities for seniors and physically disabled) [79–81]. Moreover, Q4.0 supports value creation across all dimensions of sustainability and identifies opportunities for industry through smart data-driven business model development, improving closed-loop product life cycle for creating value-based networks, cyber-physical system-based equipment, training and competence, motivation and creativity, additive manufacturing, and distributed organization focused on resource efficiency [73]. An integrated conceptual framework was developed by Martin et al. [82] with four major perspectives on quality-in-use. It was stated that societal perspective of quality integrates the value of sustainability. Improvement in quality enhances the buyer's value and minimizes environmental impact. Organizing quality work in the workplace combines both quality and sustainability in one organization or in two different but cooperating organizations [83].

By leveraging quality intelligence (big data analytics) in the manufacturing perspective, industries offer numerous prospects that deal with monitoring production process and real-time optimization, which contribute to sustainability [84].

7.4 IDENTIFICATION AND PRIORITIZATION OF Q4.0 PRACTICES

Quality 4.0 is a developing area of research, and therefore, not much literature is available that offers the evidence regarding new improvements and research opportunities from the context of the Indian economy. In this present study, we have systematically reviewed the available literature from various academic databases. We searched peer-reviewed journals and conference articles from 2017 to 2023. The search for the articles was conducted using keywords such as "Quality 4.0"; "Industry 4.0"; "Quality 4.0 and Sustainable manufacturing"; "Industry 4.0 and Quality 4.0"; "Industry 4.0 and Sustainable manufacturing and India"; and "Sustainability in manufacturing or Sustainable manufacturing". Some of the major databases linked to the research area of Industry 4.0, Quality 4.0, and sustainable manufacturing were selected for the initial search. These databases include Scopus (Elsevier), Taylor and Francis, Emerald Insight, IEEE Xplore, and Web of Science. Based on a comprehensive compilation and subsequent analysis of literature, 81 Q4.0 practices were identified. To generate distinct generic practices, those 81 practices were explored and analyzed thoroughly, with further classification of similar practices since analogy persists among the recognized ones. However, after careful analysis, those 81 practices were reduced to 56, which were further classified into four categories, namely, people, process, technology, and quality intelligence, as shown in Table 7.3. As the deployment of Q4.0 practices is in its nascent stage, all the practices identified in the literature may not be appropriate in the Indian sustainable manufacturing context. Therefore, we decided to extract the most influential/critical Q4.0 practices based on the degree of importance within the Indian context. Accordingly, a questionnaire was designed that incorporated those identified 56 Q4.0 practices, which were accompanied by a 1–5 point Likert scale, where "5" indicates very important, "4" indicates important, "3" indicates moderately important, "2" indicates degree of little importance, and "1" indicates no importance. Sixty-five respondents were identified, and questionnaires were sent accordingly; however, only 52 questionnaires were found to be complete. As such, the responses of those 52 participants were considered. The respondents/participants were mainly from the industries that include entrepreneurs from some deep-tech start-ups, senior managers from small and large-scale manufacturing firms who have already deployed or are in the process of deploying Industry 4.0 technologies, technical experts having ITIL (Information Technology Infrastructure Library) foundation certificate with experience in ITSM (information technology service management) implementation, CEOs and other IT professionals who have experience in implementing TQM in their organizations (irrespective of any magnitude), as well as few consultants with knowledge in diverse areas such as finance, marketing, operations, production, and quality control. The 56 Q4.0 practices were listed down, and the respondents were directed individually to rate the practices based on the degree of importance from the Indian context. Those Q4.0 practices that have fallen under a point value of 5 ("indicates very important") were

TABLE 7.3

Categorization of Q4.0 Practices

People	Process	Technology	Quality Intelligence
Collaboration and data presentation	e-QMS	Cloud technology/manufacturing	Deep learning
Quality culture and leadership	Compliance	Cyber-physical systems	AI and machine learning
Cross-functional team	Scalability	Internet of things	Use of smart and reliable data
Compliance	Automated inspection process	AR/VR	Big data and connectivity
ISO 9001	Quality control	Blockchain	Prescriptive analytics
Competence	Continuous monitoring	Automated data collection	Assessment and analysis
Proper training and knowledge	Appropriate operations of finished goods	Advanced robotics	Data mining
Workforce management	Quality assurance	Sensors and wireless production	Smart and reliable data
Collaboration and data presentation	Horizontal, vertical and end-to-end system integration	Virtualization	Predictive software
Quality culture and leadership	Customer focus	Service-oriented technology	-
Cross-functional team	Data availability	Advanced robotics	-
Compliance	Investment in technology	Machine-to-machine communication	-
ISO 9001	Process mapping	Smart technologies for identification and traceability	-
Clear vision and strategy	-	Cybersecurity issues	-
Customer and supplier focus	-	-	-
Change management	-	-	-
Communication	-	-	-
Time management	-	-	-
Analytical and digital skills	-	-	-
Problem-solving approach	-	-	-
Employee upskilling	-	-	-
Social and communication skills	-	-	-
Leadership commitment and support	-	-	-
Employee role	-	-	-
Cognitive skills	-	-	-

TABLE 7.4

Extracted Q4.0 Practices Based on Degree of Importance

Point Value (1–5 point scale)	Q4.0 Practices
5	Cloud technology/manufacturing
5	Cyber-physical systems
5	Data mining
5	e-quality management systems
5	Clear vision and strategy
5	Quality culture and leadership
5	AI and machine learning

further extracted for analysis. Out of 56 practices, seven have been found to have fallen under a point value of 5, and thus, it is likely that these seven Q4.0 practices are critical/relevant and could influence the sustainable manufacturing operations in India. Overall, seven critical Q4.0 practices were listed together for prioritizing using rough number-based AHP-MABAC techniques. The extracted seven critical Q4.0 practices are shown in Table 7.4.

To prioritize seven Q4.0 practices, a structured questionnaire was prepared taking all the critical Q4.0 practices. Fifteen experts' opinions were taken from the same population as previously identified during the extraction process. The next section describes the method used for prioritizing the alternatives. The filtering process that we followed in this study is a good and reliable basis for extracting relevant Q4.0 practices from a sustainable manufacturing context in the Indian economy. In this chapter, we meant Q4.0 practices as the combination of principles, digital tools/techniques, and technologies combining other aspects of QM, with reference to Industry 4.0, which can lead directly to improved quality performance and competitive advantage. Q4.0 can be considered an impetus to intelligent and sustainable manufacturing.

7.4.1 THE METHOD

Decision making is as old as human civilization. Many MCDM methods with diverse features have come up in the last 50 years to facilitate decision making in the real world. As it happens in an imprecise and inconsistent environment, the classical rough set theory as exposited by Pawlak [85] and further extended by Zhai et al. [86–87] become very useful. Realizing the potential of rough numbers to facilitate decision making in an imprecise and approximate environment, remarkable applications of this concept have been found in nontraditional machining processes selection [88], design concept evaluation [89], failure mode and effect analysis [90], assessment of renewable energy resources [91], and in many more areas. To process inconsistent and imprecise information, the use of classical rough set approximation theory can be described as follows.

Let Y be an object of universe of discourse U where exist a set, $R = \{F_1, F_2, \ldots, F_n\}$, defined in U. If it is ordered in the way of $F_1 < F_2 < \ldots < F_n$, then for, $Ci \in R$, $1 \le i \le n$, lower approximation of F_i becomes

$$\underline{Apr}(F_i) = U\{Y \in U/R(Y) \le F_i\}; \tag{1}$$

And upper approximation of F_i can be stated as

$$Apr (F_i) = U\{Y \in U/R(Y) \ge F_i\}; \tag{2}$$

the boundary region of F_i is expressed as

$$Bnd(F_i) = U\{Y \in U/R(Y) \ne F_i\};$$
$$= \{Y \in U/R(Y) > F_i\} \cup \{Y \in U/R(Y) < F_i\} \tag{3}$$

Thus Fi can be considered rough number (RN) with lower limit ($\underline{Lim}(F_i)$) and upper limit ($\overline{Lim} (F_i)$), where

$$Lim(F_i)\frac{1}{M_L}\sum R(Y) \mid Y \in Apr(F_i) \text{ and} \tag{4}$$

$$\overline{Lim}(F_i) = \frac{1}{M_U}\sum R(Y) \mid Y \in \overline{Apr}(F_i) \tag{5}$$

where M_L and M_U are the number of objects contained in lower and upper approximation of F_i.

The rough boundary interval of F_i, stated as $RBnd(F_i)$ becomes

$$RBnd(F_i) = (F_i) = (F_i) = \overline{Lim} = (F_i) - \underline{Lim}(F_i) \tag{6}$$

Thus, the vague class F_i can be defined in terms of rough number with its lower and upper limit as follows.

$$RN(F_i) = [Lim(F_i), \overline{Lim}(F_i)] = [y_{ij}^L, y_{ij}^U] \tag{7}$$

In this chapter, the relative importance of performance evaluation criteria has been determined using rough AHP while rough MABAC has been utilized to determine the performance score of alternatives on the basis of aggregation of approximate and imprecise subjective judgments of decision makers. This process takes place in two stages. In the first stage, criteria weights are determined by taking the opinion of experts followed by aggregation of the same using rough AHP. The performance score of alternative materials has been determined in the second phase using rough MABAC method.

7.4.1.1 Rough Analytic Hierarchy Process (RAHP)

Though AHP (Saaty, [92]) had been widely used, rough number-based AHP has been utilized in this chapter to determine criteria weights for its ability to deal with approximate and imprecise information. The following steps have been followed for this purpose.

Step 1: Taking the opinion of k number of experts, develop pairwise comparison matrix (PWCM) for evaluation criteria, corresponding to individual experts ($1 \le e \le k$) using Saaty's 9-point scale. Thus, the PWCM for eth expert becomes

$$F_e = [f_{ij}^e]_{nxn} = \begin{vmatrix} 1 & f_{12}^e & \cdots & f_{1n}^e \\ f_{21}^e & 1 & \cdots & f_{2n}^e \\ \cdots & \cdots & \cdots & \cdots \\ f_{n1}^e & f_{n2}^e & \cdots & 1 \end{vmatrix} \tag{8}$$

Step 2: Develop k number of PWCM. Check consistency ratio (CR) for each PWCM. If CR<0.10, accept the PWCM.

Step 3: Compute integrated PWCM by aggregating the experts' opinions. The integrated PWCM can be expressed as

$$F = [f_{ij}]_{nxn} = \begin{vmatrix} 1 & f_{12} & \cdots & f_{1n} \\ f_{21} & 1 & \cdots & f_{2n} \\ \cdots & \cdots & \cdots & \cdots \\ f_{n1} & c_{n2} & \cdots & 1 \end{vmatrix} \text{ Where } f_{ij} = \{f_{ij}^1, f_{ij}^2 \ldots \ldots f_{ij}^k\} \tag{9}$$

Step 4: Using eq. (1–7), transform the element f_{ij} into Rough Number, $RN(f_{ij}^e)$ of f_{ij} as follows.

$$RN(f_{ij}^e) = \left[f_{ij}^{eL}, f_{ij}^{eU} \right], \rightarrow (1 \le e \le k) \tag{10}$$

Step 5: Determine average rough interval $RN(f_{ij})$ as

$$\overline{(RN(f_{ij}))} = [f_{ij}^L, f_{ij}^U] \tag{11}$$

Where,
$$f_{ij}^L = \frac{\sum\limits_{e=1}^{k} f_{ij}^{eL}}{k}, c_{ij}^U = \frac{\sum\limits_{e=1}^{k} f_{ij}^{eU}}{k} \tag{12}$$

Thus, the aggregated rough PWCM becomes

$$M = [m_{ij}]_{nxn} = \begin{pmatrix} [1,1] & [f_{12}^L, f_{12}^U] & \cdots & [f_{1n}^L, f_{1n}^U] \\ [f_{21}^L, c_{21}^U] & [1,1] & \cdots & [f_{2n}^L, f_{2n}^U] \\ \cdots & \cdots & \cdots & \cdots \\ [f_{n1}^L, f_{n1}^U] & [f_{n2}^L, f_{n2}^U] & \cdots & [1,1] \end{pmatrix} \tag{13}$$

Step 6: Compute rough criteria weight W_i as the geometric mean of aggregated rough PWCM.

$$W_i = [W_i^L, W_i^U] = \left[\left(\prod_{j=1}^{n} f_{ij}^L \right)^{1/n}, \left(\prod_{j=1}^{n} f_{ij}^U \right)^{1/n} \right] \qquad (14)$$

Step 7: Normalize eq. (14) to get relative criteria weights, which are expressed as

$$w_i = [w_i^L, w_i^U] = \left[\frac{W_i^L}{\max\left(W_i^U\right)}, \frac{W_i^U}{\max\left(W_i^U\right)} \right], \text{ for all } i \qquad (15)$$

7.4.1.2 Rough MABAC

The proposed rough MABAC method works on Euclidean distance-based measure from border approximation area (BAA). The steps of the rough MABAC method are mentioned next.

Step 1: Develop a group decision matrix with the opinion of k number of experts using Saaty's 9-point scale and then convert it into aggregated rough decision matrix (D) using eq. (1–7).

$$D = [x_{ij}^L, x_{ij}^U]_{mxn} \qquad (16)$$

Where $[x_{ij}^L, x_{ij}^U]$ indicates performance of ith alternative against jth criterion in rough number.

Step 2: Compute normalized rough decision matrix $D' = [n_{ij}^L, n_{ij}^U]$ using following formula.

$$\text{(a) For higher-the-better type criteria,} \qquad (17)$$

$$\text{(b) For lower-the-better type criteria, } n_{ij}^L = \frac{x_{ij}^L - x_j^+}{x_j^- - x_j^+}, n_{ij}^U = \frac{x_{ij}^U - x_j^+}{x_j^- - x_j^+} \qquad (18)$$

$$\text{Where, } x_j^+ = \begin{cases} \max_{1 \le i \le m} \left(x_{ij}^U \right), \text{for higher and better type criteria} \\ \min_{1 \le i \le m} \left(x_{ij}^L \right), \textit{for lower and better type criteria} \end{cases}, \text{ for all } j \qquad (19)$$

$$\text{and} \qquad X_j = \begin{cases} \min_{1 \le i \le m} (x_{ij}^L, \text{for higher-the-better type criteria} \\ \max_{1 \le i \le m} (x_{ij}^L, \text{for lower-the-better type criteria} \end{cases}, \cdots \text{for all } j, \qquad (20)$$

Step 3: Determined weighted normalized matrix (D") using the following equation.

$$\begin{cases} D'' = \left[v_{ij}^L, v_{ij}^U \right]_{mxn} \\ v_{ij}^L = w_i^L \left(n_{ij}^L + 1 \right) \\ v_{ij}^U = w_i^U \left(n_{ij}^U + 1 \right) \end{cases} \qquad (21)$$

Step 4: Taking the geometric mean, compute criterion wise border approximation area (BAA) as mentioned next.

$$
\left|
\begin{aligned}
G &= \left[g_j^L, g_j^U \right], \ for \ all \ j \\
g_j^L &= \left(\prod_{i=1}^{m} v_{ij}^L \right)^{1/m} \\
g_j^U &= \left(\prod_{i=1}^{m} v_{ij}^U \right)^{1/m}
\end{aligned}
\right.
\tag{22}
$$

Step 5: Using the following formula, determine the Euclidean distance of the alternatives from BAA.

$$
s_E(v_{ij}, g_j) =
\begin{cases}
\sqrt{\left(v_{ij}^L - g_j^U \right)^2 + \left(v_{ij}^U - g_j^L \right)^2}, \text{for higher and better type criteria} \\
\sqrt{\left(v_{ij}^L - g_j^L \right)^2 + \left(v_{ij}^U - g_j^U \right)^2}, \text{for lower and better type criteria}
\end{cases}
\tag{23}
$$

Step 6: Considering the following condition, form the distance matrix U as follows:

$$
U = (u_{ij})_{mxn}
\tag{24}
$$

Where
$$
\begin{cases}
u_{ij} = \begin{cases} s_E(v_{ij}, g_j), if \ RN(v_{ij}) > RN(g_j) \\ -s_E(v_{ij}, g_j), if \ RN(v_{ij}) > RN(g_j) \end{cases}, \text{for higher and better type criteria} \\
u_{ij} = \begin{cases} -s_E(v_{ij}, g_j), if \ RN(v_{ij}) > RN(g_j) \\ s_E(v_{ij}, g_j), if \ RN(v_{ij}) > RN(g_j) \end{cases}, \text{for lower and better type criteria}
\end{cases}
$$

Step 7: Determine the performance score of alternatives by adding the row elements of distance matrix U. Thus, the score becomes

$$
S(A_i) = \sum_{j=1}^{n} u_{ij}, \text{ for all i.}
\tag{25}
$$

7.4.2 Illustrative Example

The task of prioritization of the Quality 4.0 practices for sustainable manufacturing in the Indian context has been performed in the following stages.

7.4.2.1 Selection of Alternatives and Performance Evaluation Criteria

Seven alternatives have been taken for evaluation, which includes A1 = cloud technology/manufacturing; A2 = cyber-physical systems; A3 = data mining; A4 = e-quality management system; A5 = clear vision and strategy; A6 = quality culture and leadership; and A7 = artificial intelligence and machine learning (Table 7.4).

In this study, five performance evaluation criteria have been considered, with reference to sustainability, which include F1 = increased sales; F2 = improved efficiency and productivity; F3 = compliance with regulations; F4 = enhanced reputation and image; and F5 = Better community relations [93].

7.4.2.2 Determination of Criteria Weight Using RAHP

To determine criteria weights, an expert committee has been formed with four experts (E_k, k =1 to 4). Corresponding to each expert, a pairwise comparison matrix (PWCM) has been formed using Saaty's 9-point scale. The same is presented in Table 7.5.

Now, vague and subjective linguistic judgments of individual experts corresponding to each criterion are aggregated in terms of rough numbers using eq. (1–7). The aggregated rough PWCM is shown in Table 7.6.

TABLE 7.5
Initial PWCM

$$E_1 = \begin{bmatrix} 1 & 1 & 3 & 5 & 5 \\ 1 & 1 & 5 & 7 & 5 \\ 1/3 & 1/5 & 1 & 3 & 1 \\ 1/5 & 1/7 & 1/3 & 1 & 1/3 \\ 1/5 & 1/5 & 1 & 4 & 1 \end{bmatrix} \quad E_2 = \begin{bmatrix} 1 & 2 & 3 & 4 & 5 \\ 1/2 & 1 & 3 & 5 & 5 \\ 1/3 & 1/3 & 1 & 3 & 1 \\ 1/4 & 1/5 & 1/3 & 1 & 1/3 \\ 1/5 & 1/5 & 1 & 3 & 1 \end{bmatrix}$$

$$E_3 = \begin{bmatrix} 1 & 2 & 3 & 4 & 6 \\ 1/2 & 1 & 4 & 5 & 5 \\ 1/3 & 1/4 & 1 & 3 & 3 \\ 1/4 & 1/5 & 1/3 & 1 & 1/4 \\ 1/6 & 1/5 & 1/3 & 4 & 1 \end{bmatrix} \quad E_4 = \begin{bmatrix} 1 & 1 & 4 & 4 & 6 \\ 1 & 1 & 4 & 5 & 5 \\ 1/4 & 1/4 & 1 & 3 & 3 \\ 1/4 & 1/5 & 1/3 & 1 & 1/4 \\ 1/6 & 1/5 & 1/3 & 4 & 1 \end{bmatrix}$$

Note: Saaty's 9-point scale: 1 = very low; 3 = low; 5 = moderate; 7 = high; 9 = very high.

TABLE 7.6
Aggregated Rough PWCM

	F1	F2	F3	F4	F5
F1	[1.000, 1.000]	[1.250, 1.750]	[3.063, 3.438]	[4.063, 4.438]	[5.250, 5.750]
F2	[0.625, 0.875]	[1.000, 1.000]	[3.583, 4.417]	[5.125, 5.875]	[5.000, 5.000]
F3	[0.297, 0.326]	[0.231, 0.285]	[1.000, 1.000]	[3.000, 3.000]	[1.500, 2.500]
F4	[0.228, 0.247]	[0.174, 0.196]	[0.330, 0.330]	[1.000, 1.000]	[0.271, 0.561]
F5	[0.176, 0.441]	[0.200, 0.200]	[0.499, 1.084]	[3.250, 3.750]	[1.000, 1.000]

From aggregated rough PWCM, the rough weight of each criterion has been determined using eq. (14), followed by normalization of the same using eq. (15). These normalized weights become the final rough weights of each criterion. These are presented in Table 7.7.

7.4.2.3 Determination of Criteria Weight Using RAHP

Through an exhaustive literature survey followed by expert opinion, the following seven practices have been identified as potential indicators (which are considered alternatives) of Quality 4.0 from a sustainable manufacturing perspective. Table 7.8

TABLE 7.7
Final Rough Weights

Criteria	Rough Weights	Final Rough Weights (Normalized)
F1	[2.412, 2.737]	[0.881, 1.000]
F2	[2.248, 2.576]	[0.821, 0.941]
F3	[0.790, 0.930]	[0.289, 0.340]
F4	[0.324, 0.390]	[0.118, 0.142]
F5	[0.564, 0.815]	[0.206, 0.298]

TABLE 7.8
Expert Opinion-Based Decision Matrix

Alternatives	Experts	F1	F2	F3	F4	F5
A1	1	8	7	7	6	7
	2	7	7	7	7	7
	3	7	7	6	7	7
	4	7	7	6	7	8
A2	1	6	7	7	6	7
	2	6	6	6	7	7
	3	7	5	6	6	7
	4	6	5	6	7	7
A3	1	5	7	7	6	7
	2	6	6	6	7	6
	3	6	5	6	6	7
	4	6	5	6	7	6
A4	1	7	7	7	6	7
	2	6	7	7	7	7
	3	7	5	6	7	7
	4	6	5	6	7	8

TABLE 7.8 *(Continued)*
Expert Opinion-Based Decision Matrix

Alternatives	Experts	F1	F2	F3	F4	F5
A5	1	8	9	9	8	7
	2	8	9	7	7	7
	3	8	8	7	7	7
	4	9	8	7	7	7
A6	1	9	9	8	9	7
	2	9	9	8	9	7
	3	9	7	9	7	7
	4	9	7	9	9	9
A7	1	8	8	7	7	7
	2	8	7	7	7	7
	3	7	7	7	8	8
	4	7	7	7	8	7

TABLE 7.9
Initial Rough Decision Matrix

	F1	F2	F3	F4	F5
ALT	Max	Max	Max	Max	Max
A1	[7.063, 7.438]	[7.000, 7.000]	[6.250, 6.750]	[6.250, 6.750]	[6.580, 7.415]
A2	[6.270, 7.250]	[5.270, 6.250]	[5.563, 5.938]	[6.250, 6.750]	[7.000, 7.000]
A3	[5.250, 5.750]	[5.270, 6.250]	[5.500, 6.500]	[6.250, 6.750]	[6.250, 6.750]
A4	[5.500, 6.500]	[4.750, 5.250]	[6.250, 6.750]	[6.563, 6.938]	[6.125, 6.875]
A5	[8.063, 8.438]	[8.250, 8.750]	[7.063, 7.438]	[7.063, 7.438]	[6.000, 6.000]
A6	[7.500, 8.000]	[7.500, 8.000]	[7.000, 7.000]	[7.000, 7.000]	[6.750, 7.000]
A7	[7.250, 7.750]	[7.250, 7.750]	[7.000, 7.000]	[7.250, 7.750]	[6.563, 6.938]
x+	8.438	8.750	7.438	7.750	7.415
x-	5.250	4.750	5.500	6.250	6.000

shows the list of these alternatives along with ratings of expert committee members with respect to the five criteria mentioned earlier. The judgments of experts in linguistic form are expressed in numbers using Saaty's 9-point scale.

In a similar manner, using eq. (1–7) aggregated initial rough decision matrix is constituted from these ratings of the experts. The same is presented in the following table. The ideal solutions (x_j^+, x_j^-) as identified using formula (19–20) are also shown in Table 7.9.

The following table (Table 7.10) shows the normalized rough decision matrix derived from Table 9 using formula (17–18). The last row of the table indicates criteria weights that have been determined earlier using RAHP. Subsequently, the weighted

TABLE 7.10
Normalized Rough Decision Matrix

ALT	F1	F2	F3	F4	F5
A1	[0.569, 0.686]	[0.563, 0.563]	[0.387, 0.645]	[0.000, 0.333]	[0.410, 1.000]
A2	[0.320, 0.627]	[0.130, 0.375]	[0.032, 0.226]	[0.000, 0.333]	[0.707, 0.707]
A3	[0.000, 0.157]	[0.130, 0.375]	[0.000, 0.516]	[0.000, 0.333]	[0.177, 0.530]
A4	[0.078, 0.392]	[0.000, 0.125]	[0.387, 0.645]	[0.208, 0.458]	[0.088, 0.618]
A5	[0.882, 1.000]	[0.875, 1.000]	[0.806, 1.000]	[0.542, 0.792]	[0.000, 0.000]
A6	[0.706, 0.863]	[0.688, 0.813]	[0.774, 0.774]	[0.500, 0.833]	[0.530, 0.707]
A7	[0.627, 0.784]	[0.625, 0.750]	[0.774, 0.774]	[0.667, 1.000]	[0.398, 0.663]
Weights	[0.881, 1.000]	[0.821, 0.941]	[0.289, 0.340]	[0.118, 0.142]	[0.206, 0.298]

TABLE 7.11
Weighted Normalized Rough Decision Matrix

ALT	F1	F2	F3	F4	F5
A1	[1.383, 1.686]	[1.283, 1.471]	[0.401, 0.559]	[0.118, 0.190]	[0.291, 0.595]
A2	[1.163, 1.627]	[0.928, 1.294]	[0.298, 0.417]	[0.118, 0.190]	[0.352, 0.508]
A3	[0.881, 1.157]	[0.928, 1.294]	[0.289, 0.515]	[0.118, 0.190]	[0.243, 0.455]
A4	[0.951, 1.392]	[0.821, 1.059]	[0.401, 0.559]	[0.143, 0.208]	[0.224, 0.482]
A5	[1.659, 2.000]	[1.540, 1.883]	[0.522, 0.680]	[0.182, 0.255]	[0.206, 0.298]
A6	[1.504, 1.863]	[1.386, 1.706]	[0.512, 0.603]	[0.177, 0.261]	[0.315, 0.508]
A7	[1.434, 1.784]	[1.335, 1.648]	[0.512, 0.603]	[0.197, 0.285]	[0.288, 0.495]
BAA (g)	[1.252, 1.621]	[1.145, 1.455]	[0.408, 0.557]	[0.147, 0.222]	[0.270, 0.469]

normalized rough decision matrix has been developed using formula (21) and shown in Table 11. The last row of Table 11 indicates the border approximation area, which has been computed using formula (22).

The Euclidean distance of all alternatives from BAA has been determined using formula (23). Thus, the distance matrix and final performance scores (pictorial view presented in Figure 7.1) computed for all alternatives using formula (24–25) are presented in Table 7.12. The relative ranking of alternatives is also shown in the same table.

7.5   DISCUSSIONS

The prioritized Q4.0 practices are depicted in Table 7.12. Based on ranking, it can be deciphered that "Clear vision and strategy" resembles the highest priority followed

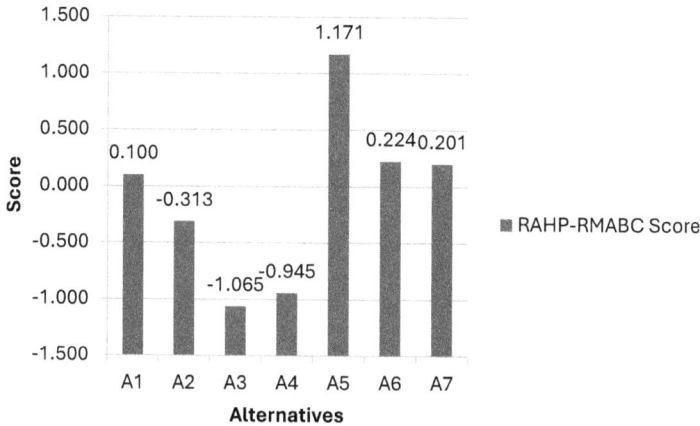

FIGURE 7.1 Performance Score of Alternatives.

TABLE 7.12
Distances from BAA, Final Score, and Ranking of Alternatives

ALT	F1	F2	F3	F4	F5	Final score S(A$_i$)	Rank
A1	0.146	−0.139	0.008	−0.044	0.128	0.100	4
A2	0.089	−0.270	−0.178	−0.044	0.091	−0.313	5
A3	−0.594	−0.270	−0.127	−0.044	−0.030	−1.065	7
A4	−0.378	−0.512	0.008	−0.015	−0.047	−0.945	6
A5	0.556	0.582	0.167	0.048	−0.182	1.171	1
A6	0.349	−0.348	0.114	0.049	0.060	0.224	2
A7	0.245	−0.270	0.114	0.080	0.032	0.201	3

by "quality culture and leadership", "AI and machine learning", and "cloud technology/manufacturing". These four prioritized practices are more relevant and critical based on the context of the study. The detailed discussions of the critical practices are given next:

Clear vision and strategy: The digital data extracted from the organizations that use emerging technologies should be addressed strategically. Based on the respondents' perception toward this indicator, it may be envisaged that proper planning is essential for managing organizations in Industry 4.0 to create better quality products by offering advantage of price value over the competitors. To implement Q4.0, organizations should build a data strategy for establishing linkage between information technology and operational technology to improve reliability of quality inputs into new product introduction.

Quality culture and leadership: Quality culture influences the organizations' members in terms of behavior, performance outcome, and external environment [94], whereas leadership is the talent that inspires through systematic activities to attain organizational goals [95]. Based on the respondents' opinion, it is found that organizations need to focus on the process of innovation and learning as the core concept of Q4.0 in order to bring in QM practices with reference to Industry 4.0. From a quality culture perspective, it is thus recommended that organizations that are in the process of implementing Q4.0 must exhibit an adhocratic culture. From a leadership perspective, organizations need to adopt knowledge-oriented leadership with a focus on innovation and learning. Leadership in Q4.0 should emphasize the use of emerging technologies/tools for quality planning, quality control, and quality improvement with employee training in a strategic manner.

AI and machine learning: Artificial intelligence (AI) is one of the most disruptive technologies that have transformed management and business models. Its key applications toward Q4.0 implementation continuously improve the quality of production. AI techniques such as machine learning (ML), if properly implemented, can have significant positive effects in terms of return on investment, which is in line with Deming's Chain Reaction. The results indicate that AI/ML could be used to predict nonconforming products. Based on priority, AI/ML can be considered critical for Q4.0 environment in a sustainable manufacturing context, and therefore, quality managers need to comprehend the types of data that could be used and how data collection could be automated using information systems and other allied software. AI/ML can also be used by the manufacturing firm to perform data analytics that could predict unwanted signals that are detrimental to the environment and organizations as a whole. By combining AI/ML approaches, managers in organizations can make processes more efficient and robust by eliminating non-value adding activities from the process by reengineering.

Cloud technology/manufacturing: Cloud technology/manufacturing can enable AI/ML approaches with reference to Industry 4.0 [96]. It enables manufacturing network to carry out operational activities. By deploying advanced cloud manufacturing, data can be shared across various levels of organizations that results in flexibility and agility [97], which contributes to sustainable operations in manufacturing.

Overall findings indicate that technology and quality intelligence are not only the main drivers of Q4.0, but it is the fusion of people, process, technology, and quality intelligence that could lead to a successful Q4.0 implementation for sustainable manufacturing operations explicitly in the Indian context. It can be envisaged that organizations need to devise strategic initiatives while dealing with "change management and culture" taking the importance of human factor in Q4.0 implementation. AI/ML can be used to reduce waste with predictive maintenance and intelligent supply chain. Manufacturing organizations can better enhance the energy efficiency by investing in renewable and low-carbon energy systems with the help of AI. The major goal of quality intelligence is to contribute to sustainability, which includes digital tools/techniques that help organizations generate, collect, track, and analyze the vast amount of data. It is evident from the findings that the process of digital transformation (Quality 4.0) must be continuous and adaptive to organizations' culture. The results also indicate that deployments of Q4.0 practices not only relate to automation process but also that concepts of human thinking and people management are very crucial. Finally, the most important point to highlight is the role of managers who implement strategies to determine value proposition by altering the business model.

7.6 CONTRIBUTION TO THE STUDY

Q4.0 is an emerging research area, and this study attempts to systematically and methodically explore and identify its critical practices. This study contributes to advance Q4.0 research by proposing a novel framework of Quality 4.0 for sustainable manufacturing in the context of the Indian economy. The framework addresses the research objective by identifying the relevant Q4.0 practices through a multi-step filtering approach and further categorizing the practices into four dimensions, namely, people, process, technology, and quality intelligence. The dimension named "quality intelligence" has not been reported earlier in the said context. Moreover, prioritizing critical Q4.0 practices using rough number-based AHP-MABAC approach to decipher the pattern and relative importance of Q4.0 practices is another major contribution of this study. Hardly any literature is available that has adopted this approach of identifying and prioritizing critical Q4.0 practices explicitly from the Indian context, and therefore, the study may be deemed novel in this regard.

7.7 CONCLUSIONS

The framework proposed in this study would help organizations assess the status of readiness of Q4.0 for sustainable manufacturing that would stand as an interface between academia and industry, both of which can benefit from it. To address the challenges of Q4.0 implementation, "human skills and leadership orientation" are essential requirements for developing an efficient transitional plan. Moreover, to enable sustainable manufacturing through Q4.0 practices, it is imperative to integrate digitalization of quality initiatives into their strategic plans so that the organizations' leadership can be embraced and committed to Q4.0. The identification and prioritization of Q4.0 practices using a multi-step filtering approach and MCDM technique, respectively, would serve as a partial guideline for the Indian manufacturing organizations of any magnitude for managing intelligent systems. Small and medium-scale businesses can take input from this study to decipher which Quality 4.0 practices should be considered for its implementation.

This study may be extended using a greater number of respondents in service sector organizations. A longitudinal case study taking the identified critical Q4.0 practices within the organizations would be valuable in understanding the influence of such practices. It is anticipated that industry leaders, academicians, and policy makers can benefit from the findings of this study, and accordingly, an action plan may be formulated. Future studies should emphasize the framework focusing on

Q4.0 practices for achieving sustainability in both large and small-to-medium-scale organizations. In the future, an in-depth exploration on the role of AI/ML for intelligent and sustainable manufacturing system may be pursued.

REFERENCES

1. Yamada, T. T., Poltronieri, C. F., do NascimentoGambi, L., & Gerolamo, M. C. (2013). Why does the implementation of quality management practices fail? A qualitative study of barriers in Brazilian companies. *Procedia-Social and Behavioral Sciences*, 81, 366–370.
2. Goh, T. N. (2011). Six Sigma in industry: Some observations after twenty five years. *Quality and Reliability Engineering International, 27*(2), 221–227.
3. Sony, M., Antony, J., & Douglas, J. A. (2020). Essential ingredients for the implementation of Quality 4.0: A narrative review of literature and future directions for research. *The TQM Journal*, 32(4), 779–793
4. De Sousa Jabbour, A. B. L., Jabbour, C. J. C., Foropon, C., & Filho, M. G. (2018). When Titans meet—can industry 4.0 revolutionise the environmentally-sustainable manufacturing wave? The role of critical success factors. *Technological Forecasting and Social Change*, 132, 18–25.
5. ASQ (2016). *Global State of Quality Research—Discoveries*. Milwaukee, WI.
6. ASQ (2015). *Future of Quality Report*. Milwaukee, WI.
7. Hanover Research (2020). *Quality Management Trends*. Pleasanton, CA.
8. Sureshcander, G. S. (2023). Quality 4.0—a measurement model using the confirmatory factor analysis (CFA) approach. *International Journal of Quality and Reliability Management, 40*(1), 280–303.
9. Aleksandrova, S. V., Vasiliev, V. A., & Alexandrov, M. N. (2019, September). Integration of quality management and digital technologies. In *2019 International Conference "Quality Management, Transport and Information Security, Information Technologies" (IT&QM&IS)* (pp. 20–22). IEEE.
10. Carvalho, A. V., Enrique, D. V., Chouchene, A., & Charrua-Santos, F. (2021). Quality 4.0—an overview. *Procedia Computer Science*, 181, 341–346.
11. Chiarini, A., & Kumar, M. (2021). What is quality 4.0? An exploratory sequential mixed methods study of Italian manufacturing companies. *International Journal of Production Research*, 0, 1–21. https://doi.org/10.1080/00207543.2021.1942285
12. Antony, J., McDermott, O., Sony, M., Toner, A., Bhat, S., Cudney, E. A., & Doulatabadi, M. (2023). Benefits, challenges, critical success factors and motivations of Quality 4.0—A qualitative global study. *Total Quality Management & Business Excellence*. Vol 34 (7-8), pp. 827-846 https://doi.org/10.1080/14783363.2022.2113737
13. Radziwill, N. M. (2018). *Let's Get Digital/ASQ [WWWDocument]*. ASQ.org. Available at: https://asq.org/quality-progress/articles/lets-get-digital?id=526b64168f1f4f-2c80648300336bad1a (accessed 24 June 2021).
14. ASQ (2021). Quality glossary. *Quality Resources*. Available at: https://asq.org/quality-resources/quality-glossary/q (accessed 1 November 2021).
15. Antony, J., & Sony, M. (2021). An empirical study into qualifications and skills of quality management practitioners in contemporary organizations: Results from a global survey and agenda for future research. *IEEE Transactions on Engineering Management*, 1–17. https://doi.org/10.1109/TEM.2021.3050460
16. Dias, A. M., Carvalho, A. M., & Sampaio, P. (2021). Quality 4.0: Literature review analysis, definition and impacts of the digital transformation process on quality. *International Journal of Quality & Reliability Management*, ahead-of-print No. ahead-of-print. https://doi.org/10.1108/IJQRM-07–2021–0247
17. Radziwill, N. M. (2020). *Connected, Intelligent, Automated: The Definitive Guide to Digital Transformation and Q4.0*. Quality Press.

18. Aldag, M. C., & Eker, B. (2018). What is Quality 4.0 in the era of industry 4.0? In *3rd International Conference on Quality of Life*, November 2018. University of Kragujevac.
19. Jacob, D. (2017a). *Quality 4.0 Impact and Strategy Handbook*. LNS Research, MaterControl.
20. Jacob, D. (2017b). *What Is Quality 4.0?* LNS Research, MaterControl. https://blog.lns-research.com/quality4.0
21. Allcock, A. (2018). Nikon talks quality 4.0. *Machinery*, 176(4276), 49–50. Available at: www.machinery.co.uk/machinery-features/nikon-talksquality-4-0-industry-4-0 (accessed 30 September 2019).
22. Schönreiter, I. (2017). Significance of quality 4.0 in post merger process harmonization. *Lecture Notes in Business Information Processing*, 285, 123–134. https://doi.org/10.1007/978-3-319-58801-8_11
23. Sader, S., Husti, I., & Daroczi, M. (2022) A review of quality 4.0: Definitions, features, technologies, applications, and challenges. *Total Quality Management & Business Excellence*, 33(9–10), 1164–1182. https://doi.org/10.1080/14783363.2021.1944082
24. Nyendick, M. (2017). Qualität 4.0—IT-Rückgratfüreinenfertigungsintegriertenqualitätsmotor. *ZWF ZeitschriftFürWirtschaftlichenFabrikbetrieb*, 111(4), 167–168. https://doi.org/10.3139/104.111506
25. Radziwill, N. M. (2018). Quality 4.0: Let's get digital—the many ways the fourth industrial revolution is reshaping the way we think about quality. *arXiv:1810.07829*, October, 10. https://arxiv.org/abs/1810.07829
26. Nenadal, J. (2020). The new EFQM model: What is really new and could be considered as a suitable tool with respect to quality 4.0 concept? *Quality Innovation Prosperity*, 24(1), 17–28. https://doi.org/10.12776/qip.v24i1.1415
27. Foidl, H., & Felderer, M. (2015). Research challenges of industry 4.0 for quality management. In *International Conference on Enterprise Resource Planning Systems* (pp. 121–137). Springer.
28. Chiarini, A., & Kumar, M. (2022). What is Quality 4.0? An exploratory sequential mixed methods study of Italian manufacturing companies. *International Journal of Production Research*, 60(16), 4890–4910. https://doi.org/10.1080/00207543.2021.1942285
29. Gunasekaran, A., Subramanian, N., & Ngai, W. T. E. (2019). Quality management in the 21st century enterprises research pathway towards industry4.0. *International Journal of Production Economics*, 207(1), 125–129.
30. Zhong, R. Y., Xu, X., & Wang, L. (2017). IoT-enabled smart factory visibility and traceability using laser-scanners. *Procedia Manufacturing*, 10(2), 1–14.
31. Javaid, M., Haleem, A., Singh, R. P., & Suman, R. (2021). Significance of quality 4.0 towards comprehensive enhancement in manufacturing sector. *Sensors International*, 2, 100109.
32. Bag, S., Gupta, S., & Kumar, S. (2021). Industry 4.0 adoption and 10R advance manufacturing capabilities for sustainable development. *International Journal of Production Economics*, 231, 107844.
33. Fatorachian, H., & Kazemi, H. (2021). Impact of industry 4.0 on supply chain performance. *Production Planning and Control*, 32(1), 63–81.
34. Savelyeva, M., & Shumakova, N. (2020). Innovative approach to training industry 4.0 experts. *Journal of Physics: Conference Series*, 1515 (3), 32065 (IOP Publishing).
35. de Sousa Jabbour, A. B. L., Jabbour, C. J. C., GodinhoFilho, M., & Roubaud, D. (2018). Industry 4.0 and the circular economy: A proposed research agenda and original roadmap for sustainable operations. *Annals of Operations Research*, 270(1), 273–286.
36. Sudrajat, D., Achdisty, M., Kurniasih, N., Mulyati, S., Purnomo, A., & Sallu, S. (2019). The implementation of innovation in educational technology to improve the quality of website learning in industrial revolution era 4.0 using waterfall method. *Journal of Physics: Conference Series*, 1364(1), 12044 (IOP Publishing).
37. Fettermann, D. C., Cavalcante, C. G. S., Almeida, T. D. D., & Tortorella, G. L. (2018). How does Industry 4.0 contribute to operations management? *Journal of Industrial and Production Engineering*, 35(4), 255–268.

38. Villalba-Diez, J., Schmidt, D., Gevers, R., Ordieres-Mer, J., Buchwitz, E. M., & Wellbrock, W. (2019). Deep learning for industrial computer vision quality control in the printing industry 4.0. *Sensors*, 19(18), 3987.

39. Vrchota, J., Rehor, P., Maríková, M., & Pech, M. (2021). Critical success factors of the project management in relation to industry 4.0 for sustainability of projects. *Sustainability*, 13(1), 281.

40. Cugno, M., Castagnoli, R., & Büchi, G. (2021). Openness to Industry 4.0 and performance: The impact of barriers and incentives. *Technological Forecasting and Social Change*, 168, 120756.

41. Hidayatno, A., Destyanto, R., & Hulu, C. A. (2019). Industry 4.0 technology implementation impact to industrial sustainable energy in Indonesia: A model conceptualization. *Energy Procedia*, 156, 227–233.

42. Kurnikova, M., Bolgova, E., Bolgov, S., Khaitbaev, V., & Dodorina, I. (2020, March). Industry 4.0: Theoretical foundations and the strategic priorities of Russian regions. In *New Silk Road: Business Cooperation and Prospective of Economic Development (NSRBCPED 2019)* (pp. 195–200). Atlantis Press.

43. Hollowell, J. C., Kollar, B., Vrbka, J., & Kovalova, E. (2019). Cognitive decision-making algorithms for sustainable manufacturing processes in Industry 4.0: Networked, smart, and responsive devices. *Economics, Management, and Financial Markets*, 14(4), 9–15.

44. Tortorella, G. L., Silva, E., & Vargas, D. (2018). An empirical analysis of total quality management and total productive maintenance in industry 4.0. In *Proceedings of the International Conference on Industrial Engineering and Operations Management (IEOM)* (pp. 742–753).

45. da Silva, V. L., Kovaleski, J. L., & Pagani, R. N. (2019). Technology transfer in the supply chain oriented to industry 4.0: A literature review. *Technology Analysis & Strategic Management*, 31(5), 546–562.

46. Marciniak, R., Moricz, P., & Baksa, M. (2019). Towards business services 4.0-digital transformation of business services at a global technology company. In *International Workshop on Global Sourcing of Information Technology and Business Processes* (pp. 124–144). Springer.

47. Bigliardi, B., Bottani, E., & Casella, G. (2020). Enabling technologies, application areas and impact of industry 4.0: A bibliographic analysis. *Procedia Manufacturing*, 42, 322–326.

48. Ancarani, A., Di Mauro, C., & Mascali, F. (2019). Backshoring strategy and the adoption of industry 4.0: Evidence from Europe. *Journal of World Business*, 54(4), 360–371.

49. Alaloul, W. S., Liew, M. S., Zawawi, N. A. W. A., & Kennedy, I. B. (2020). Industrial Revolution 4.0 in the construction industry: Challenges and opportunities for stakeholders. *Ainshams Engineering Journal*, 11(1), 225–230.

50. Dachs, B., Kinkel, S., & Jäger, A. (2019). Bringing it all back home? Backshoring of manufacturing activities and the adoption of Industry 4.0 technologies. *Journal of World Business*, 54(6), 101017.

51. Dalmarco, G., Ramalho, F. R., Barros, A. C., & Soares, A. L. (2019). Providing industry 4.0 technologies: The case of a production technology cluster. *The Journal of High Technology Management Research*, 30(2), 100355

52. Mitra, A. (2021). On the capabilities of cellular automata-based MapReduce model in industry 4.0. *Journal of Industrial Information Integration*, 21, 100195.

53. Li, D., Fast-Berglund, A., & Paulin, D. (2019). Current and future Industry 4.0 capabilities for information and knowledge sharing. *The International Journal of Advanced Manufacturing Technology*, 105(9), 3951–3963.

54. Zonnenshain, A., & Kenett, R. S. (2020). Quality 4.0—the challenging future of quality engineering. *Quality Engineering*, 32(4), 614–626.

55. Zulfiqar, M., Antony, J., Swarnakar, V., Sony, M., Jayaraman, R., & McDermott, O. (2023). A readiness assessment of Quality 4.0 in packaging companies: An empirical investigation, Vol 34 (11-12), pp. 1334 - 1352 *Total Quality Management & Business Excellence*. https://doi.org/10.1080/14783363.2023.2170223

56. Sader, S., Husti, I., & Daroczi, M. (2021). A review of quality 4.0: Definitions, features, technologies, applications, and challenges. *Total Quality Management & Business Excellence*, 1, 1164–1182.

57. Tsai, C. W., Lai, C. F., Chao, H. C., & Vasilakos, A. V. (2015). Big data analytics: A survey. *Journal of Big Data*, 2(1), 1–32.

58. Johnson, S. (2019). Quality 4.0: A trend within a trend. *Quality*, 58(2), 21–23.

59. Sahu, C. K., Young, C., & Rai, R. (2021). Artificial Intelligence (AI) in augmented reality (AR)-assisted manufacturing applications: A review. *International Journal of Production Research*, Vol. 59 (16), pp. 4903 - 4959https://doi.org/10.1080/00207543.2020.1859636.

60. Velandia, D. M. S., Kaur, N., Whittow, W. G., Conway, P. P., & West, A. A. (2016). Towards industrial internet of things: Crankshaft monitoring, traceability and tracking using RFID. *Robotics and Computer-Integrated Manufacturing*, 41(1), 66–77.

61. Zhong, R. Y., Xu, X., & Wang, L. (2017). IoT-enabled smart factory visibility and traceability using laser-scanners. *Procedia Manufacturing*, 10(2), 1–14.

62. Gilchrist, A. (2016). *Industry4.0: The Industrial Internet of Things*. Apress.

63. Funk, M., Kosch, T., Kettner, R., Korn, O., & Schmidt, A. (2016). Motion EAP: An overview of 4 years of combining industrial assembly with augmented reality for industry4.0. In *Proceedings of the 16th International Conference on Knowledge Technologies and Data Driven Business*. Hochschule.

64. Romero, D., Stahre, J., Wuest, T., Noran, O., Bernus, P., FastBerglund, Å., & Gorecky, D. (2016). Towards an operator 4.0 typology: A human-centric perspective on the fourth industrial revolution technologies. In *Proceedings of the International Conference on Computers and Industrial Engineering (CIE46)* (pp. 29–31), Elsevier.

65. Bortolini, M., Ferrari, E., Gamberi, M., Pilati, F., & Faccio, M. (2017). Assembly system design in the industry 4.0 era: A general framework. *IFAC-Papers On-Line*, 50(1), 5700–5705.

66. Satoglu, S., Ustundag, A., Cevikcan, E., & Durmusoglu, M. B. (2018). Lean transformation integrated with industry 4.0 implementation methodology. In *Industrial Engineering in the Industry 4.0 Era* (pp. 97–107). Springer.

67. Tortorella, G. L., Cauchick-Miguel, P. A., Li, W., Staines, J., & McFarlane, D. (2022). What does operational excellence mean in the fourth industrial revolution era? *International Journal of Production Research*, 60(9), 2901–2917 https://doi.org/10.1080/00207 543.2021.1905903.

68. Srinivasan, R., Kumar, M., & Narayanan, S. (2020). Human resource management in an industry 4.0 era: A supply chain management perspective. In T. Y. Choi et al. (Eds.), *The Oxford Handbook of Supply Chain Management*. Oxford University Press. https://doi. org/10.1093/oxfordhb/9780190066727.013.39.

69. Bittencourt, V., Alves, A. C., & Leao, C. (2021). Industry 4.0 triggered by lean thinking: Insights from a systematic literature review. *International Journal of Production Research*, 59(4), 1–15.

70. Miladin Stefanović, M., Dorđević, A., Puškarić, H., & Petronijević, M. (2019). Web based cloud solution for support of quality management 4.0 in the concept of industry 4.0. *Quality Festival 2019 Conference*, 29th May–1st June (pp. 443–448).

71. Zavadska, Z., & Zavadsky, J. (2018). Quality managers and their future technological expectations related to Industry 4.0. *Total Quality Management and Business Excellence*, 1–25.

72. Jabbour, C. J. C., & Renwick, D. W. S. (2018). The soft side of environmentally sustainable organizations. *RAUSP Management Journal*, 53(4), 622–627.

73. Machado, C. G., Winroth, M. P., & da Silva, E. H. D. R. (2020). Sustainable manufacturing in Industry 4.0: An emerging research agenda. *International Journal of Production Research*, 58(5), 1462–1484. https://doi.org/10.1080/00207543.2019.1652777

74. Gunasekaran, A., & Spalanzani, A. (2012). Sustainability of manufacturing and services: Investigations for research and applications. *International Journal of Production Economics*, 140(1), 35–47. https://doi.org/10.1016/j.ijpe.2011.05.011.

75. Bonvoisin, J., Stark, R., & Seliger, G. (2017). Field of research in sustainable manufacturing. In R. Stark, G. Seliger, & J. Bonvoisin (Eds.), *Sustainable Manufacturing: Challenges, Solutions and Implementation*. Springer International Publishing. Sustainable Production, Life Cycle Engineering and Management. https://doi.org/10.1007/978-3-319-48514-0.

76. Teknikföretagen. (2017). *Made in Sweden 2030—Strategic Agenda for Innovation in Production*. www.teknikforetagen.se/globalassets/i-debatten/publikationer/produktion/made-in-sweden-2030-engelsk.pdf.

77. Waibel, M. W., Steenkamp, L. P., Moloko, N., & Oosthuizen, G. A. (2017). Investigating the effects of smart production systems on sustainability elements. *Procedia Manufacturing*, 8, 731–737. https://doi.org/10.1016/j.promfg.2017.02.094.

78. Duarte, S., & Cruz-Machado, V. (2017). Exploring linkages between lean and green supply chain and the industry 4.0. In *Proceedings of the Eleventh International Conference on Management Science and Engineering Management* (pp. 1242–1252).

79. Hermann, M., Pentek, T., & Otto, B. (2016). Design principles for industrie 4.0 scenarios. In *2016 49th Hawaii International Conference on System Sciences (HICSS)* (pp. 3928–3937). IEEE. https://doi.org/10.1109/HICSS.2016.488.

80. Kiel, D., Muller, J., Arnold, C., & Voigt, K.-I. (2017). Sustainable industrial value creation: Benefits and challenges of industry 4.0. *International Journal of Innovation Management*, 21(8), 1740015. https://doi.org/10.1142/S1363919617400151.

81. Waibel, M. W., Steenkamp, L. P., Moloko, N., & Oosthuizen, G. A. (2017). Investigating the effects of smart production systems on sustainability elements. *Procedia Manufacturing*, 8, 731–737. https://doi.org/10.1016/j.promfg.2017.02.094.

82. Martin, J., Elg, M. H., & Gremyr, I. (2020). The many meanings of quality: Towards a definition in support of sustainable operations. *Total Quality Management & Business Excellence*. https://doi.org/10.1080/14783363.2020.1844564

83. Siva, V., Gremyr, I., & Halldórsson, Á. (2018). Organising sustainability competencies through quality management: Integration or specialisation. *Sustainability*, 10(5), 1326. https://doi.org/10.3390/su10051326

84. Jamwal, A., Agarwal, R., Sharma, M., & Giallanza, A. (2021). Industry 4.0 technologies for manufacturing sustainability: A systematic review and future research directions. *Applied Sciences*, 11, 5725. https://doi.org/10.3390/app11125725

85. Pawlak, Z. (1982). Rough sets. *International Journal of Computer & Information Sciences*, 11(5), 341–356.

86. Zhai, L. Y., Khoo, L. P., & Zhong, Z. W. (2008). A rough set enhanced fuzzy approach to quality function deployment. *The International Journal of Advanced Manufacturing Technology*, 37(5), 613–624.

87. Zhai, L. Y., Khoo, L. P., & Zhong, Z. W. (2009). A rough set based QFD approach to the management of imprecise design information in product development. *Advanced Engineering Informatics*, 23(2), 222–228.

88. Chakraborty, S., Dandge, S. S., & Agarwal, S. (2020). Non-traditional machining processes selection and evaluation: A rough multi-attributive border approximation area comparison approach. *Computers & Industrial Engineering*, 139, 106201.

89. Zhu, G. N., Hu, J., & Ren, H. (2020). A fuzzy rough number-based AHP-TOPSIS for design concept evaluation under uncertain environments. *Applied Soft Computing*, 91, 106228.

90. Huang, G., Xiao, L., & Zhang, G. (2020). Improved failure mode and effect analysis with interval-valued intuitionistic fuzzy rough number theory. *Engineering Applications of Artificial Intelligence*, 95, 103856.
91. Ecer, F., Pamucar, D., Mardani, A., & Alrasheedi, M. (2021). Assessment of renewable energy resources using new interval rough number extension of the level based weight assessment and combinative distance-based assessment. *Renewable Energy*, 170, 1156–1177.
92. Saaty, T. L. (1988). *What Is the Analytic Hierarchy Process?* (pp. 109–121). Springer Berlin Heidelberg.
93. Alayón, C. (2016). *Exploring Sustainable Manufacturing Principles and Practices* (Doctoral dissertation). Jonkoping: Jönköping University, School of Engineering.
94. De Long, D. W., & Fahey, L. (2000). Diagnosing cultural barriers to knowledge management. *Academy of Management Perspectives*, 14(4), 113–127.
95. Waddell, D., Devine, J., Jones, G. R., & George, J. M. (2007). *Contemporary Management*. McGraw Hill Irwin.
96. Xu, X. From cloud computing to cloud manufacturing. *Robotics and Computer-Integrated Manufacturing*, 28(1), 75–86.
97. Liu, Y., Wang, L., Wang, X. V., Xu, X., & Zhang, L. (2019). Scheduling in cloud manufacturing: State-of-the-art and research challenges. *International Journal of Production Research*, 57, 4854–4879.

8 Intelligent Manufacturing in the Context of Industry 4.0 in Belarus
Overview and Perspectives

Yuliya Pranuza
Francisk Skorina Gomel State University, Gomel, Belarus

8.1 INTRODUCTION

Digital transformation in manufacturing and adopting Industry 4.0 tools are significant factors in ensuring national economic competitiveness. Digital technologies change the economy, public policy, governance, industry, and manufacturing. Digital transformation is a new driving force accelerating innovation in manufacturing.

International experience confirms the importance of digital transformation and intelligent manufacturing. Digital transformation is used by many countries to create industrial competitiveness and to solve economic problems. Digital transformation is promoted through government policies. Many countries all of world aim to implement strategies to facilitate the transition to a digital economy. To develop intelligent manufacturing in Belarus, it's important to study the foreign economy experience and improve the tools of government support.

The Digital Strategy 2025 in Germany is a comprehensive governmental plan which aimed to make the national economy competitive under the conditions of digitalization. The strategy aims to promote digital innovation, increase connectivity, and enhance digital skills due to education. The digital strategy in Germany seeks to smart manufacturing using Industry 4.0 online platform. There are five working groups in Platform Industry 4.0 covering the following topics: standards and norms, research and development (R&D) and innovation, security of network systems, legal frameworks, and education and training [1].

In the United States, the Industrial Internet solution has been developed. The Industrial Internet is a network of connected devices, machines, and systems that collect and share data to improve industrial processes and operations. The Industrial Internet is being used to optimize production processes, reduce downtime, improve quality control in manufacturing facilities, and transform manufacturing.

DOI: 10.1201/9781003405870-8

The Digitising European Industry Strategy was adopted in the European Union in order to achieve considerable progress in the digital economy. Digital transformation is accomplished by governments, and the strategy is oriented toward digitalization of small and medium enterprises.

Japan is pushing digitalization by increasing R&D investments in manufacturing under the government. Japan's intelligent manufacturing is a concept that focuses on the integrated use of artificial intelligence, the Internet of Things (IoT), and robotics in the manufacturing process. The ultimate goal is the creation of more efficient manufacturing that can compete globally. One of the key features of Japan's intelligent manufacturing is the use of data analytics to optimize production processes and improve product quality. This involves data-collecting sensors and other sources throughout the manufacturing process.

Korea is pushing to create an ecosystem for innovative growth and digitalized industries. The government is supporting manufacturing innovation using its "smart factories" policy.

Belarus has created conditions for intelligent manufacturing development. At the same time, the practice of its implementation has revealed a number of problems and new practical tasks.

This chapter aims to summarize the experience of Belarus in the field of intelligent manufacturing, research the potential perspectives for digital transformation of the industry, and define directions for a digital transformation of the economy.

8.2   DIGITAL TRANSFORMATION OF MANUFACTURING

Modern digital technologies allow manufacturers to automate repetitive tasks and optimize production processes. Virtual copies of the physical world (cloud systems, IoT, AI, robotics, big data technology, blockchain, the formation of smart factories, etc.) are created as a result of digital transformation and the application of Industry 4.0 technologies.

Intelligent manufacturing means the integration of advanced technologies into the production processes for improving it. Intelligent manufacturing involves the use of data analytics, machine learning, and automation to optimize production processes, reduce waste and errors, and enhance decision-making. As a result, modern digital manufacturing technologies are transforming the industry and are key to increasing industrial efficiency.

Industry 4.0 is the ongoing automation of conventional manufacturing and industrial processes using modern smart technologies. Industry 4.0 leads to "smart anything", such as smart grids, smart energy, smart logistics, smart devices, smart buildings, smart services, smart manufacturing, smart factories, smart cities, and so on [2].

Some of the key benefits of intelligent manufacturing include the following aspects.

First, intelligent manufacturing has ability to increase productivity and added value by optimizing production processes.

Second, enhanced quality. The use of advanced technologies such as sensors and data analytics help to prevent defects and errors early in the production process, leading to higher-quality products.

Third, greater flexibility. Intelligent manufacturing allows rapid reconfiguration of production lines in order to adapt to changes in the demand or market conditions.

Next, cost savings. By improving efficiency and enhancing quality, intelligent manufacturing can help manufacturers reduce costs and increase profitability.

Intelligent manufacturing enables manufacturers to create smart factories that are more flexible, agile, and responsive to customer needs. It also allows real-time monitoring and control of production processes, which helps to identify and resolve issues quickly.

A smart factory is a "people-oriented high-tech intelligent factory that integrates all production processes from product planning to sales with information and communication technology to produce customized products at minimum cost and time" [3].

A smart factory combines the manufacturing processes with IoT, accumulates big data generated from production processes through sensors, analyzes them with artificial intelligence, and optimizes the process intelligently in virtual space.

Overall, intelligent manufacturing is transforming the manufacturing itself by enabling manufacturers to create more efficient and flexible production processes that can be adapted to the changes of the customer needs and market conditions.

8.3 INNOVATION POLICY IN BELARUS

In Belarus, great attention is paid to the development of innovative activity. There is a broad government recognition of the importance of innovation for the future growth and competitiveness of industry and the national economy.

The aim of the state innovation policy is establishing a favorable environment for innovative development in Belarus and raising the competitiveness of the national economy. The highest level of government and all relevant public bodies are responsible for this policy.[1]

One of the key objectives of innovation policies is the transition of Belarus towards a knowledge economy and intelligent manufacturing.

Belarus realizes its innovation policy. The national innovation policy is managed by numerous governmental documents. The main objective of social and economic development determines the priority and importance of the innovative development for the economy of Belarus. These documents include the program of social and economic development of Belarus for 2021–2025,[2] the National Strategy for Sustainable Development of Belarus until 2035,[3] the strategy Science and Technology: 2018–2040,[4] and the State Program for Innovative Development of Belarus for 2021–2025.[5]

The State Programme for Social and Economic Development includes digital transformation among five key priorities for development.

The main document defining the frameworks for the development of science and technology is the Law on Innovation Policy and Innovation Activities. The law provides a legal framework for the implementation of innovation policies. This document unifies strategic documents, formulates measures for accelerating research and innovation activities, and determines the functions of public authorities. It has an objective to create a favorable environment for Industry 4.0. By providing a legal framework for innovation policies and activities, the document aims to promote the development of new technologies and products.

The list of priorities for single digital transformation activities that will manage science and innovation starting in 2021 is formulated for Belarus:

- biological, pharmaceutical and chemical technologies;
- digital transformation;
- energy, construction and environmental management;
- agriculture and food technologies;
- engineering, machine building and materials science;
- social well-being and national defense.[6]

Belarus has elaborated a well-developed system of public bodies and organizations in support of innovation activity, which forms the backbone of the national innovation system. These bodies in the innovation system have defined functions and roles in the innovation process.

The State Committee on Science and Technology of Belarus[7] is the republican body of state administration, the aim of which is to conduct state policy and implement the function of state management in innovative activities. The State Committee on Science and Technology focuses on innovative developments and the latest technologies. The State Committee on Science and Technology coordinates innovation policy and technology development, controls state science and R&D programs, and provides the state scientific and technical expertise.[8]

The implementation of the national innovation policy is realized on the "project-based" principle. The planning of innovation policy measures is realized through the adoption of innovative programs. The State Programmes for Innovative Development are the means for controlling innovation policy. The program is a tool for planning and monitoring the innovation process in Belarus and represents the combination of programs and projects focused on the innovative development of Belarus.

Belarus has been implementing state programs of innovative development since 2007. The first program was adopted for the period 2007–2010, the second one for 2011–2015, and the third one for 2016–2020.

The State Programme for Innovative Development of Belarus for the period 2021–2025 is a comprehensive plan that aims to promote innovation and technological advances in various sectors of the Belarusian economy. The state program aims to define the priorities of economic development of Belarus for 2021–2025 and accelerate development of innovative sectors of the economy and is the main instrument for the implementation of the most important state innovation policy.

The program is designed to foster innovation and entrepreneurship, to enhance the competitiveness of local businesses, and to promote sustainable economic growth.

The drafts of the digital transformation policy were included in the State Programme for Innovative Development of Belarus for the period 2021–2025. The program is introducing innovative projects for starting new industries based on the Fourth Industrial Revolution.

The objective of the State Programme for Innovative Development of Belarus for 2021–2025 is "to achieve the level of innovative development of the leading countries of Eastern Europe through the intellectual potential of the Belarusian nation".[9]

To achieve this goal, efforts will be put into the following main tasks:

- to create the best conditions in the Eastern European region for boosting innovative activity by the implementation of the best world practices;
- to ensure the innovative development of traditional sectors of the national economy through increasing the scientific intensity of production;
- to start new and accelerate the development of existing high-tech economies;
- to increase the presence and consolidation of the position of Belarus in the world markets of science-intensive and high-tech products.

The State Programme for Innovative Development of Belarus for 2021–2025 is expected to considerably contribute to the modernization and diversification of the Belarusian economy, as well as to enhance its competitiveness in the global market.

The Programme for Innovative Development includes a list of target indicators (Table 8.1).

Table 8.1 shows that in 2022, the level of manufacturing companies' innovative activity reached 27.8%. The government have set a target of reaching the share of enterprises active in innovations in the total number of manufacturing companies 30.5% by 2035. The share of the total volume of shipped manufacturing products rose from 13.1% in 2015 to 17.7% in 2022. It is planned to increase the share of shipped innovative products in the total volume of shipped industrial products to 21.0% in 2025.

Table 8.2 contains indicators of innovation at the level of firms. In 2022, 27.8% of the industrial enterprises were considered active in innovation (compared with 15.4% in 2010) [4].

In 2022, the indicator of the specific weight of shipped innovative products reached 17.7%. Thus, in 2010–2022, it was possible to significantly increase the share of shipped innovative products.

TABLE 8.1

Target Indicators of the State Program for Innovative Development of Belarus for 2021–2025

		Forecast		
Indicators	2022	2023	2024	2025
Share of companies active in innovations in the total number of manufacturing companies, percent	27.8	30.0	30.2	30.5
Share of shipped innovative output in total industrial output shipped, percent	17.7	20.4	20.6	21.0

Source: State Program for Innovative Development of Belarus for 2021–2025.

TABLE 8.2

Indicators of Innovation Activity in Belarus for 2010–2022

Indicators	2010	2013	2014	2015	2016	2017	2018	2019	2020	2021	2022
Share of companies active in innovations in the total number of manufacturing companies, %	15.4	21.7	20.9	19.6	20.4	21.0	23.3	24.5	27.1	27.5	27.8
Share of shipped innovative output in total industrial output shipped, %	14.5	17.8	13.9	13.1	16.3	17.4	18.6	16.6	17.9	19.8	17.7

Source: National Statistic Committee of Belarus [4].

The modern structure of industry in Belarus is characterized by a high percentage of manufacturing industries. Manufacturing in Belarus is diverse, with a range of industries including machinery, electronics, chemicals, textile and metal production, and food processing.

The country has a well-developed machinery industry, producing a range of products including tractors, trucks, and construction equipment. Belarus also has a developed electronics industry, producing a range of consumer goods and electronic components.

Chemical production is another important industry in Belarus, with the country producing a range of chemicals including fertilizers, plastics, and pharmaceuticals. The textile industry is also significant, with Belarus producing a range of clothing and textiles for both domestic and export markets.

The food processing industry is also an important sector in Belarus, with the country producing a range of food products, including meat and dairy products and processed foods. The sector has grown significantly in recent years, with a focus on increasing exports to new markets.

Belarus has achieved significant success and is confidently moving along the path of innovative development. Following the 2022 results, the Global Innovation Index (GII) for Belarus is the 77th position out of 132 world's economies (the 72nd position in 2019) [5].

The level of industrial development in Belarus can be compared with the help of the Competitive Industrial Performance Index (CIP). The index is a measure of a country's ability to compete in the global market. It takes into account factors such as productivity, innovation, infrastructure, and human capital. The index is calculated based on a number of indicators, including the share of high-tech exports, research and development expenditures, and quality of infrastructure. Countries with a high CIP index are generally considered to have a strong industrial base.

Belarus has not significantly changed its positions in the CIP rank for 2010–2020. Belarus took the 53th place among 154 countries ranked in the world in 2021 (upper middle-income industrial economy) (Table 8.3) [6].

TABLE 8.3

Belarus CIP Rank for 2010–2021

Year	2010	2011	2012	2013	2014	2015	2016	2017	2018	2019	2020	2021
CIP rank	44	41	39	44	43	49	50	46	46	46	47	53

Source: Competitive Industrial Performance Index [6].

TABLE 8.4

Individual Indicators of Development of Manufacturing Industry in Belarus

Indicators	2015	2016	2017	2018	2019	2020	2021
Manufacturing value added as a proportion of GDP, %	20.6	21.1	22.0	22.4	22.4	22.6	23.4
Manufacturing employment as a proportion of total employment, %	20.4	18.1	18.4	17.6	17.4	17.8	17.6
Proportion of medium and high-tech industry value added in total value added, %	41.1	38.8	40.0	42.2	42.2	37.4	40.4
Research and development expenditure as a proportion of GDP, %	0.5	0.5	0.58	0.6	0.58	0.54	0.47
Medium- and high-technology product exports, %	30.3	32.7	30.6	30.7	32.1	33.9	31.7
High-technology exports (% of manufactured exports)	4.5	4.9	4.4	4.1	4.3	4.8	5.6
Sales of new-to-market and new-to-enterprise, %	12.3	15.3	16.2	17.3	15.3	15.7	18.0

Source: [4, 7].

Manufacturing is the leading industry of the Belarusian economy, contributing 20–23% of gross domestic product (GDP) in 2015–2021 (Table 4). The highest share (18.0% in 2021) of employees in Belarus is employed in the manufacturing sector. This is a significant proportion of the country's workforce, and it reflects the importance of manufacturing to the Belarusian economy.

One of the important indicators in the analysis of the level of manufacturing is gross value added, which serves as a source of economic growth and the formation of state income. More than one fifth of the value added of the GDP in Belarus is created in manufacturing (23.0% in 2021). For 2015–2021, there was a slight increase in this indicator (2.1 percentage points). Moreover, medium and high-tech industry value added is at a level of around 40% of total value added.

Research and development expenditures were no more than 1% of GDP over 2015–2021. In 2021, Belarus spent 0.47% of GDP on research and development activity. This is a relatively low percentage compared to other countries, especially developed nations, which spend around 2–4% of their GDP annually on research and development activity. The share of public expenditure on research and development will be raised by 2030 by the government to 2.5% of GDP.

Manufacturing enterprises implement innovation activity. In 2021 in Belarus, 18% of sales were new-to-market and new-to-enterprise.

By the end of 2021, the share of medium- and high-tech product exports in total exports reached 31.7%, or up 1.8 percentage points in comparison with 2018.

Products with high R&D intensity, such as scientific instruments, pharmaceuticals, and electrical machinery, are high-technology exports. Their share within the total export volume made up 5.6% in 2021 (4.1% in 2018).

According to the OECD, the exists a four-way classification of exports: high, medium-high, medium-low, and low technology. This classification is based on the relevance of expenditures on research and development in relation to the gross output and value added of different types of industries producing goods for export.

The evolution of the structure of production of manufacturing in Belarus according to technological intensity for 2016–2021 was analyzed. The results are represented in Table 8.5.

The analysis of the structure of the manufacturing industries for 2016–2021 allows us to determine that the medium-low technology industries dominate (37.3% in 2021) and increase (+2.3 percentage points for 2016–2021) in the structure of production of manufacturing in Belarus. The medium-low technology industries include the production of coke and refined petroleum products (+2.5 percentage points for 2016–2021) and the manufacture of fabricated metal products (+1.0 percentage points for 2016–2021). The share of the high-technology industries did not change during the research period (for 2016–2021).

The larger share of production of manufacturing industries in Belarus belongs to industries with medium-low technology and low technology (73.3% in 2021). The government has been actively promoting the development of high-tech manufacturing

TABLE 8.5
Structure of Manufacturing in Belarus According to Technological Intensity, 2011–2021, %

Manufacturing Industry	2011	2016	2017	2018	2019	2020	2021
High technology	1.7	2.9	3.0	2.8	2.8	3.3	3.1
Medium high technology	25.9	20.0	21.1	22.4	22.8	21.9	23.4
Medium low technology	35.7	27.7	29.5	30.5	29.2	26.3	28.0
Low technology	26.7	35.1	34.5	32.9	33.9	37.0	34.8

Source: [4].

sectors. At the same time, the share of high-technology industry is extremely low (3.3% in 2021).

On the whole, Belarus is among the world's scientifically and technologically advanced nations and has considerable potential for innovation-driven economy development.

Belarus has been making significant strides in manufacturing innovation in recent years, with a focus on advanced technologies. To confirm this point, one can mention a high proportion of high-tech exports in total volume of trade and medium-and high-tech production.

However, the manufacturing sector in Belarus faces a number of challenges. These include a lack of investments in research and development, outdated equipment and technology, and reliance on exports to Russia.

At the same time, innovative development in manufacturing is one of the priorities in Belarus. While Belarus has made progress in improving its industrial competitiveness in recent years, there is still room for improvement in such areas as digital transformation and intelligent manufacturing.

8.4   EXPERIENCE OF BELARUS IN INTELLIGENT MANUFACTURING

In recent years, Belarus has accomplished considerable progress in the innovation area in the manufacturing industry, with particular emphasis on advanced technologies.

The Belarusian government has accepted digital transformation for intelligent manufacturing as a key task of economic development. The digital transformation of the economy is an important factor of national development.

Belarus has been focusing on digitalization in manufacturing to improve efficiency and competitiveness.

First, Belarus is actively developing its intelligent manufacturing capabilities. The country's government has launched several initiatives to support the promotion and adoption of advanced technologies in the manufacturing sector.

The Program for Social and Economic Development of Belarus (for 2021–2025) has a separate chapter devoted to digital transformation (chapter 7). This program provides the introduction of intelligent manufacturing based on the concept of Industry 4.0, the formation of smart industry.[10]

The Strategy for Science and Technology (2018–2040) provides "transformation of traditional industries and the transition to digital production".[11]

The National Strategy for Sustainable Development (until the year 2035) provides a regulatory framework for supporting the digital transformation of the economy (chapter 6). The basic part of the strategy is devoted to digitalization of the economy. The development will aim at increasing its competitiveness through the implementation of the concept of Industry 4.0, digitalization of manufacturing. A lot of attention in the strategy is given to the "formation of digital platforms for the interaction of industrial organizations, the creation of smart factories, the digitalization of traditional activities, the development of consulting in the field of digital transformation".[12]

Projects for the digitalization of the mechanical, pharmaceutical, petrochemical, energy, and transportation industries are being designed and implemented through the National Program for Innovative Development 2021–2025.[13]

The State Program for Digital Development of Belarus for 2021–2025 includes the sub-program Digital Development for Branches of Economy.[14]

Second, the Council for Digital Development of the economy was established in 2019. The council determines the aims of digital transformation of the national economy and sets priority tasks for the introduction of digital technologies in industries. The council has the competence to create and develop digital infrastructure.

The Ministry of Communications and Informatization of Belarus has additional powers in digital development [8]. The Digital Development Centre and the Centre for Prospective Research have been set up to provide practical support to digital development issues. The Institute of Digital Offices was formed.[15]

The Digital Development Center is the leading organization for the development of activities in the field of digital transformation [9]. Digital development of manufacturing is also one of the main sectors of its work.

Third, the State Committee for Standardization of Belarus adopted the state standard on digital transformation in which the basic definitions of digital transformation in manufacturing were given (digital transformation, digital technology, digital infrastructure, digital twins, digitalization, digital platform, etc.) [10].

One of the key initiatives is the creation of a national innovation platform for intelligent manufacturing. The country's national innovative platform for intelligent manufacturing aims to bring industry, academia, and government together to drive innovation and collaboration in this field.

Belarus is also investing in the development of advanced technologies such as robotics, artificial intelligence, and the IoT. The country has established several research centers and innovation hubs to support the development of these technologies and promote their adoption in the manufacturing sector.

EnCata (Engineering Catalyst) is a fully equipped factory focused on product development, design engineering, and prototype manufacturing [11]. EnCata helps companies supervise a wide variety a number of challenges in production. EnCata provides digital transformation services for all forms of business. EnCata has all the necessary stack technology for industrial transformation in manufacturing.

To share South Korea's development experience and knowledge in the digital transformation in the machine building industry, cooperation between South Korea and Belarus was examined in detail. The Korea Development Institute embarked on a knowledge sharing program with Belarus in 2018 under the overarching topic "Digital Transformation of the National Economy of Belarus" [12]. As a part of the project Intellectual Support for the Organization of Digital Transformation of the Belarusian Industry, the program Development of a Digital Transformation of Manufacturing Process Model in the Machine Building Sector ("Digital Factory" Creation) was implemented (2021). The objective of this project was developing guidelines for digital transformation in industry in Belarus. The goal of this project is to share Korean knowledge and development experience of digital transformation in the machine building sector by using specific examples of Korean corporations.

The experts of South Korea and Belarus suggested recommendations for digital transformation and a roadmap for implementation and a program of transformation in industrial sectors (petrochemical industry, pharmaceutics, mechanical engineering).

A pilot manufacturing demo center for development of smart industry in the technologies of the Industry 4.0 concept was set up in the Brest region in Belarus in 2023 with assistance from the European Union (through an international technical assistance project with UNIDO) [13].

The establishment of demonstration center for collective use of equipment for smart manufacturing promote learning intelligent manufacturing. The center will provide Industry 4.0 technological learning, innovation, and smart manufacturing and support capacity-building services in the Brest region. The demonstration center will be open to students and professionals. The prepared training programs will allow students and specialists to get acquainted with the technologies of Industry 4.0 and introduce modern technologies at their enterprises.

The project is to ensure a smooth transformation of Belarus to Industry 4.0 by meeting challenges such as lack of information on modern technological solutions in manufacturing. Establishing demonstration and innovation centers for Industry 4.0 technologies promotes innovation and smart manufacturing in Belarus. This is the first such center in Belarus. It is planned to scale this pilot project and launch a similar project in other regions of Belarus (Mogilev and Vitebsk regions).

A new statistical form, "Questionnaire on the Use of Digital Technologies in an Organization" has been adopted and put into effect starting from the state statistical observation of January 1, 2023 [14].

The country is promoting the use of digital technologies in manufacturing through training programs and workshops to enhance the skills of the workforce and improve efficiency and productivity.

In addition, Belarus is actively promoting the use of digital technologies in manufacturing through various training programs and workshops. These programs aim to enhance the skills of the workforce and help manufacturers leverage digital technologies to improve efficiency and productivity.

With its focus on innovation and technology, strategic location, and skilled workforce, Belarus is well positioned to become a leader in intelligent manufacturing in the region.

Belarus is also actively promoting innovation in other sectors, such as information technology (IT) and bio-technology. The country has established several innovation centers and technology parks to support startups and entrepreneurs, providing them with access to funding, mentoring, and networking opportunities. The government is also offering tax incentives and other benefits to companies engaged in R&D activities.

Belarus has a strong IT industry, with a large pool of skilled software developers and engineers, and is home to several successful tech startups. The country is leveraging this expertise to develop new technologies and solutions in areas such as cybersecurity, fintech, and e-commerce. In the bio-tech sector, Belarus is investing in research and development of new drugs and medical devices, as well as promoting the growth of local companies in this field.

In addition, Belarus has the capacity to build AI systems [15].

Conferences on intelligent manufacturing were held in Belarus: Smart Industry Expo [16], Digital Economy Forum [17], and others.

All in all, Belarus is well positioned to become a leader in intelligent manufacturing in the region. The attention paid to innovation and technology, combined with the country's strategic location and skilled workforce, make it an attractive area for manufacturers looking to adopt advanced technologies and improve their competitiveness.

Belarus is committed to fostering a culture of innovation and entrepreneurship, with the aim of driving economic growth and creating new opportunities for its citizens. Digitalization is a key driver of innovation and competitiveness in Belarusian manufacturing, and the government and companies are investing heavily in this area.

8.5 MAJOR OBSTACLES REGARDING INTELLIGENT MANUFACTURING IN BELARUS

For Belarus, despite recent measures, the processes of digital transformation and intelligent manufacturing are at an early stage of development.

At present, the digital transformation in Belarus is not developing properly. In the manufacturing enterprises, there is a negative trend concerning the dynamics of implementing technological innovations.

The following problems can be distinguished in the field of digital transformation in manufacturing in Belarus.

1. Limited investment in research and development and advanced technologies. Government financial support is not sufficient. Investment and research in intelligent manufacturing area are very weak. Intelligent manufacturing requires significant investment in research and development to develop new technologies and processes based on advanced technologies.
2. Limited access to advanced technology and intelligent manufacturing technology. Belarus does not have access to the latest manufacturing technologies due to trade restrictions.
3. Lack of efficient tools of state support for intelligent manufacturing.
4. Inadequate infrastructure. Intelligent manufacturing requires a robust digital infrastructure, including high-speed internet connectivity and secure data storage facilities.
5. Lack of bases of technological solutions for intelligent manufacturing in industrial enterprises. Many enterprises are not ready for digitalization. There are just a few examples of organizations using intelligent manufacturing (Minsk Motor Plant, Metallopolymer, Optron, Adani, Polymaster, etc.)
6. Shortage of skilled workers with expertise in intelligent manufacturing areas.
7. Lack of established business models of smart factories targeting different types of industrial enterprises.
8. Lack of collaboration between industry and academia.

8.6 RECOMMENDATIONS FOR IMPLEMENTATION OF INTELLIGENT MANUFACTURING IN BELARUS

The government of Belarus is working to develop the adoption of intelligent technologies in manufacturing in the country. This involves providing incentives for companies to invest in these technologies, promoting research and development in this area.

In order to develop intelligent manufacturing in Belarus, it is essential to pay attention to the following recommendations.

1. Researching the tools of state support of intelligent manufacturing initiatives existing in the world and selection and establishment of various support tools relevant for Belarus.

It is possible to receive government support (government innovation fund provision, loans from financial institutions, and tax reductions) for intelligent manufacturing enterprises and reduction of construction costs.

2. Selection of priority areas of intelligent manufacturing, concentrating on the nation's resources and capabilities.

It is important to determine directions and develop strategies for intensive investment and determine areas that have large spillover effects.

3. Preparation and implementation of a government intelligent manufacturing plan.
4. Implementation of the diagnostics of the readiness of enterprises for intelligent manufacturing (to assess their level of digital maturity), setting the digitalization recognition index as a tool of measurement index, sharing the results of measurement, and establishing the means of improvement.
5. Conversion of representative companies in each manufacturing industry into advanced smart factories (demonstrating pilot projects) with the IoT, cyber physical systems, big data, and so on. These factories can be used by other companies as hubs for research and test beds. After introducing pilot projects of smart factories, the results of the introduction can be available to other companies.
6. Designing a digitalization-applied project for the basic industries of Belarus and determining the key element technologies for each industry. The expansion and introduction of smart factories is based on the evaluation of best practices.

It is important to guarantee partial fund provision by the government for the needed cost and tax reductions for new smart factories.

7. Creating an intelligent manufacturing information platform for demonstration of solutions and results of the introduction of Industry 4.0 technologies in manufacturing.

For example, in Korea, a manufacturing platform which provides a high-quality AI-based data analytics services (free high-performance cloud infrastructure, data storage, simple manufacturing data analysis tools, manufacturing datasets and guides) is established [18].

8. Creation of a library of implemented projects of smart factories and the best practices of intelligent manufacturing by industries.
9. Creation of a digital twin data bank (production, products, services).
10. Providing a competent workforce for intelligent manufacturing.

A qualified workforce is the one of the most important factors in digitalization in the economy. Training technical personnel support enterprises' digital transformation. To do this, it is essential to examine and motivate personnel and improve the educational system, taking into account the necessity of digitalization.

11. Promotion of the necessity of digitalization in the manufacturing industries.

Belarusian companies need to strengthen awareness of digital transformation. Digitalization promotion can be carried out by the Ministry of Information and Communication, Ministry of Industry, and Ministry of Economy.

8.7   CONCLUSION AND FUTURE SCOPE

Digitalization and digital transformation are the new driving forces, accelerating manufacturing innovation. To secure the competitiveness of the manufacturing industry in Belarus globally, it is essential to implement smart manufacturing's efficiency of production.

We have shown that there are lots of problems that hinder accelerated development of intelligent manufacturing in Belarus. Among these should be highlighted: limited investment in the intelligent manufacturing area, lack of government tools to support intelligent manufacturing, inadequate digital infrastructure, lack of skilled workers in intelligent manufacturing areas, lack of smart factories, and the existence of single examples of the organization of digital transformation of industrial enterprises.

Recommendations for intelligent manufacturing transformation suitable for Belarus were proposed: establish support policy tools for intelligent manufacturing, select the priority areas of intelligent manufacturing, elaborate government intelligent manufacturing plans, implement diagnostics for the digital maturity of enterprises, implement pilot projects of smart factories in each industry in manufacturing and expand this practice, create an intelligent manufacturing information platform, and ensure a competent workforce for intelligent manufacturing.

This chapter helps to identify the current level of digitalization in Belarus and presents recommendations for intelligent manufacturing transformation in the national economy. The results of this study can be used by the government to clarify measures for the acceleration of digitalization in manufacturing in Belarus.

Intelligent manufacturing in Belarus will continue to evolve. Further research should strive to lead this change. There are a few areas for future research: (1) improving the understanding of the practical tools of intelligent manufacturing (AI,

IoT, big data technology, virtual copies, etc.) and their application; (2) adaptation of the international experience of intelligent manufacturing to economic conditions in Belarus; 3) addition of new government tools to motivate the introduction intelligent manufacturing, design measures, and tools of new state economic policy aimed to develop intelligent manufacturing in Belarus.

Successful digitalization in the national economy is important for increasing the competitiveness of manufacturing and promoting sustainable growth in Belarus. It is important the future research to be focused on adapting the international experience to the economic situation in Belarus. This may constitute the object of future studies.

8.8 ACKNOWLEDGMENTS

This work was given support by the Department of Economics and Management of the Francisk Skorina Gomel State University. My sincere thanks to Valeriy Sialitski, my scientific mentor, for valuable recommendations while this work was being written.

NOTES

1. State Innovation Policy and Innovation Activity in Belarus, Law of the Republic of Belarus No 425-Z (2012).
2. Program for Social and Economic Development of Belarus for 2021–2025, Decree of the President of the Republic of Belarus No 292 (2021).
3. National Strategy of Sustainable Development until 2035, Decision of the Council of Ministers of the Republic of Belarus (2020).
4. Strategy Science and Technology for 2018–2040, Decision of the National Academy of Sciences of Belarus No 17 (2018).
5. State Program for Innovative Development of Belarus for 2021–2025, Decree of the President of the Republic of Belarus No 348 (2022).
6. Priorities of Scientific and Innovative Activities for 2021–2025, Decree of the President of the Republic of Belarus No 166 (2015).
7. State Committee on Science and Technology. Accessed March 24, 2023. https://gknt. gov.by/en/.
8. Priorities of Scientific and Innovative Activities for 2021–2025. Decree of the President of the Republic of Belarus No 166 (2015).
9. State Program for Innovative Development of Belarus for 2021–2025, Decree of the President of the Republic of Belarus No 348 (2022).
10. Program for Social and Economic Development of Belarus for 2021–2025, Decree of the President of the Republic of Belarus No 292 (2021).
11. Strategy Science and Technology for 2018–2040, Decision of the National Academy of Sciences of Belarus No 17 (2018).
12. National Strategy of Sustainable Development until 2035, Decision of the Council of Ministers of the Republic of Belarus (2020).
13. State Program for Innovative Development of Belarus for 2021–2025, Decree of the President of the Republic of Belarus No 348 (2022).
14. State Program Digital Development of Belarus for 2021–2025. Resolution of the Council of Ministers No 66 (2022).
15. The public administration body in the field of digital development and informatization issues, Decree of the President of the Republic of Belarus No 136 (2022).

REFERENCES

[1] German Federal Ministry of Education and Research. "Platform Industry 4.0." Accessed March 19, 2023. www.plattform-i40.de/IP/Navigation/EN/Home/home.html/.

[2] Kumar, Ajay, Hari Singh, Parveen Kumar, and Bandar AlMangour, eds. *Handbook of Smart Manufacturing: Forecasting the Future of Industry 4.0*. CRC Press, 2023. doi: 10.1201/9781003333760.

[3] Korean Smart Manufacturing Office (KOSMO). "Smart Factory in Korea." Accessed March 23, 2023. www.smart-factory.kr/eng/orgct?menuId=02/.

[4] National Statistic Committee of Belarus. Accessed March 17, 2023. www.belstat.gov.by/en/.

[5] World Intellectual Property Organization. "Global Innovation Index." Accessed March 17, 2023. www.globalinnovationindex.org/Home/.

[6] United Nations Industrial Development Organization (UNIDO). "Competitive Industrial Performance Index." Accessed March 29, 2023. https://stat.unido.org/database/CIP%20-%20Competitive%20Industrial%20Performance%20Index/.

[7] United Nations Industrial Development Organization (UNIDO). "Monitoring the SDG 9." Accessed March 10, 2023. https://stat.unido.org/sdg/BLR/.

[8] Ministry of Communications and Informatization of Belarus. Accessed March 9, 2023. www.mpt.gov.by/en/.

[9] Ministry of Communications and Informatization of Belarus. "The Digital Development Center." Accessed March 18, 2023. https://ipps.by/about-us/.

[10] State Committee for Standardization of Belarus, STB 2583: 2021: Digital transformation. Terminology (Minsk: State Committee for Standardization of Belarus, 2021).

[11] EnCata. Accessed March 18, 2023. www.encata.net/.

[12] Knowledge Sharing Program (Korea). "Digital Transformation of the National Economy of the Republic of Belarus." Accessed March 24, 2023. www.ksp.go.kr/english/pageView/info-eng/676/.

[13] United Nations Industrial Development Organization (UNIDO). "Regular Program of Technical Cooperation." Accessed March 28, 2023. https://open.unido.org/projects/BY/donors/400390/.

[14] National Statistic Committee of Belarus. "Questionnaire on the Use of Digital Technologies in an Organization." Accessed March 11, 2023. https://pravo.by/document/?guid=3961&p0=T22205159p/.

[15] Pranuza, Y. 2021. "Capacity to Build Artificial Intelligence Systems for Nuclear Energy Security and Sustainability: Experience of Belarus," *IFIP Advances in Information and Communication Technology*, vol. 637, pp. 128–140, 2022. doi: 10.1007/978-3-030-96592-1_10.

[16] Robotics and AI Association. "Smart Industry Expo." Accessed March 8, 2023. https://smartexpo.pro/.

[17] Technics and Communications. "Digital Economy Forum." Accessed March 8, 2023. http://de.tibo.by/.

[18] Korea Smart Manufacturing Office (KOSMO). "Smart Factory Digital Library." Accessed March 11, 2023. https://library.smart-factory.kr/SDLP/main-en/main/.

9 Investigation of the Chip Reduction Coefficient of X-625 Using Coated Tools

Neeraj Sharma[1] and Manjeet Bohat[2]

1 Engineering Department, UIET, Kurukshetra University, Kurukshetra, Haryana, India

2 Department of Mechanical Engineering Maharishi Markandeshwar (Deemed to be University), Mullana Ambala, India

9.1 INTRODUCTION

There is a huge demand for super alloys due to their mechanical, corrosive, and heat resistant characteristics. There are a number of alloys, such as Hastelloy, Waspaloy, and Rene alloys, under the category of super alloy, and nickel alloy is one of them. Nickel alloys are in high demand due to their good strength at high temperatures, even up to 700 °C; high resistance to corrosion; and high strength in harsh environments [1]. Nickel alloys have several significant applications in turbines, both gas and steam; medical instruments; nuclear reactors; and aircraft. The main challenge with nickel alloys is machining with conventional methods. A huge amount of heat is created in processing nickel alloys, so the cutter must have good properties to withstand the generated heat. In this study, a TiAlN-coated carbide tool is cast off to practice the turning operation on an X-625 nickel alloy on a lathe machine. The turning process is used to remove the extra material in the desired shape. Productivity may be increased by increasing the values of cutting parameters, but this reduces the surface quality and life of the tool [2]. During the turning operation, chips are produced as (a) discontinuous, (b) continuous with build-up, and (c) continuous. The chip reduction coefficient is the ratio of chip thickness to before uncut thickness. It is necessary to optimize the chip thickness so less power or energy is required to run the machine [3].

Product quality may be enhanced from the optimization of process parameters. There is a relationship between the input value and output value within the process parameters to reach the optimum machining state [4]. For smooth machining, lubrication may be used to reduce the friction between the workpiece and tool; it also ventilates the heat generated during machining [5]. The use of lubricating fluids has harmful effects on the health of laborers and on the surroundings. Minimum quality

DOI: 10.1201/9781003405870-9

lubrication is used to lessen tool wear and surface roughness by reducing the cutting sector heat [6].

Tool condition worsens due to an increase in friction among tools and chips; increased twinning deformation leads to tensile and compressive chip formation. The authors investigated the result of tool wear on the formation of chips throughout the machining of Ti-6Al-4V in dry machining. [7]. It is observed that shear strain, cutting force, friction, and length of contact of tool chips are reduced at the shear zone during the turning of Ti-6Al-4V material [8]. Zeqiri et al. conducted 38 experiments on 42CrMo4 hardened steel and found that chip reduction coefficient is reduced by increasing the cutting angle and cutting speed. They also found that the chip reduction coefficient increases as the cut gets deeper [9].

Khamel et al. used L27 orthogonal array for making design of experiment and trials were directed on AISI (52100) steel with carbon boron nitride (CBN) tool. They chose the three input parameters, that is, speed of cut, rate of feed, and penetration of cut with two outputs, that is, surface quality and cutting force. They observed that the cutting speed affected the life of the tool by 59.14%. They also detected that the rate of feed disturbed the surface quality by 64.09% [10]. Sahoo et al. used the Taguchi method for "smaller is better" and conducted the experiment to find out the results of various process factors on chip reduction coefficient and cutting force. Literature explains that rate of feed and depth of cut effect the cutting force parameter. It is also noted that the depth of cut and the rate of feed had an influence on the CRC [11].

Thakur et al. analyzed the wear (tool) and chip characteristics while turning on Nimonic C263 nickel super alloy by coated and uncoated cutting tools. It found that the CRC is reduced by the increase in time and speed of cutting [12]. Das et al. conducted an experiment on chromium moly alloy steel using Taguchi orthogonal array L27. There are four responses, namely, wear (flank), surface roughness (SR), chip reduction coefficient, and power consumption with three input parameters, DoC, speed (cutting), and feed (rate). They observed that the feed, speed, and depth of cut have an effect on reducing SR, wear (tool), power intake, and CRC [13, 14]. The authors adopted the response surface methodology to plan the number of experiments with input factors such as cutting speed, axial cut to generate a model to find out the cutting force using a carbide tool to machine aluminum alloy Al7075-T6. Genetic algorithms were applied to determine the optimized input parameters with minimum cutting force applied [15].

Chip temperature, cutting force, and chip morphology are compared in two cases, namely, cryogenic compressed air and dry machining. The authors observed that chip temperature is higher while machining, using compressed air during machining (dry) as compared to cryogenic compressed air. They also found that more force (cutting) is requisite by way of cryogenic compressed air as associated with machining (dry). It is noted that minimum flank wear is observed in both cases as previously discussed [16].

Titanium alloy is known for its mechanical properties and the ability to withstand high temperatures. As such, it is used for biomedical, aerospace, military, and nuclear plant purposes. The authors explained that the cutting speed, cutting temperature, and cutting force are responsible for obtaining the good surface quality with minimum tool insert wear. They suggested that minimum cutting force is required with high speed and low feed rate combination. They mentioned that the effect of coating on insert is desirable; it lowers the cutting force and temperature and enhances surface quality [17].

9.2 EXPERIMENTAL DETAILS

Here, turning operation is executed on semi-automatic lathe machine (Figure 9.1) using X-625 nickel alloy. The chemical structure of the X-625 round bar is specified in Table 9.1. TiAlNi coated tool insert (make ZCC.CT, YBG 102) is used for turning operation as shown in Figure 9.2. Specimen of the workpiece with the same diameter (40mm) and same length (120mm) is used as shown in Figure 9.3.

In this study, three input parameters are considered, namely, tool rotational speed, feed, and cut (depth) and one response, that is, chip reduction coefficient. Chip reduction coefficient is the fraction of chip thickness to uncut chip thickness. Values of the CRC are observed and noted from a Mitutoyo micrometer with least count of 0.001 mm.

9.3 RESULTS

Six experiments are conducted at different parametric settings and their response values are shown in Table 9.1. Various graphs are drawn between CRC vs rotational

FIGURE 9.1 Lathe Machine Setup.

TABLE 9.1

Chemical Structure of X-625

Name of Element	Percentage
Molybdenum	8 to 10
Iron	5
Niobium with tantalum	4
Carbon	0.15
Chromium	22
Manganese	0.51
Nickel	58
Silicon	0.55
Phosphorus	0.016
Cobalt	1
Aluminum	0.3
Sulphur	0.015
Titanium	0.5

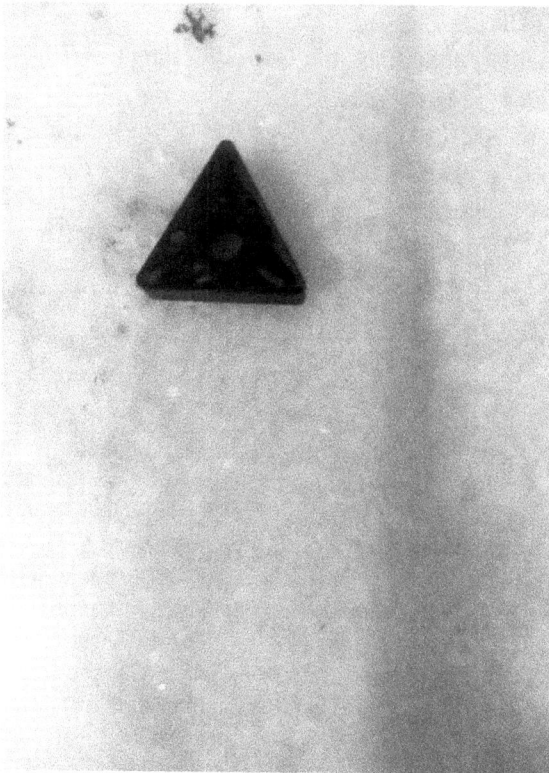

FIGURE 9.2 Tool Insert.

FIGURE 9.3   Workpiece.

TABLE 9.2
Different Settings and Corresponding Responses

RS	DoC	F	CRC
800	0.2	7	1.675
800	0.4	11	1.934
800	0.3	15	1.628
1600	0.4	7	1.431
1600	0.3	11	1.256
1600	0.2	15	1.005

speed, CRC vs feed, and CRC vs DoC as presented in Figures 9.1, 9.2, and 9.3. From Figure 9.1, it is observed that the value of CRC goes down from 1.745 to 1.23 as rotational speed increases from 800rpm to 1600 rpm. The reason may be because of the increase in strain hardening near the shear zone and because of the temperature increase in that area. If the CRC values go down, it is a desirable condition. Figure 9.2 shows that CRC gradually decreases from 1.553 to 1.3165, a desirable condition, when feed increases from 7 to 15 mm/rev. because of the increase in strain hardening and brittleness in workpiece leads to lower CRC and von Mises stress. Figure 9.3 shows the increase of CRC, an undesirable condition, from 1.34 to 1.682 when DoC increases from 0.02 to 0.04 mm. Machining area is softened due to generation of high temperature that leads thick chips formation (higher CRC) with increase in DoC.

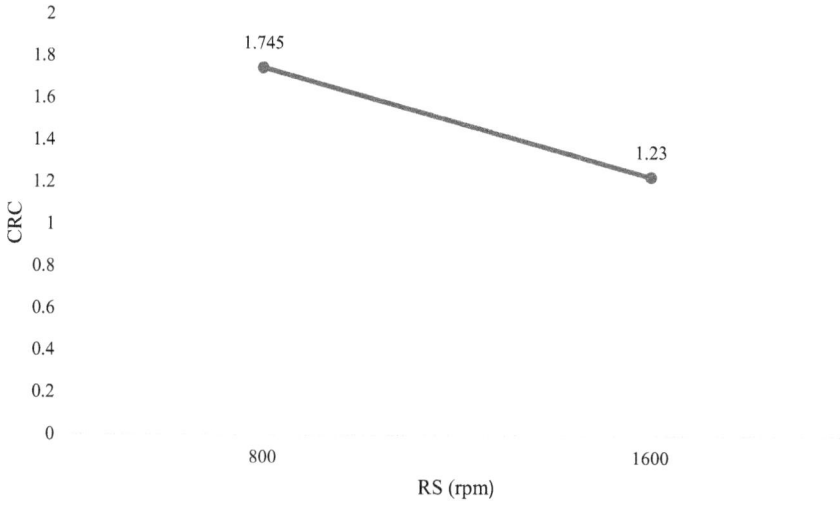

FIGURE 9.4 Variation of CRC vs RS.

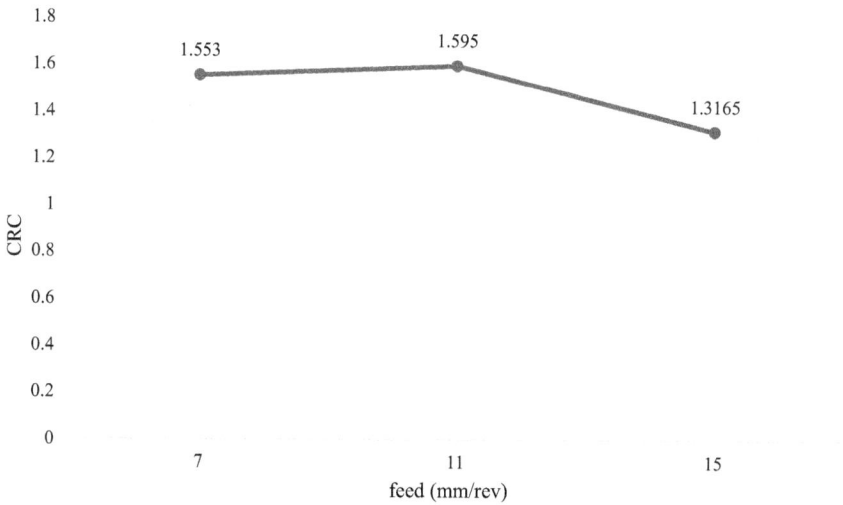

FIGURE 9.5 Variation of CRC vs Feed.

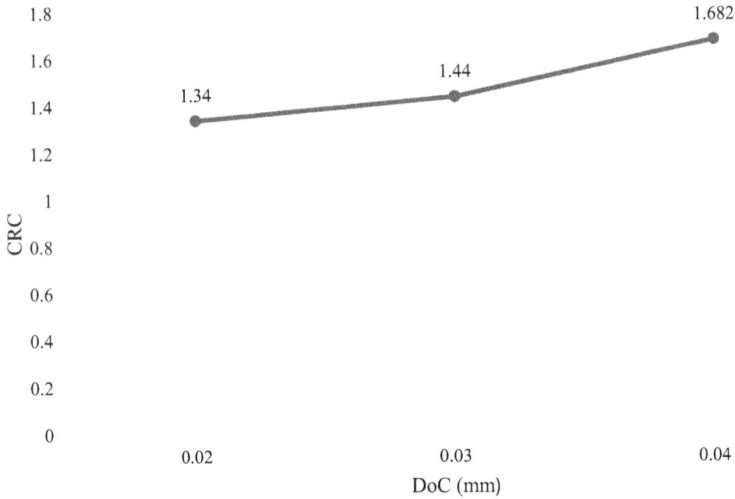

FIGURE 9.6 Variation of CRC vs DoC.

9.4 CONCLUSIONS

The following points are observed:

1. Tool rotational speed (TRS) is indirectly proportional to chip reduction coefficient (CRC). If tool rotational speed increases, then the CRC will decrease. In this study, if the TRS increases from 800 to 1600 rpm, then the CRC will decrease from 1.745 to 1.23.
2. When feed rate increases from 7 to 15 mm/rev, the CRC will drop from 1.553 to 1.3165.
3. When DoC increases from 0.02 to 0.04, the CRC also increases from 1.34 to 1.682.
4. It is evident from the results that tool rotational speed has a major effect of lowering the CRC and trailing by feed and cut (depth).

REFERENCES

[1] Bohat, M. and Sharma, N., Some studies of the CRC for X-750 nickel alloy at different parametric setting. *Materials Today: Proceedings*, vol. 63, pp. 46–48, 2022.
[2] Sharma, V. S., Dogra, M. and Suri, N. M., Advances in the turning process for productivity improvement—A review. *Proceedings of the Institution of Mechanical Engineers, Part B: Journal of Engineering Manufacture*, vol. 222, no. 11, 1417–1442, 2008.
[3] Sahoo, A. K. and Mohanty, T., Optimization of multiple performance characteristics in turning using Taguchi's quality loss function: An experimental investigation. *International Journal of Industrial Engineering Computations*, vol. 4, pp. 325–336, 2013.
[4] Das, S. R. and Nayak, R. P., Optimization of cutting parameters on tool wear and workpiece surface temperature in turning of AISI D2 steel. *International Journal of Lean Thinking*, vol. 3, pp. 140–156, 2013.

[5] Yan, P., Rong, Y. and Wang, G., The effect of cutting fluids applied in metal cutting process. *Proceedings of the Institution of Mechanical Engineers, Part B: Journal of Engineering Manufacture*, vol. 230, no. 1, pp. 19–37, 2015.

[6] Dhar, N. R., Kamruzzaman, M. and Mahiuddin, A., Effect of minimum quantity lubrication (MQL) on tool wear and surface roughness in turning AISI-4340 steel. *Journal of Materials Processing Technology*, vol. 172, pp. 299304, 2006.

[7] Daragusch, M., Sun, S., Kim, J., Li, T., Trimby, P. and Cairney, J., Effect of tool wear evolution on chip formation during dry machining of Ti-6Al-4V alloy. *International Journal of Machine Tools and Manufacture*, vol. 126, pp. 13–17, 2018.

[8] Bai, W., Sun, R., Roy, A. and Silberschmidt, V., Improved analytical prediction of chip formation in orthogonal cutting of titanium alloy Ti6Al4V. *International Journal of Mechanical Sciences*, vol. 133, pp. 357–367, 2017.

[9] Zeqiri, H., Salihu, A., Bunjaku, A., Osmani, H., Qehaja, N. and Zeqiri, F., Chip deformation and its morphology in orthogonal cutting of hardened steel 42CrMo4. *14th International Research/Expert Conference, TMT 2010*, pp. 653–656, September 2010.

[10] Khamel, S., Ouelea, N. and Bouacha, K., Analysis and prediction of tool wear, surface roughness and cutting forces in hard turning with CBN tool. *Journal of Mechanical Science and Technology*, pp. 3605–3616, November 2012.

[11] Sahoo, A. K. and Mohanty, T., Optimization of multiple performance characteristics in turning using Taguchi's quality loss function: An experimental investigation. *International Journal of Industrial Engineering computations*, pp. 325336, 2013.

[12] Thakur, A., Gangopadhyay, S., Mohanty, A. and Maity, K. P., Performance evaluation of CVD multilayer coating on tool wear characteristics during dry machining of nimonic c-263. *5th International and 26th All India Manufacturing, Design and Research Conference*, pp. 124.1–124.5, December 2014.

[13] Das, A., Khan, A. and Mohanty, S., Optimization of machining parameter using Taguchi approach during hard turning of alloy steel with uncoated carbide under dry cutting environment. *International Journal of Lean Thinking*, vol. 6, no. 2, pp. 1–15, December 2015.

[14] Subramanian, M., Sakthivel, M., Sooryaprakash, K. and Sudhakaran, R., Optimization of cutting parameters for cutting force in shoulder milling of Al7075-t6 using response surface methodology and genetic algorithm. *International Conference on Design and Manufacturing*, vol. 64, pp. 690–700, 2013.

[15] Pickering, B., Ikeda, S., Choudhary, R. and Ooka, R., Comparison of metaheuristic and linear programming models for the purpose of optimising building energy supply operation schedule. *CLIMA 2016: 12th REHVA World Congress*, Aalborg, Denmark, 2016, vol. 6, 2020.

[16] Aktel, A., Yagmahan, B., Özcan, T., Yenisey, M. M. and Sansarcı, E., The comparison of the metaheuristic algorithms performances on airport gate assignment problem. *Transportation Research Procedia*, vol. 22, pp. 469–478, 2017.

[17] Raborn, A. W., Leite, W. L. and Marcoulides, K. M., A comparison of metaheuristic optimization algorithms for scale short-form development. *Sage Journal*, vol. 80, no. 5, pp. 910–931, 2020.

10 Effect of ECM, Hard Turning, and Deep Cryogenic on Properties of AISI S-1 Tool Steel

Yogesh Dubey[1], Pankaj Sharma[1], and Mahendra Pratap Singh[1]
1 Department of Mechanical Engineering, JECRC University, Jaipur, Rajasthan, India

10.1 INTRODUCTION

Welding is a permanent joint and is used in many manufacturing sectors [1]. There has been a growing demand for welding in recent days because of its use in fabrication such as structures or joining of fractured and damaged machinery parts. The type of welding utilized in industries is friction stir welding (FSW), tungsten inert gas (TIG), laser welding (LW), metal inert gas (MIG), brazing, and soldering [2]. The properties of low-alloying metals have been controlled by the addition of alloying elements [3]. The prime elements utilized in the microstructure of steel are ferrite, nitride, carbide, and pearlite. The quantity of carbide present is very small in steel [4]. Ferrite available in low-alloy steel can be improved by utilizing available niobium, vanadium, titanium, and molybdenum [5]. The mixing of alloying elements in the steel increases the heat affected zone (HAZ). This also improves the properties of the parent specimen toward brittleness and corrosion. Generally, the elements present in low-alloy steel such as Mo, Ni, V, and Cr with 0.05% to 0.3% carbon content produce remarkable weldability. Furthermore, the low-alloy steels are commonly welded by various types of welding methods. The different welding approaches can be friction stir welding (FSW), gas metal arc welding (GMAW), tungsten inert gas welding (TIG), etc. [5, 6]. Many researchers prescribed the suggestions on working circumstances on processes after reviewing the feasibility to the work conditions of high-strength low-alloy (HSLA) steel. Many researchers have also reviewed the behavior of the oxidation and workability of low-carbon steel [7–11]. Welding aluminum and magnesium alloys takes part in fulfilling the demand for low emissive structures, more fuel efficiency, and lightweight structures [12]. Large quantities of aluminum, magnesium, silicon, steel, and other alloys have been in great demand because of their strength, and they are widely used in aerospace, defense,

DOI: 10.1201/9781003405870-10

infrastructure, electronics, electrical, and for many other purposes. Smart manufacturing with Industry 4.0 is processing for smart materials and recovery of wastes to achieve sustainable development goals in additive manufacturing using artificial intelligence and bio-manufacturing [13–18].

The outcome of hard turning, deep cryogenic, and electrochemical machining (ECM) on S-1 tool steel has been studied. Tools steels are often used to get favorable toughness and are also heat-treated to increase hardness. Despite this, the actual relationship between toughness and hardness of tool steel is not properly quantified, and the required heat treatment is to increase the hardness and toughness. Tool steel has superior properties and uses hard turning as a secondary method of grinding operation. Cost, lower manufacturing power, good surface finish without cooling machining, and shot machining time are some of the good properties of hard turning. Hard turning also provides accuracy in dimensions, good quality of surface, and good manufacturing flexibility in complex geometries [19–21]. The feed rate and power consumption during surface roughness process of AISI 2040 steel were the main factors for consideration during machining [22]. The high speed of cutting and low feed rate provide good surface quality on AISI D2 tool steel, but higher cutting speed also results in tool flank wear [23]. The affecting properties for AISI S-1 tool steel were soaking and tempering time when the material was tempered at 200 °C and DCT at –196 °C. Further, heating rates, cooling rates, austenitizing time, and temperature were constant at 900 °C for 60 minutes at deep cryogenic treatment [24]. Machining of alloys and tool steels is difficult in traditional machining processes. Thus, ECM is a cost-effective option for alloys and tool steels and will be extremely fundamental in the future [25].

10.2 PRINCIPLE OF WORK

10.2.1 ELECTROCHEMICAL MACHINING (ECM)

The main choice for S-1 tool steel machining is electrochemical machining practice, which is a nontraditional manufacturing process [25]. The operator must have good technological experience for operating ECM due to the various ranges together with a convenient collection of parameters of machining [26]. The operator must select the machining variables sensibly and not randomly for experiments [27]. The major chemical and physical systems of ECM with other variations have been recognized for years [28, 29], but demonstrations exhibiting them have been primarily centered on the attainable precisions in geometry [30]. Recent studies abstracted the multi-physics interdisciplinary modeling to incorporate temperature, local chemical dissolution, fluid dynamics, and electrical field aspects [31, 32]. ECM principally provides good surface bonding cause of the contactless mechanical working principle [33]. It consists of a DC power supply, spindle, cathode, electrolyte pump, and tank [34].

Electrochemical machining comprises disciplined specimens in an anodic dissolution of electrochemical with the properly structured tool of the cathode as mentioned by Faraday's laws of electrolysis [35–37]. In between the anode specimen and cathode tool, the solution of electrolyte is permitted to stream with a high velocity. ECM is the converse of material deposition as an electroplating practice and is solely used in electrically conductive materials [38].

10.2.2 HARD TURNING

The hard turning practice can be adopted against the grinding operation during the finishing phase for hardened steel (HRC 45–70), utilizing a one-point tool for cutting of ceramics and cubic boron nitride (CBN) [39–42]. The finished surface quality formed by hard turning is like the quality produced through grinding [43]. Numerous and dissimilar studies were conducted to analyze the output of process parameters such as quality of surface, material removal rate (MRR), feed rate (FR), cutting speed (CS), depth of cut (DoC), and further, the influence of process parameters on the integrity of surface with novel methods in the manufacturing processes [44–48]. The roughness surface quality is 0.42 μm through hard turning. Similarly, the profile of surface, form, roughness, waviness, and microroughness was investigated by several authors [49, 50]. Reduction in time of processing and lubrication is possible to eliminate in hard turning compared to grinding operation. Therefore, productivity improvement is achieved with a reduction in machining time [51, 52].

10.2.3 DEEP CRYOGENIC

"Cryogenic" comes from the Greek words "kryo", which means "frost", and "genic", which means "to produce". Based on this etymology, the cryogenic method involves a temperature below 0 °C. [54, 55].

Deep cryogenic treatment (DCT) is usually accomplished before tempering and after quenching. DCT is useful once the material properties like resistance in wear, material toughness, and hardness are boosted by −196 °C low temperature, and high-speed steel (HSS) is the second listed material used for DCT to attain upgraded properties [57–60]. The different names of cryogenic treatment in many works of literature are subzero, cryo, low-temperature treatment, cryo treatment, low-temperature treatment, cryogenics, ultra-low treatment, cold treatment, cryogenating, cryogenically treated (CT), cryogenic hardening, cryogenic stress relief (CSR), cryogenic thermal cycling (CTC), and cryo processing [61–69]. HSS is generally required by many industries to produce dies, punches, and cutting tools because of its exceptional hardness property due to a combination of cobalt, chromium, molybdenum, vanadium, tungsten, etc. [70–72]. DCT makes the transformation of grains from austenite to martensite, reduction in-site and redistribution of carbon along with free energy dropping in the crystal structure [73, 74], precipitation of secondary-tertiary and clear sub-microscopic carbides [75–84].

10.3 LITERATURE REVIEW

The effect of different manufacturing processes (e.g., ECM, cryogenic, hard turning, etc.) on the tool steel material properties has been studied broadly. Table 2.1 summarizes the studies done by various researchers in this field.

TABLE 10.1

ECM, Hard Turning, and Deep Cryogenic Literature on Various Tool Steels

Authors	Material	Method	Comment
J. Janardhanan et al. (2017) [27]	SKD-12 tool steel	ECM	For enhancing the responses of MRR, the different variables studied are electrolyte concentration, voltage, and current.
J.Y. Huang et al. (2003) [85]	M2 tool steel	Cryogenic	The resistance on the wear of steel has been enhanced after cryogenic treatment by the growth in density of carbide with heat treatment, and cryogenic also enables the development of carbon clustering.
A. Akhbarizadeh et al. (2009) [86]	D6 tool steel	Cryogenic	The hardness resistance and the wear resistance were upgraded as a result of the transformation of austenite into martensite by increments in holding time at cryogenic temperature.
Kamran Amini et al. (2012) [87]	1.2080 tool steel	Cryogenic	The particle size, hardness, carbide-sharing homogeneity, carbide percentage, and microhardness attain the stable value and never change when holding period increases over optimal holding period.
Young-Mok Rhyim et al. (2006) [88]	D2 tool steel	Cryogenic	The wear resistance affects the wetness of very minute fine particles of carbides and hardness. Further, the growth shown in carbide were improved using the process of deep cryogenic.
Shaohong Li et al. (2010) [89]	Cold work die steel	Cryogenic	Austenite is available between laths of the martensite. The lengthy soaking time of liquid nitrogen does not affect the stability of martensite.
M.A. Shalaby et al. (2014) [90]	D2 tool steel	Hard turning	Tool-chip interface was enhanced by the development of chromium-oxygen tribo-films on the tool surface due to contact of environment and work material. Result shows the outstanding performance of all types of PCBN at varying criteria of machinability.
M.K. Das et al. (2014) [91]	EN31 tool steel	ECM	The finest combination for the least surface roughness (SR) and extreme MRR is the 10 V voltage, 0.2 mm electrode gap, 0.25 mm/min FR, and 10% electrolyte concentration.
M.K. Das et al. (2017) [92]	EN31 tool steel	ECM	The best grouping of progression parameters is attained as the lowermost level of the electrolyte concentration and at the mid-level of voltage (V), feed rate, and internal electrode gap (A1B2C2D2) for extreme MRR and least surface roughness.
G.S. Prayogo et al. (2019) [93]	AISI D2 tool steel	ECM	With a 1 mm gap width, 150 g/l electrolyte concentration, and 48 V, the MRR is maximum at 0.153 g/min due to the influence of electrolyte concentration.
S.S. Bosheh et al. (2006) [94]	H13 tool steel	Hard turning	The processed samples were examined through electron micro probe, EDAX software-enabled SEM, and microhardness tester. Hardness was increased at processed surface compared to the majority of the materials. White layer hardness and depth were reduced despite the increase in the cutting speed of tool wear.

(Continued)

TABLE 10.1 (*Continued*)
ECM, Hard Turning, and Deep Cryogenic Literature on Various Tool Steels

Authors	Material	Method	Comment
G.K. Dosbaeva et al. (2015) [95]	D2 tool steel	Hard turning	The hard turning of D2 tool steel by using polycrystalline cubic boron nitride (PCBN) and coated carbide (CVD) tools. Wear patterns by cutting tools were studied through EDS/SEM, and X-ray photoelectron spectroscopy (XPS) was utilized to explore the tribo-films. Improved tool life was observed due to lubricious Cr-O tribo-films.
Sanjeev Saini et al. (2012) [96]	AISI H11 Tool Steel	Hard turning	The period of tool life decreases when nose radius and speed of cutting increases with constant cutting depth of 0.30 mm and feed of 0.13 mm/rev.
M. Nayak et al. (2021) [97]	AISI D6 tool steel	Hard turning	The o/p parameters of responses were the cutting temperature, the thrust force, and the surface roughness; the main cutting force with the respective fluctuations in feed rate and speed gives the optimal mathematical model.
W.B. Rashid et al. (2016) [98]	AISI 4340 tool steel	Hard turning	The findings show great wear in tool due to lower feed rate. Further, the quality and cost considerations show the important role of selecting machining parameters.
V.G. Navas et al. (2008) [99]	AISI O1 tool steel	Hard turning	The representation of quenching, auto tempering, and austenitizing was established by the white layer (WL) induction by EDM, which shows the distinctive representation of thermal source beneath a fine melted and re-solidified layer.
L. Tang et al. (2017) [100]	AISI D2 tool steel	Hard turning	The strain rate and brittleness are increased by the cutting. Further, the increase in quenching hardness also raises the brittleness of the sample and further affects the shear band with great damage.
A. Elsadek et al. (2020) [101]	AISI H13 tool steel	Hard turning	The hardness of sample greatly influenced the temperature of cutting during dry cutting. The cutting temperature increases when the depth of cut is increased.
J.S. Dureja et al. (2014) [102]	D3 tool steel	Hard turning	The tool wear was largely affected by the rate of feed and the speed of cutting. Further, the roughness of the surface was mostly influenced by the feed rate.

10.4   WORK DONE ON S-1 TOOL STEEL

Sahinoglu and Rafighi [103] conducted research on the AISI S-1 tool steel hard turning process using CBN to find the outcome of i/p parameters such as the feed rate, the cutting speed, and the cutting depth against output parameters such as surface roughness, power consumption, and sound. The authors selected the AISI S-1 tool steel bar of Ø50 mm with a length of 200 mm. The length–diameter ratio to uphold the stiffness of the cylindrical sample, cutting tool, and chuck is four. To obtain 60 HRC hardness and remove residual stresses, the sample was heat-treated for two hours at 860°C; further sample was oil quenched for half an hour; and last, the sample was tempered at 200°C for one hour. The measuring period was two minutes for surface roughness and one minute for each pass for machining. The CNC lathe TAKSAN TTC-630 with the top speed of 4000 rpm and 20 kW power was used for turning tests under dry cutting conditions. Further CBN inserts were used for cutting with the thickness of 3.969 mm and tool nose radius of 0.40 mm. To measure in-process turning operation sound, the LUTRON SL-401 device was used, and to measure machine current, the UNI-T UT201 device was employed. The portable device Mitutoyo SJ 201 was employed to measure Ra surface roughness. Further, the response surface methodology (RSM) was utilized to detect relationships among response variables and cutting parameters.

The findings of this research were obtained by using Minitab 19, RSM, analysis of variance (ANOVA), and regression equation on L27, considering the feed rate, the depth of cut, and the cutting speed are analyzed.

The primary expression to assume responses according to the parameters of cutting is given by

$$Y = b_0 + b_1a + b_2V + b_3f + b_{11}a^2 + b_{22}V^2 + b_{33}f^2 + b_{12}a \times V + b_{13}a \times f + b_{23}V \times f$$

Here Y (output parameter) and b_0, b_1, . . ., b_{33} is to be resolved using simple linear regression coefficient for every o/p parameter.

Using ANOVA, the research found that 91.77% of the surface roughness is damaged by the feed rate. The radius of nose does not have a direct relation with feed rate as compared to surface roughness. So, cutting forces, chatters, and vibration increase as the increment of feed rate rises, which directly affects the quality of the workpiece. Further, surface quality is not affected by the cutting speed and the depth of cut. Using ANOVA, results showed 43.61% FR, 30.71% CS, and 19.78% DoC.

L.K. Sharma et al. [104] finished their work on a 15 x 25 x 2 S-1 tool steel and used $NaNO_3$ as an electrolyte. The control valve is used to control the electrolyte circulation rate, and the stepper motor is used to control the machining gap. The authors used .16 mm electrode of copper. Voltage, inter-electrode gap (IEG), electrolyte concentration, and flow rate are machining parameters, and low, intermediate, and high are different levels of experiments. Further, response calculation for the higher the better is applied.

The finding is that after increasing the concentration of electrolyte, electrolytes diverge from the original way due to a rise in resistance of the solution because of saturation. The flow rate, concentration, and IEG of electrolyte are important factors for MRR after practical voltage. The excellent machining parameters for MRR are 200 g/ltr electrolyte concentration, 4.09 l/min flow rate, 0.5 mm IEG, and 30 V voltage. The best machining parameters for radial overcut (ROC) are 150 g/l electrolyte concentration, 2.18 l/min flow rate, 1.5 mm IEG, and 20 V voltage.

L.K. Sharma et al. [105] did their experiment on S-1 tool steel pieces of 15 X 25 X 2 mm as specimens and used 0.16 mm electrode of copper diameter as the material of tool. The EC, V, FR, and IEG have opted for variables for research. ANOVA was used to check the effectiveness of the RSM to get the optimal results of electrochemical machining (ECM) parameters utilizing S-1 tool steel.

Results show that the selected procedure gives a statistically good model as the R^2 value of ROC is 94.63%. ROC increases at the increase in voltage and decreases at the decrease in voltage, which directly affects the cuts. ROC decreases when there is an increase in the gap of the electrode due to an increment in resistance between sample and tool, which gives low ROC as stray current density is decreasing. Further, the concentration of the electrolyte is maximum on ROC and then IEG applied voltage and flow rate of electrolyte.

Vahdat and Niaki [106] used 12 sets of AISI S-1 samples for testing, where one set of samples goes through a hardening process and the other samples go through full treatment cycles. Further, the samples were treated after proper machining. The test was continually repeated three times.

Results show no change in the tensile strength (TS) and the yield strength (YS). The period of soaking was reduced for extreme hardness after the rise of tempering time and with the rise of the time in soaking, the extreme hardness was lowering the tempering time. The scanning transmission electron microscope method discovers fresh carbides of Cr3C2, M6C, and Fe3C, which were not noticed by the scanning electron microscope (SEM).

Niaki and Vahdat [107] finalized an experiment on AISI S-1 tool steel, where 12 sets of samples were set out to experience the traditional method of hardening [108] and other samples undergo complete treatment. DCT activity used a cryogenic programmable processor. The authors recorded the heating and cooling rates by thermocouple and controlled them by using a check valve.

Results show the maximum hardness was reduced with an increment in the time of tempering for the required time of soaking, whereas increasing soaking time gives maximum hardness required for lesser tempering time. For samples 48–2 and 36–1, greater tensile toughness compared to the traditional procedure showed a higher DCT period required for a larger tempering time to get greater tensile toughness.

10.5 CONCLUSION

In the current review, research related to ECM done on SKD-12 tool steel, EN13 tool steel, D2 tool steel etc. has been compiled. The effect on material properties of different parameters such as the concentration of electrolyte, current, feed rate, voltage, and electrode gap was investigated.

In the ECM process, after the electrolyte concentration, the voltage and current are affecting extra on MRR of the SKD tool steel. Further, in AISI D2 tool steel, the MRR response does not get affected by the gap of machining but has substantially been affected by electrolyte concentration and voltage. The grades of EN31 tool steel enhance the gray relational grades by 48% using optimal parametric condition. The optimum parameters used were 10 V voltage, 15% electrolyte concentration, 0.2 mm IEG, and 0.25 mm/min feed rate.

In the hard turning, the speed of cutting and feed has played an important role in affecting the cutting temperature of the samples. Both parameters have contributed less than one-third of sample hardness and greatly affects the result. The CBN tool is used during the AISI D6 tool steel hard turning. The report shows that the dominant factor is feed and that it significantly affects the Ra with 52.45% contribution with interruption. Further, the effect of feed is quadratic. According to heat treatment in cryogenic behavior, the foremost types of precipitate of the carbides in the D2 cold work tool steel are M7C3 and M23C6. The growth in tempering temperature decreases the hardness with improvement in wear resistance. Further, the resistance in wear is concerned with optimal heat treatment and evolution of carbide. Cryogenic treatment can generate extra homogeneous spreading of carbides.

10.6   SCOPE OF FURTHER WORK

The literature review shows that the research studies are limited to wear analysis. There are very few research articles based on the study of the fusion zone (FZ) and the heat affected zone (HAZ) by welding on S-1 tool steel. The welding processes were TIG, MIG, and Arc welding for determination of the most appropriate metal used in different industrial applications. Welding has a wide range of uses, such as in houses, railways, offices, bridges, space, and aerospace [109].

Metallurgical properties after welding represent the dominant factor that influences the mechanical properties of welded joints. Segregation of samples play a very important role and affect the joint's mechanical properties.

REFERENCES

[1] Rahul Madhusudan, S. P. Shivprakasam, B. R. Vishnu, K. R. Balasubramanian and Mohan Sreejith, "Health issue owing to exposure with welding fumes and their control strategies at the source–A review," *Materials Today: Proceeding* 46, no. 19 (2021): 9239–9245.

[2] Scott A. Civjan, T. Guihan and Kara Peterman, "Testing of oxyacetylene weld strength," *Journal of Constructional Steel Research* 168. (2020): 105921.

[3] Niyamat U. Khan, Sunil K. Rajput, Vertika Gupta, Vijay Verma and Tarun Soota, "To study mechanical properties and microstructures of MIG welded high strength low alloy steel," *Materials Today: Proceedings* 18 (2019): 2550–2555.

[4] Ramachandran Oyyaravelu, Palaniyandi Kuppan and Natarajan Arivazhagan, "Comparative study on metallurgical and mechanical properties of laser and laser-arc-hybrid welding of HSLA steel," *Journal of Advanced Research* 5, no. 5 (2016): 463–472.

[5] S. Ramanathan, V. Balasubramanian, S. Malarvizhi and A. Gourav Rao, "Effect of welding processes on mechanical and microstructural characteristics of high strength low alloy naval grade steel joints," *Defence Technology* 11, no. 3 (2015): 308–317.

[6] Bijay Kumar Show, Rowthu Veerababu, Ramalingam Balamuralikrishnan and G. Malakondaiah, "Effect of vanadium and titanium modification on the microstructure and mechanical properties of a microalloyed HSLA steel," *Materials Science and Engineering: A* 527, no. 6 (2010): 1595–1604.

[7] Sunil K. Rajput, Martina Dikovits, Gajanan Chaudhari, Maria C. Poletti, Fernando War-chomicka, Vivek Pancholi and Sumeer K. Nath, "Physical simulation of hot deformation and microstructural evolution of AISI 1016 steel using processing maps," *Materials Science and Engineering: A* 587 (2013): 291–300.

[8] Sunil K. Rajput, Gajanan Chaudhari and Sumeer K. Nath, "Physical Simulation of hot deformation of low-carbon Ti-Nb micro alloyed steel and microstructural studies," *Journal of Materials Engineering and Performance* 23, no. 8 (2014): 2930–2942.

[9] Sunil K. Rajput, Gajanan Chaudhari and Sumeer K. Nath, "Characterization of hot deformation behavior of a low carbon steel using processing maps, constitutive equations and Zener-Holloman parameter," *Journal of Materials Processing Technology*237 (2016): 113–125.

[10] Anjneya Sharma, Sunil Kumar Rajput and Shashee Kant Soni, "Cyclic high temperature oxidation behavior of bare and NiCr coated mild steel and low alloyed steel," *Materials Today: Proceedings* 5 (2018): 18433–18441.

[11] Lei Shi and Chuan Song Wu, "Transient model of heat transfer and material flow at different stages of friction stir welding process," *Journal of Manufacturing Processes* 25 (2017): 323–339.

[12] Abidin Sahinoglu and Mohammad Rafighi, "Optimization of cutting parameters with respect to roughness for machining of hardened AISI 1040 steel," *Materials Testing* 62, no. 1 (2020): 85–95.

[13] Love Kumar, Ajay, Rajiv Kumar Sharma and Parveen, "Smart manufacturing and industry 4.0: State-of-the-art review," *CRC* (2023): 1–28.

[14] Hari Singh Ajay, Parveen and Bandar AlMangour, *Handbook of Smart Manufacturing: Forecasting the Future of Industry 4.0*. CRC Press (2023).

[15] Ravi K. Mittal, *Incremental Sheet Forming Technologies: Principles, Merits, Limitations, and Applications*. CRC Press (2020).

[16] Ajay Kumar, Praveen Kumar and Ashish K. Srivastava, *Modeling, Characterization, and Processing of Smart Materials*. IGI Global (2023).

[17] Ajay, Parveen and Ashwini Kumar, *Waste Recovery and Management: An Approach Toward Sustainable Development Goals*. CRC Press (2023).

[18] Ajay Kumar, Ravi Kant Mittal and Abid Haleem, *Advances in Additive Manufacturing: Artificial Intelligence, Nature-Inspired, and Biomanufacturing*. Elsevier (2022).

[19] Hans Kurt Toenshoff, Ronald C. Arnett and R. Ben Amor, "Cutting of hardened steel," *CIRP Annals* 49, no. 2 (2000): 547–566.

[20] Mozammel Mia and Nikhil Ranjan Dhar, "Prediction of surface roughness in hard turning under high pressure coolant using Artificial Neural Network," *Measurement* 92 (2016): 464–474.

[21] Yildirim Cagri Vakkas, "Investigation of hard turning performance of eco-friendly cooling strategies: Cryogenic cooling and nanofluid based MQL," *Tribology International* 144 (2020): 106127.

[22] Abidin Sahinoglu and Mohammad Rafighi, "Optimization of cutting parameters with respect to roughness for machining of hardened AISI 1040 steel," *Materials Testing* 62, no. 1 (2020): 85–95.

[23] J.G. Lima, R.F. Avila, A.M. Abrao, M. Faustino and J. Paulo Davim, "Hard turning: AISI 4340 high strength low alloy steel and AISI D2 cold work tool steel," *Journal of Materials Processing Technology* 169, no. 3 (2005): 388–395.

[24] Junwan Li, Leilei Tang, Shaohong Li and Xiaochun Wu, "Finite element simulation of deep cryogenic treatment incorporating transformation kinetics," *Materials & Design* 47 (2013): 653–666.

[25] Love Kishore Sharma, Dilip Gehlot, Anil Kumar Sharma and Bhupendra Verma, "RSM application for optimization of ECMM parameter using S1 Tool steel," *International Journal of Advance Research, Ideas and Innovations in Technology* 4, no. 4 (2018): 364–369.

[26] P Asokan, R Ravi Kumar, R Jeyapaul and Maddela Santhi, "Development of multi-objective optimization models for electrochemical machining process," *The International Journal of Advanced Manufacturing Technology* 39 (2008): 55–63.

[27] Jeykrishnan Janardhanan, Vijaya Ramnath, C Elanchezhian and Sivaswamy Akilesh, "Parametric analysis on Electro-chemical machining of SKD-12 tool steel," *Materials Today: Proceedings* 4 (2017): 3760–3766.

[28] Joseph A. McGeough, *Principles of Electrochemical Machining.* Halsted Press Division, Wiley (1974).

[29] Kamlakar P. Rajurkar, Jerzy Kozak, Binqiang Wei and Joseph A. McGeough, "Study of pulse electrochemical machining characteristics," *CIRP Annals Manufacturing Technology* 42 (1993): 231–234.

[30] Srichand Hinduja and Masanori Kunieda, "Modelling of ECM and EDM processes," *CIRP Annals Manufacturing Technology* 62, no. 2 (2013): 775–797.

[31] Fritz Klocke, Markus Zeis and Andreas Klink, "Interdisciplinary modelling of the electrochemical machining process for engine blades. Processes," *CIRP Annals Manufacturing Technology* 64, no.-1 (2015): 217–220.

[32] Markus Zeis, "Modeling of the removal process of the electrochemical sinking machining of engine blades," *Dissertation RWTH Aachen University* (2015).

[33] Thomas Bergsa and Simon Harst, "Development of a process signature for electrochemical machining," *CIRP Annals—Manufacturing Technology* 69 (2020): 153–156.

[34] L. I. N. Guomin and C. A. I. Hongzhuan, "Electrochemical machining technology and its latest applications," *Advanced Materials Research* 472, no. 475 (2012): 875–878.

[35] Neelesh Kumar Jain and Vijay Kumar Jain, "Optimization of Electro-chemical machining process parameters using genetic algorithms," *Machining Science and Technology* 11 (2007): 235–258.

[36] Alexey Dmitrievich Davydov, Tatyana B. Kabanova and Vladimir Volgin, "Electrochemical Machining of Titanium," *Russian Journal of Electrochemistry* 53, no. 9 (2017): 941–965.

[37] Joseph A. McGeough, *Principles of Electrochemical Machining.* Chapman and Hall (1974).

[38] N. Rajesh Jesudoss Hynesa and R. Kumar, "Electrochemical machining of aluminium metal matrix composites," *Surface Engineering and Applied Electrochemistry* 54, no. 4 (2018): 367–373.

[39] K. Seveen and K. Muniswaran, "Experimental study on hard turning of hardened tool steel with coated carbide cutting tools," *Masters thesis, Universiti Teknologi Malaysia, Faculty of Mechanical Engineering* (2007).

[40] Dilbag Singh and Venkateswara Rao Paruchuri, "A surface roughness prediction model for hard turning process," *The International Journal of Advanced Manufacturing Technology* 32 (2007): 1115–1124.

[41] Yusuf Sahin and Ali Riza Motorcu, "Surface roughness model for machining mild steel with coated carbide tool," *Journal of Materials Design* 26 (2005): 321–326.

[42] Hamza Bensouilah, Hamdi Aouici, Ikhlas Meddour, Mohamed Athmane Yallese, Tarek Mabrouki and Francois Girardin, "Performance of coated and uncoated mixed ceramic tools in hard turning process," *Measurement* 82 (2016): 1–18.

[43] M. C. Gosiger, "Fundamentals of hard turning-An depth look at the process" (2012). https://cdn2.hubspot.net/hub/139128/file-17761415-pdf/docs/gos_wp_hardturning_f.pdf

[44] Miroslav Piska and Michal Forejt, "Theory of machining, forming and cutting tools," *UST FSI VUT, Brno, Czech Republic* (2006).

[45] Toth Tibor, Janos Kundrak and Karoly Gyani, "The removal rate as a parameter of qualification for hard turning and grinding," *Tools and Methods of Competitive Engineering* 1, no. 2 (2004): 629–639.

[46] Janos Kundrak, Bernhard Karpuschewski, Gyani K and Bana V, "Accuracy of Hard Turning," *Journal of Materials Processing Technology* 202 (2008): 328–338.

[47] Alexandre M. Abrao and David K. Aspinwall, "The surface integrity of turned and ground hardened bearing steel," *Wear* 196 (1996): 279–284.

[48] Youngsik Cho, "A comprehensive study of residual stress distribution induced by hard machining versus grinding," *Tribological Letters* 36 (2009): 277–284.

[49] Diptikanta Das, Ashok Kumar Sahoo, Ratanakar Das and Bharat Routara, "Investigations on hard turning using coated carbide insert: Grey based Taguchi and regression methodology," *Procedia Materials Science* 6 (2014): 1351–1358.

[50] Zawada-Tomkiewicz A, "Analysis of surface roughness parameters achieved by hard turning with the use of PCBN tools," *Estonian Journal of Engineering* 17, no. 1 (2011): 88–99.

[51] Bharat Routara, A. Bandyopadhyay and Prasanta Sahoo, "Roughness modeling and optimization in CNC and milling response surface method: Effect of workpiece material variation," *The International Journal of Advanced Manufacturing Technology* 40 (2009): 1166–1180.

[52] Aouici Hamdi, Mohamed Athmane Yallese, Brahim Fnides and Tarek Mabrouki, "Machinability investigation in hard turning of AISI H11 hot work steel with CBN tool," *Mechanika* 6, no. 86 (2010): 71–77.

[53] M. S. Cheema and A. Batish, "Investigation of effects of process parameters and insert geometry on hard turning of steels," *Materials Science* (2011): 139592814.

[54] Patricia Jovicevic-Klug and Bojan Podgornik, "Review on the effect of deep cryogenic treatment of metallic materials in automotive applications," *Metals* 10, no. 434 (2020): 1–12.

[55] Chitrang A. Dumasia, V. A. Kulkarni and Kunal Sonar, "A review on the effect of cryogenic treatment on metals," *International Research Journal of Engineering and Technology*, no. 4 (2017): 2402–2406.

[56] Vishnu Vardhan Mukkoti, G. Sankaraiah and M. Yohan, "Effect of cryogenic treatment of tungsten carbide tools on cutting force and power consumption in CNC milling process," *Production & Manufacturing Research* 6, no. 1 (2011): 149–170.

[57] P. Baldissera and C. Delprete, "Deep cryogenic treatment: A bibliographic review," *The Open Mechanical Engineering Journal*, no. 2 (2008): 1–11,

[58] Emmanuel Ogu, Daniel El Chami, Patricia Jovicevic-Klug and Bojan Podgornik, "Deep cryogenic treatment of metallic materials," *Encyclopedia* (2020): 1–11.

[59] D. Senthilkumar, "Cryogenic treatment: Shallow and deep," in: G. E. Totten and R. Colas, editors. *Encyclopedia of Iron, Steel, and Their Alloys*. Taylor and Francis (2016): 995–1007.

[60] Patricia Jovicevic-Klug and Bojan Podgornik, "Review on the effect of deep cryogenic treatment of metallic materials in automotive applications," *Metals*, no. 10 (2020): 434.

[61] Bojan Podgornik, Irena Paulin, Bostjan Zajec, Staffan Jacobson and Vojteh Leskovsek, "Deep cryogenic treatment of tool steels," *Journal of Materials Processing Technology* no. 229 (2016): 398–406.

[62] V. Leskovsek and B. Podgornik, "Vacuum heat treatment, deep cryogenic treatment and simultaneous pulse plasma nitriding and tempering of P/M S390MC steel," *Materials Science and Engineering: A* 531 (2012): 119–129.

[63] J. Indumathi, Jayashree Bijwe, Anup Kumar Ghosh, M. Fahim and N. Krishnaraj, "Wear of Cryo-treated engineering polymers and composites," *Wear* 225, no. 229 (1999): 343–353.

[64] C. L. Gogte, Ajay Likhite, D. R. Peshwe, Aniruddha Bhokarikar and Rahul Shetty, "Effect of cryogenic processing on surface roughness of age hardenable AA6061 alloy," *Materials and Manufacturing Processes* 29 (2014): 710–714.

[65] Simranpreet Singh Gill, Harpreet Singh, Rupinder Singh and Jagdev Singh, "Cryoprocessing of cutting tool materials-A review," *The International Journal of Advanced Manufacturing Technology* 48 (2010): 175–192.

[66] Albert Bensely, A. Prabhakaran, D. Mohan Lal and Govindan Nagarajan, "Enhancing the wear resistance of case carburized steel (En 353) by cryogenic treatment," *Cryogenics* 45 (2006): 747–754.

[67] Silvio Gobbi, Vagner Joao Gobbi and Gustavo Reinke, "Ultra low temperature process effects on micro-scale abrasion of tool steel AISI D2," *Materials Science and Technology* 35 (2019): 1355–1364.

[68] Wayne Reitz and John Pendray, "Cryoprocessing of materials: A review of current status. *Materials and Manufacturing Processes* 16 (2001): 829–840.

[69] Adem Cicek, Turgay Kivak, Ilyas Uygur, Ergun Ekici and Yakup Turgut, "Performance of cryogenically treated M35 HSS drills in drilling of austenitic stainless steels," *The International Journal of Advanced Manufacturing Technology* 60 (2012): 65–73.

[70] Anil Kumar Singla, Jagtar Singh and Vishal S. Sharma, "Processing of materials at cryogenic temperature and its implications in manufacturing: A review," *Materials and Manufacturing Processes* 33 (2018): 1–38.

[71] Vengatesh. M, Srivignesh. R, Pradeep. T and Karthik N. R, "Review on cryogenic treatment of steels," *International Journal of Innovative Research in Technology, Science & Engineering* 3 (2016): 417–22.

[72] Alan M. Bayer and Bruce A. Becherer, "High-speed tool steels. In: ASM handbook," *ASM International* 16 (1989): 51–59.

[73] D. Senthil Kumar, "Cryogenic treatment: Shallow and deep," in: *Encyclopedia of Iron, Steel, and Their Alloys*. CRC (2016): 995–1007.

[74] F. Diekman, "Cold and cryogenic treatment of steel," in: *Steel Heat Treating Fundamentals and Processes*. ASM International (2013): 382–386.

[75] S. Paul and A. B. Chattopadhyay, "Environmentally conscious machining and grinding with cryogenic cooling," *Machine Learning: Science and Technology* 10 (2007): 87–131.

[76] N. Govindaraju, Shakeel Ahmed Liyakhath and M. Pradeep Kumar, "Experimental investigations on cryogenic cooling in the drilling of AISI 1045 Steel," *Materials and Manufacturing Processes* 29 (2014): 1417–1421.

[77] Marcos Perez, Cristina Rodriguez and F. J. Belzunce, "The use of cryogenic thermal treatments to increase the fracture toughness of a hot work tool steel used to make forging dies," *Procedia Materials Science* 3 (2014): 604–609.

[78] Timmerhaus, K.D, "Superconductivity," in: *Advances in Cryogenic Engineering*. Springer (1960): 145–148.

[79] Chen Jer Ming, "Cryogenic Treatment of Music Wire. Master's Thesis," *Department of Mechanical Engineering, National University of Singapore* (2004).

[80] Jesse Jones and Chris Rogers, "The acoustic effect of cryogenically treating," *The Journal of the Acoustical Society of America* 114 (2013): 2349.

[81] D. Senthilkumar, "Cryogenic treatment: Shallow and deep," in: *Encyclopedia of Iron, Steel, and Their Alloys*. CRC (2016): 995–1007

[82] Tushar Sonar, Sachin Lomte and Chandrashekhar Gogte, "Cryogenic treatment of metal—A review," *Materials Today* 5 (2018): 25219–25228.

[83] Narendra Dhokey, Ashutosh Hake, V. T. Thavale, Rahul Gite and R. Batheja, "Microstructure and mechanical properties of cryotreated SAE8620 and D3 steels," *Current Advances in Materials Sciences Research (CAMSR) Microstructure* 1 (2014): 23–27.

[84] Narendra Dhokey, Janhavi Dandawate and Raja Rawat, "Effect of cryosoaking time on transition in wear mechanism of M2 tool steel," *ISRN Tribology* (2013): 1–6.

[85] Jianyu Huang, Yuntian T. Zhu, Xiaozhou Liao, Irene J. Beyerlein, Mark A. M. Bourke and Terrence E. Mitchell, "Microstructure of cryogenic treated M2 tool steel," *Materials Science and Engineering* A339 (2003): 241–244.

[86] Amin Akhbarizadeh, Ali Shafyei and M. A. Golozar, "Effects of cryogenic treatment on wear behavior of D6 tool steel," *Materials and Design* 30 (2009): 3259–3264.

[87] Kamran Amini, Amin Akhbarizadeh and Sirus Javadpour, "Investigating the effect of holding duration on the microstructure of 1.2080 tool steel during the deep cryogenic heat treatment," *Vacuum* 86 (2012): 1534–1540.

[88] Young-Mok Rhyim, Sang-Ho Han, Young-Sang Na and Jong-Hoon Lee, "Effect of deep cryogenic treatment on carbide precipitation and mechanical properties of tool steel," *Solid State Phenomena* 118 (2006): 9–14.

[89] Shaohong Li, Lihui Deng, Xiaochun Wu, Yong'an Min and Hongbin Wang, "Influence of deep cryogenic treatment on microstructure and evaluation by internal friction of a tool steel," *Cryogenics* 50 (2010): 754–758.

[90] Mohamed Shalaby, Mohamed Abd El Hakim, Magdy M. Abdelhameed, James E. Krzanowski, Stephen C. Veldhuis and Goulnara K. Dosbaeva, "Wear mechanisms of several cutting tool materials in hard turning of high carbon–chromium tool steel," *Tribology International* 70 (2014): 148–154.

[91] Milan Kumar Das, Kaushik Kumar, Tapan Kr. Barman and Prasanta Sahoo, "Optimization of surface roughness and MRR in electrochemical machining of EN31 tool steel using grey-Taguchi approach," *Procedia Materials Science* 6 (2014): 729–740.

[92] Milan Kumar Das, Kaushik Kumar, Tapan Kr. Barman and Prasanta Sahoo, "Effect of process parameters on MRR and surface roughness in ECM of EN 31 tool steel using WPCA," *International Journal of Materials Forming and Machining Processes* 4, no. 2 (2017): 45–63.

[93] Galang Sandy Prayogo and Nuraini Lusi, "Determining the effect of machining parameters on material removal rate of AISI D2 tool steel in electrochemical machining process using the Taguchi method," *IOP Conference Series: Materials Science and Engineering* 494 (2019): 012055.

[94] S. S. Bosheh and Paul Tarisai Mativenga, "White layer formation in hard turning of H13 tool steel at high cutting speeds using CBN tooling," *International Journal of Machine Tools & Manufacture* 46 (2006): 225–233.

[95] George Karakostas Dosbaeva, Marwan El-Hakim, Mohamed Shalaby, James Krzanowski and S.C. Veldhuis, "Cutting temperature effect on PCBN and CVD coated carbide tools in hard turning of D2 tool steel," *International Journal of Refractory Metals and Hard Materials* 50 (2015): 1–8.

[96] Sanjeev Saini, Inderpreet Singh Ahuja and Vishal S. Sharma, "Influence of cutting parameters on tool wear and surface roughness in hard turning of AISI H11 tool steel using ceramic tools," *International Journal of Precision Engineering and Manufacturing* 13, no. 8 (2012): 1295–1302.

[97] Manoj Nayak, Rakesh Sehgal and Rajender Kumar, "Investigating machinability of AISI D6 tool steel using CBN tools during hard turning," *Materials Today: Proceedings* 47, no. 13 (2021): 3960–3965.

[98] Waleed Bin Rashid, Saurav Goel, J. Paulo Davim and Shrikrishna N. Joshi, "Parametric design optimization of hard turning of AISI 4340 steel (69 HRC)," *The International Journal of Advanced Manufacturing Technology* 82 (2016):451–462.

[99] Virginia Garcia Navas, Imanol Ferreres, Jose Anged Maranon, Carmen Garcia-Rosales and Javier Gil Sevillano, "White layers generated in AISI O1 tool steel by hard turning or by EDM," *International Journal of Machining and Machinability of Materials* 4, no. 4 (2008).

[100] Linhu Tang, Jun Yin, Yongji Sun, Hao Shen and Chengxiu Gao, "Chip formation mechanism in dry hard high-speed orthogonal turning of hardened AISI D2 tool steel with different hardness levels," *The International Journal of Advanced Manufacturing Technology*, 93, no. 5–8 (2017): 2341–2356.

[101] Ahmed A. Elsadek, Ahmed M. Gaafer, Samah Samir Mohamed and A.A. Mohamed, "Prediction and optimization of cutting temperature on hard-turning of AISI H13 hot work steel," *SN Applied Sciences* 2 (2020): 540.

[102] Jasminder Singh Dureja, Rupinder Singh and Manpreet S. Bhatti, "Optimizing flank wear and surface roughness during hard turning of AISI D3 steel by Taguchi and RSM methods," *Production & Manufacturing Research: An Open Access Journal* 2, no. 1 (2014): 767–783.

[103] Abidin Sahinoglu and Mohammad Rafighi, "Machinability of hardened AISI S1 cold work tool steel using cubic boron nitride," *Scientica Iranica* 28, no. 5 (2021): 2655–2670.

[104] Love K. Sharma, Sumit Sharma, Yogesh Dubey and Lajwanti Parwani, "Taguchi method approach for multi factor optimization of S1 tool steel in electrochemical machining," *IJRAR* 6, no. 2 (2019): 403–409.

[105] Love K. Sharma, Dilip Gehlot, Anil Kumar Sharma and Bhupendra Verma, "RSM application for optimization of ECMM parameter using S1 Tool steel," *International Journal of Advance Research, Ideas and Innovations in Technology* 4, no. 4 (2018): 364–369.

[106] Seyed Ebrahim Vahdat and Keyvan Niaki, "XRD, STEM, and tensile properties of AISI S1 tool steel after deep cryogenic treatment," *Advanced Materials Research* 1088 (2015): 195–199.

[107] Keyvan Niaki and Seyed Ebrahim Vahdat, "Optimization of Tensile Properties of AISI S1 Tool Steel," *Transactions of the Indian Institute of Metals* 68, no. 5 (2015): 777–781.

[108] C. W. Wegst, *Key to Steel*. Verlag Stahlschlussel Wegst GMBH (1989).

[109] Ario Sunar Baskoro, Mohammad Azwar Amat and Muhammad Fikri Arifardi, "Investigation effect of ECR's thickness and initial value of resistance spot welding simulation using 2-dimensional thermo-electric coupled," *Joint Journal of Novel Carbon Resource Sciences & Green Asia Strategy* 8, no. 4 (2021): 821–828.

11 A Study on Adoption of Information and Communication Tools in Emergency Disaster Management

Sandeep Chhillar¹, Ranbir Singh², and Pankaj Sharma¹

1 Department of Mechanical Engineering, JECRC University, Jaipur, Rajasthan, India

2 School of Engineering & Technology, BML Munjal University, Sidhrawali, Gurgaon, Haryana, India

11.1 INTRODUCTION

In the setting of disaster risk management, avoiding or decreasing the incidence of fatalities among those impacted is of the utmost importance [1]. Emergency disaster management includes numerous domains that integrates health in the organizational objectives to actively handle the disaster-related activities. A bottom-up strategy that integrates localized methods of coping, acknowledging, and strengthening region capacities is crucial for disaster risk reduction [2]. All catastrophes have nationwide impacts [2].

A disaster is an occurrence or sequence of occurrences that endangers and disturbs people's lives and livelihoods. These threats and disruptions can be triggered by both natural and non-natural elements as well as human behaviors, and they can result in individual casualties, damage to the surroundings, and losses. It is crucial to strengthen various sectors to reduce the risk of disasters, and this work is performed continually. Different types of disasters are listed as follows:

- Occurrences that are triggered by natural catastrophes or a succession of natural occurrences such as earthquakes, tsunamis, volcanic eruptions, slides, and flooding.
- A sequence of occurrences that are not caused by nature, including failures that are the result of modernization and technology, epidemiology, and illnesses that are brought about by epidemics.

DOI: 10.1201/9781003405870-11

- Events that are produced by humans are referred to as "human-made disasters", and this includes conflicts between groups of people.
- Tragic events are referred as natural disasters when they are documented according to events and places. Different types of catastrophes causes different types of damages.
- An earthquake is a sequence of tremors or shocks that occur on the surface of the planet and are generated by the collision of the earth's tectonic plates, faults that are active, erupting volcanoes, or avalanches.
- The term "eruption" refers to the overall volcanic event that includes volcanic eruptions. Hot cloud and lahars are potential hazards from volcanic eruptions.
- The word "tsunami" is originated from Japanese, meaning "harbor wave". Tsumani is caused by an earthquake that originates in the ocean.
- A landslide is the name given to the movement of earth and rock that occurs when unstable soil causes dirt and rock to slide down a slope.
- A flood occurs with a sudden rise in water levels that causes the land to submerge.
- High water discharge with obstacles in the river flow can result in flash floods.
- There is a large need for water, but there is also a shortage of water; despite this, the demand for water can still influence the agricultural industry, which may have a knock-on effect on the economic component of selling plants or flowers that are traded.
- A house or settlement that is engulfed in flames is the starting point for a chain of fires that might result in deaths or property damage.
- Areas of forests and land that have been consumed by fire, causing harm to both the forest and the land, may result in monetary losses or reductions in environmental value. Smoke catastrophes are frequently caused by forest and land fires, which can have a negative impact on both the activities and health of the community that is immediately adjacent.
- The appearance of extremely powerful winds with circular movements that resembles a spiral and can reach speeds of up to 150 kilometers per hour is referred as a tornado, and it will dissipate within three to five minutes [3].

Disaster management depends on timely information, timely and actively response to the information, recovery, and its mitigation. Figure 11.1 demonstrates the concept of a disaster management system. Information, response, recovery, and mitigation are four important aspects of any disaster management system.

Information and communication systems make life simpler and easier. It supports decision-making system, and is the most effective framework for disaster managers. Industry 4.0 technologies are developing rapidly. In general, disaster is a major catastrophe that abruptly disrupts human existence. According to their origins, disasters can be categorized into two categories: those resulting from natural causes and those caused by humans, including technological components. The factors influencing disaster management can be biological, geophysical, hydrological, and meteorological. Disasters significantly impact human life across the globe. With the evolution of Internet technologies, more robust and efficient solutions for disaster management

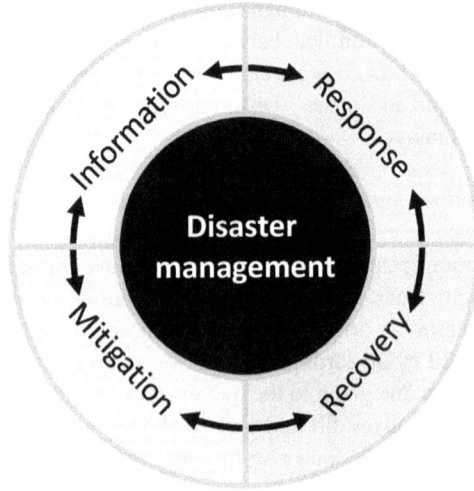

FIGURE 11.1   Concept of disaster management system.

are possible. Disaster management attempts to enhance measurements associated with disaster avoidance, prevention, recovery, crisis management, and emergency preparedness through the systematic observation and analysis of disasters [4]. A disaster management system is a four-phase planning process for disasters [5]: relief, prevention, response, and mitigation. This chapter aims to analyze the application of information and communication tools in disaster management.

11.2   LITERATURE REVIEW

Wireless sensor networks (WSNs) have been extensively utilized in the tracking of natural disasters in recent times. It is common knowledge that natural catastrophes can be meticulously tracked by using various sorts of sensors that have been upgraded [6]. Depending on the problem type, for example, game theory, the system for disaster management may be separated into different types using the available research [7]. These classifications can be found in the academic literature. Geographic information systems (GISs) are another emerging tool for disaster management [8]. Wireless telecommunication technologies will revolutionize the future of disaster management [9]. Social media has a significant impact on people's lives [10]. In past, researchers have developed the Disaster Notification and Resource Allocation System (DNRAS) and Alert Management System (AMS) [11], disaster information systems (DISs) and disaster operations management (DOM) [12], and the information system SIRENE based on spatial data infrastructure (SDI) [13] for disaster management. Information on previous disasters is essential for learning from mistakes made in the past, investigating the causes of disasters, remembering what caused them, and working on projects related to disaster planning [9]. A disaster logistics information system can be of assistance to survivors in the aftermath of a disaster [14, 15]. Data mining techniques are used to define disaster risk, including financial loss [16]. Disaster relief strategies can be developed based on operational research methodologies and geographic information systems to assist decision-making during disasters. Researchers have used two-stage stochastic programming [17], probabilistic routing of vehicles [18], multifaceted and multi-commodity distribution models [19], and probabilistic network models [20–22] to solve the problem of dynamic vehicle routing during a disaster. Medical commodities are critical in a crisis [23]. Information regarding disasters should be categorized in line with risk management strategy [24]. Emergency data transfer rates may vary in the area affected by the disaster [25].

Multi-agent systems for emergency response management can incorporate real-time data from various sources. This type of system can be designed to help emergency responders make informed decisions quickly during a crisis. "Multi-Agent Systems for Disaster Management" focused on the advantages and challenges of using these systems in real-world scenarios [26]. "An Agent-Based Simulation Framework for Disaster Response Operations" was based on a simulation framework for disaster response operations to incorporate factors such as resource allocation and communication among responders to test different response strategies and evaluate their effectiveness [27]. Rui-na et al. developed "A Multi-Agent System for Emergency Management in Large-Scale Disasters" for emergency management in large-scale disasters that integrates various sources of information and decision-making processes to improve coordination among emergency responders and reduce response time [28]. "Multi-Agent Systems for Emergency Response and Disaster Management" provides an overview of the use of multi-agent systems in emergency response and disaster management; focusing on the advantages and challenges of using these systems in real-world scenarios is also proposed by researchers [29]. Various applications of multi-agent systems in disaster management have been tested. These systems can help improve coordination among responders, reduce response time, and facilitate data sharing and decision-making. However, there are also challenges associated with implementing these systems in real-world scenarios, such as the need for accurate and timely data and effective communication among agents. Disaster management is critical in e-waste management and smart manufacturing [30, 31]. Figure 11.2 presents the word cloud of information and communication tools

FIGURE 11.2 Word cloud of information and communication tools in emergency disaster management.

in emergency disaster management, developed by a bibliometric analysis of research articles from 1997 to 2023.

A bibliometric analysis was conducted by the authors for the time span of 1997 to 2023. A total of 380 articles were published in 211 sources with the keywords information, communication, artificial intelligence, machine learning, Internet of Things, Industry 4.0, and data analytics with disaster management in the Scopus database. Table 11.1 presents the details of the bibliometric data on information and communication tools in emergency disaster management. The bibliometric analysis shows that the annual growth rate of the related articles is 13.83%, with 20.49 average citations per document.

The open source Bibliometrix package is used for graphical illustrations. The annual scientific production of research articles on information and communication tools in emergency disaster management is presented in Figure 11.3. An average annual growth rate of 13.83% was witnessed during 1997 to 2023, with rapid record growth in the last five years. Figure 11.4 presents a tree map of the most used/cited words.

TABLE 11.1

Summary of Bibliometric Data on Information and Communication Tools in Emergency Disaster Management

Description	Results
Timespan	1997:2023
Sources (journals, books, etc.)	211
Documents	380
Annual growth rate, %	13.83
Document average age (in years)	5.94
Average citations per document	20.49

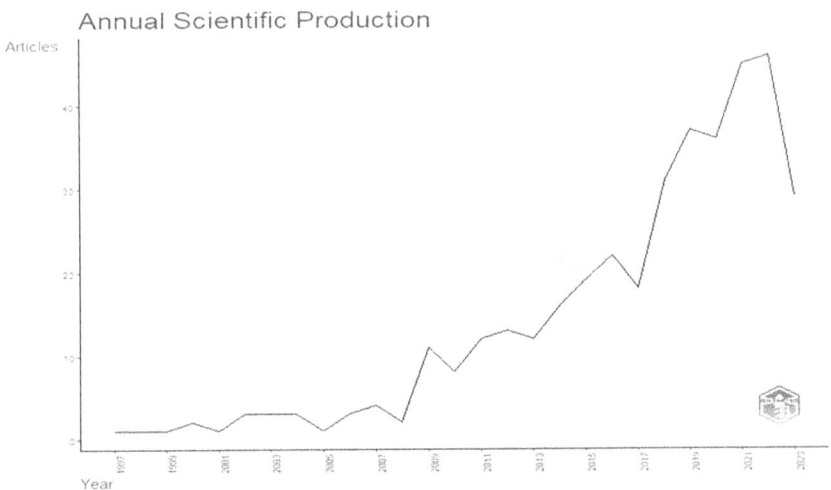

FIGURE 11.3 Annual scientific production of research articles on information and communication tools in emergency disaster management.

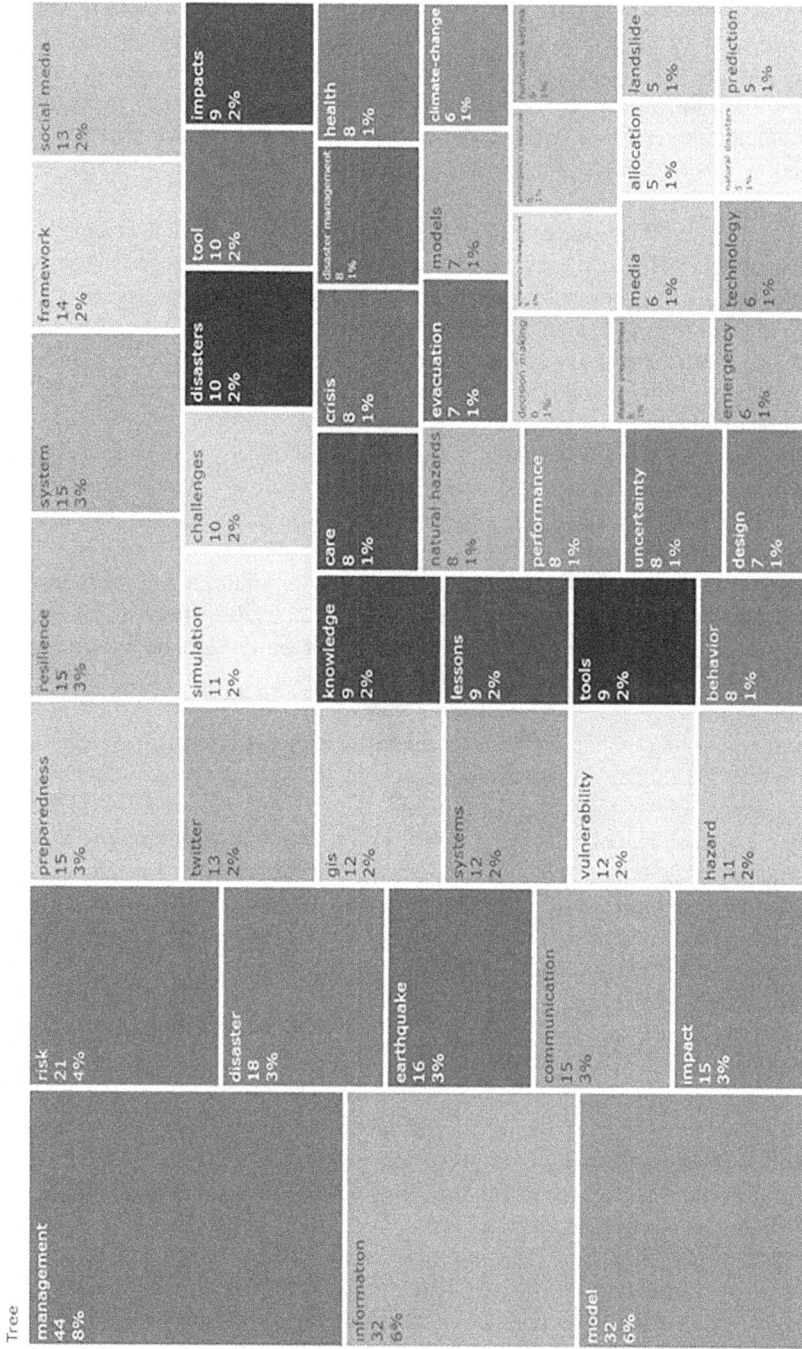

FIGURE 11.4   Tree map of keywords on information and communication tools in emergency disaster management.

11.3 RESULTS AND DISCUSSION

Disaster management and emergency management are two interrelated fields that involve disasters of all kinds which can be prevented, dealt with, and recovered from. Lessening the effects of disasters on individuals, property, and the natural world is the aim of both areas. Effective disaster management and emergency management have resulted in several positive outcomes. The major benefits of an efficient disaster management system include, but are not limited to, the following:

- Saving Lives: The primary goal of disaster management and emergency management is to save lives. Effective planning and response can reduce the number of casualties and injuries during a disaster.
- Reduced Property Damage: Effective disaster management and emergency management can also reduce property damage. Early warning systems, evacuation plans, and proper infrastructure can help to protect buildings and infrastructure from damage during a disaster.
- Improved Resilience: Disaster management and emergency management can help communities become more resilient to future disasters. By learning from past experiences, communities can develop better disaster preparedness plans and response strategies.
- Enhanced Cooperation: Disasters often require multiple agencies and organizations to work together to respond effectively. Effective disaster management and emergency management can promote cooperation and coordination between these organizations.

Despite the positive outcomes of disaster management and emergency management, there are several challenges that must be addressed. Resource limitations mean disaster management and emergency management require significant resources, including funding, personnel, and equipment. Limited resources can hamper effective disaster preparedness and response, and coordination and communication regarding disasters often require multiple agencies and organizations to work together, which can be challenging. Coordination and communication can be hindered by organizational silos, different communication protocols, and limited resources.

11.4 CONCLUSION

In conclusion, early prediction and strategies has been a significant step towards life and property savings. Disaster management has made it possible to provide timely and accurate information, improve communication and coordination among various response agencies, and enhance decision-making processes. Emergency management has been found to be effective in reducing the impact of disasters by enabling the detection of early warning signs and facilitating rapid response actions. Incorporating cutting-edge technology like artificial intelligence, machine learning, and data analytics can further improve the capabilities of disaster management systems. Figure 11.5 presents trending words in the research field of information and communication tools in emergency disaster management.

Trend Topics

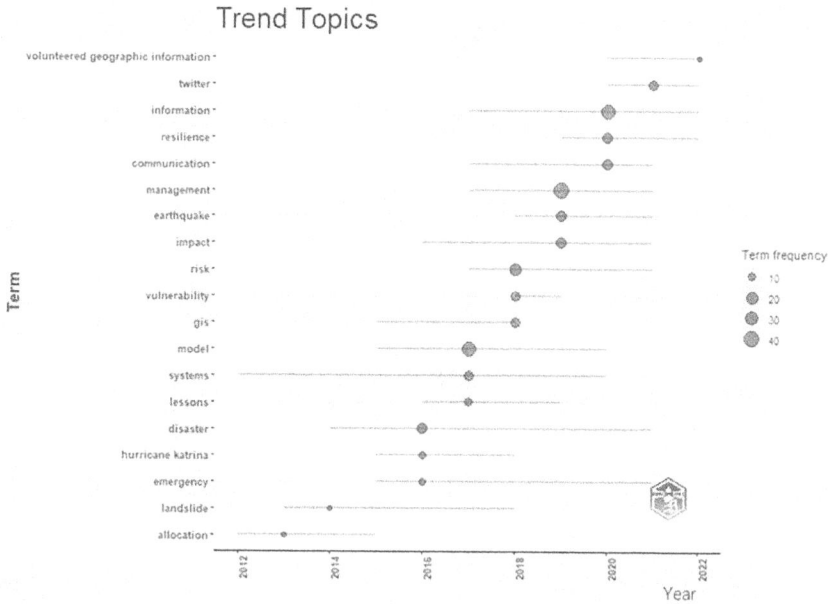

FIGURE 11.5   Trending words in the research field of information and communication tools in emergency disaster management.

Despite the numerous benefits of disaster management, there are still some challenges that need to be addressed, such as interoperability issues among various response agencies, limited access to real-time data, and the need for more extensive training programs for emergency responders. Overall, the development of disaster management is a promising approach that has the potential to save lives, reduce damage, and enhance resilience to disasters. Ongoing research and development efforts in this field will undoubtedly lead to even more advanced disaster management systems in the future. There is a need for integration of emerging technologies such as artificial intelligence, machine learning, and the Internet of Things in disaster management systems. Also, there is a need to investigate the influence of these technologies in early warning systems and decision support systems. The latest technological tools like blockchain, 5G technology, data analytics, and social media have emerged as life-saving kits for disaster management in real time. This future scope will surely offer great opportunities for research and innovation in the field of emergency disaster management, thereby ensuring more effective and efficient response efforts in the face of disasters.

REFERENCES

[1] Her Nugrahandika, Widyasari, and Ramadhani Nuresa Putri. "Sustainable Community-Based Disaster Management (CBDM) in Wonolelo Village, Bantul Regency, Special Region of Yogyakarta, Indonesia." *IOP Conference Series: Earth and Environmental Science* 764, no. 1 (2021): 012036. doi:10.1088/1755-1315/764/1/012036.

[2] Pandey, Chandra Lal. "Making Communities Disaster Resilient." *Disaster Prevention and Management: An International Journal* 28, no. 1 (2018): 106–18. doi:10.1108/dpm-05-2018-0156.

[3] BNPB. *Guidelines for the Quick Handling of Covid-19 Medical and Public Health in Indonesia.* Center for Health Policy and Management, Faculty of Medicine, Public Health and Nursing, Universitas Gadjah Mada, 2020.

[4] Carter, W. Nick. *Disaster Management: A Disaster Manager's Handbook.* ADB, 1991.

[5] Altay, Nezih, and Walter G. Green. "Or/Ms Research in Disaster Operations Management." *European Journal of Operational Research* 175, no. 1 (2006): 475–93. doi:10.1016/j.ejor.2005.05.016.

[6] Chen, Dan, Zhixin Liu, Lizhe Wang, Minggang Dou, Jingying Chen, and Hui Li. "Natural Disaster Monitoring with Wireless Sensor Networks: A Case Study of Data-Intensive Applications upon Low-Cost Scalable Systems." *Mobile Networks and Applications* 18, no. 5 (2013): 651–63. doi:10.1007/s11036-013-0456-9

[7] Seaberg, Daniel, Laura Devine, and Jun Zhuang. "A Review of Game Theory Applications in Natural Disaster Management Research." *Natural Hazards* 89, no. 3 (2017): 1461–83. doi:10.1007/s11069-017-3033-x

[8] Assilzadeha, H., and S. B. Mansora. "Natural Disaster Data and Information Management System." University Putra Malaysia: Institute of advanced Technology (ITMA) (2004).

[9] Chu, Yuechun, and Aura Ganz. "WISTA: A Wireless Telemedicine System for Disaster Patient Care." *Mobile Networks and Applications* 12, no. 2–3 (2007): 201–14. doi:10.1007/s11036-007-0012-6

[10] Ghosh, Saptarshi, Kripabandhu Ghosh, Debasis Ganguly, Tanmoy Chakraborty, Gareth J. Jones, Marie-Francine Moens, and Muhammad Imran. "Exploitation of Social Media for Emergency Relief and Preparedness: Recent Research and Trends." *Information Systems Frontiers* 20, no. 5 (2018): 901–7. doi:10.1007/s10796-018-9878-z.

[11] Chiu, Dickson K., Drake T. Lin, Eleanna Kafeza, Minhong Wang, Haiyang Hu, Hua Hu, and Yi Zhuang. "Alert Based Disaster Notification and Resource Allocation." *Information Systems Frontiers* 12, no. 1 (2009): 29–47. doi:10.1007/s10796-009-9165-0.

[12] Lee, J., and T. Bui. "A Template-Based Methodology for Disaster Management Information Systems." *Proceedings of the 33rd Annual Hawaii International Conference on System Sciences*, Maui, HI, USA, vol. 2, p. 7 (2000). doi:10.1109/hicss.2000.926635.

[13] Takahagi, Kazuhiro, Tomoyuki Ishida, Akira Sakuraba, Kaoru Sugita, Noriki Uchida, and Yoshitaka Shibat. "Proposal of the Disaster Information Transmission Common Infrastructure System Intended to Rapid Sharing of Information in a Time of Mega Disaster." *2015 18th International Conference on Network-Based Information Systems*, pp. 505–510 (2015). doi:10.1109/nbis.2015.75.

[14] Levy, Yair, and Timothy J. Ellis. "A Systems Approach to Conduct an Effective Literature Review in Support of Information Systems Research." *Informing Science: The International Journal of an Emerging Transdiscipline* 9 (2006): 181–212. doi:10.28945/479.

[15] Aydin, Can, Cigdem Tarhan, Ahmet Selcuk Ozgur, and Vahap Tecim. "Improving Disaster Resilience Using Mobile Based Disaster Management System." *Procedia Technology* 22 (2016): 382–90. doi:10.1016/j.protcy.2016.01.027.

[16] Palen, Leysia, and Amanda L. Hughes. "Social Media in Disaster Communication." *Handbook of Disaster Research* (2017): 497–518. doi:10.1007/978-3-319-63254-4_24

[17] Tomoyasu, Kota, Yuanyuan Wang, Reo Kimura, and Kazutoshi Sumiya. "Will Mail-Based Disaster Information Systems Work Well? A Case Study in Japan.". In *Smart Digital Futures 2014*, pp. 293–303, edited by R. Neves-Silva et al. IOS Press, 2014. doi:10.3233/978-1-61499-405-3-293

[18] Ahmadi, Morteza, Abbas Seifi, and Behnam Tootooni. "A Humanitarian Logistics Model for Disaster Relief Operation Considering Network Failure and Standard Relief Time: A Case Study on San Francisco District." *Transportation Research Part E: Logistics and Transportation Review* 75 (2015): 145–63. doi:10.1016/j.tre.2015.01.008.

[19] Otomo, Masaki, Goshi Sato, and Yoshitaka Shibata. "In-Vehicle Cloudlet Computing Based on Delay Tolerant Network Protocol for Disaster Information System." *Advances on Broad-Band Wireless Computing, Communication and Applications* (2016); 255–66. doi:10.1007/978-3-319-49106-6_24.

[20] Zhou, Qifeng, Ruifeng Yuan, and Tao Li. "An Improved Textual Storyline Generating Framework for Disaster Information Management." *2017 12th International Conference on Intelligent Systems and Knowledge Engineering (ISKE)*. Nanjing, China, pp. 1–8 (2017). doi:10.1109/iske.2017.8258738.

[21] Alem, Douglas, Alistair Clark, and Alfredo Moreno. "Stochastic Network Models for Logistics Planning in Disaster Relief." *European Journal of Operational Research* 255, no. 1 (2016): 187–206. doi:10.1016/j.ejor.2016.04.041.

[22] McLoughlin, David. "A Framework for Integrated Emergency Management." *Public Administration Review* 45 (1985): 165–172. doi:10.2307/3135011.

[23] Bandyopadhyay, Ayan, Debasis Ganguly, Mandar Mitra, Sanjoy Kumar Saha, and Gareth J.F. Jones. "An Embedding Based IR Model for Disaster Situations." *Information Systems Frontiers* 20, no. 5 (2018): 925–32. doi:10.1007/s10796-018-9847-6

[24] Liu, Sophia B., Leysia Palen, and Elisa Giaccardi. "Heritage Matters in Crisis Informatics: How Information and Communication Technology Can Support Legacies of Crisis Events." *Crisis Information Management* (2012): 65–86. doi:10.1016/b978-1-84334-647-0.50004-7.

[25] Besiou, Maria, Alfonso J. Pedraza-Martinez, and Luk N. Van Wassenhove. "Or Applied to Humanitarian Operations." *European Journal of Operational Research* 269, no. 2 (2018): 397–405. doi:10.1016/j.ejor.2018.02.046.

[26] Sharmeen, Zahra, Ana Maria Martinez-Enriquez, Muhammad Aslam, Afraz Zahra Syed, and Talha Waheed. "Multi Agent System Based Interface for Natural Disaster." *Active Media Technology* (2014); 299–310. doi:10.1007/978-3-319-09912-5_25.

[27] Yang, Weilong, Yue Hu, Cong Hu, and Mei Yang. "An Agent-Based Simulation of Deep Foundation Pit Emergency Evacuation Modeling in the Presence of Collapse Disaster." *Symmetry* 10, no. 11 (2018): 581. doi:10.3390/sym10110581.

[28] Rui-na, L. I. U., C. H. E. N. Jin-hua, C. H. E. N. Xi, and C. A. O. Wen. "Risk Assessment of Spring Frost Damage to Tea Plant in Anhui Province." *Chinese Journal of Agrometeorology* 42, no. 10 (2021): 870–879.

[29] Zahra, Kiran, Rahul Deb Das, Frank O. Ostermann, and Ross S. Purves. "Towards an Automated Information Extraction Model from Twitter Threads during Disasters." *Proceedings of the 19th International Conference on Information Systems for Crisis Response and Management*, Tarbes, France, pp. 637–653 (2022). ISCRAM Digital Library.

[30] Ajay, Ashwini Kumar, Parveen, Rajesh Goel, and Ravi Kant Mittal. "Waste Recovery and Management." In *An Approach Toward Sustainable Development Goals,* edited by P. Ajay, A. Kumar, R. Kant Mittal, and R. Goel. CRC Press, 2023. doi:10.1201/9781003359784.

[31] Kumar, Love, Ajay Kumar, Rajiv Kumar Sharma, and Parveen Kumar. "Smart Manufacturing and Industry 4.0." *Handbook of Smart Manufacturing* (2023): 1–28. doi:10.1201/9781003333760-1.

12 Optimization of Process Parameters of Counterbore Hole Made on Workpiece of Al-6061 Using DOE and MCDM Techniques

Rajesh Choudhary[1], Pankaj Sharma[1], and Gaurav Tiwari[1]

1 Department of Mechanical Engineering, JECRC University, Jaipur, Rajasthan, India

12.1 INTRODUCTION

Material removal process, which in general is also called "machining", removes undesirable portions of metal workpieces in the form of chips to acquire a machined product having good quality with final shape and size optimization. Like other conventional machining processes, the drilling machining process is also the most reliable and important method to generate holes in a product.

The drilling operation is a process in which the metal is removed by applying feed rate to the workpiece using a rotating multi-toothed cutting tool. This tool rotates by machine assembly that then removes the material, which is called "metal removal rate" (MRR). An examination was done on drilling activity [1] on an AISI 1020 steel sheet. The limited component reproduction was completed using ANSYS software, which involved the utilization of feed rate and shaft speed. Bilgin et al. [2] conducted experiments on AISI 1015 steel using a tungsten carbide tool and analyzed the effect of the feed rate. Analysis of variance (ANOVA) strategy was utilized to examine the importance of each parameter. Chow et al. [3] broke down the sintered carbide drill for drilling the sheet made of AISI 304 steel. Information parameters were drilling speed, friction contact region proportion, friction angle, and feed rate; yield parameter was surface roughness. Results showed that high hardness with fine grain structure was found in the zone close to the penetrated gap. Navanth et al. [4] did the experiments on AI 2014 combination, utilizing HSS

DOI: 10.1201/9781003405870-12

turned drill. Shaft speed, feed rate, point, and helix angle were under consideration for finding improvement in processes. A mix of a feed pace of 0.36 mm/ fire up, angle and Helix angle of 900/150, and shaft speed of 200 rpm was ideal, which created the ideal consequences of opening width. Sumesh et al. [5] did an improvement concentrate to advance parameters of drilling activity to acquire least surface roughness. The material utilized was solid metal, and the boring apparatus was HSS. Results showed that 4 mm measurement, 0.1 mm/fire up feed, and 80 rpm speed was the ideal blend. Yahya Hypman Çelik [6] did an examination on process parameter impacts on drilling of Ti-6Al-4V compound. Heisel et al. [7] completed an investigation on the angle of drill point while drilling on carbon fiber strengthened plastic. The result indicated that an increment in point angle and feed improves the nature of the drill opening at entrance yet the quality declines at exit. El-Bahloul et al. [8] led drilling experiments on AISI 304 tempered steel, and to enhance the process of drilling, the Taguchi technique and fluffy rationale approach were utilized. Somasundaram et al. [9] completed drilling activity on aluminum network composite sheets utilizing fast steel. Yogendra et al. [10] directed drilling experiments utilizing L27 symmetrical cluster on mellow steel. Lower cutting rate and feed rate alongside medium point angle was the ideal mix to limit the surface roughness. Sharma et al. [11] completed CNC drilling experiments on tempered steel AISI 304 and upgraded the process parameters of drilling activity by Taguchi strategy (L-16) symmetrical exhibit. The parameters were surface roughness and ovality of the bored gap. Results demonstrated that with an increment in feed rate and depth of cut, the surface roughness increases. Ovality increases with increase in depth of cut.

Patel et al. [12] led the drilling experiments on EN8, EN24, and EN31 materials, having the drill made of cobalt composite steel whose point angle was 1350 and a helix angle of 300. ANOVA was completed to decide the ideal parameter mix. Sundeep et al. [13] examined the drilling of AISI 316 tempered steel and utilized the Taguchi system with L-9 symmetrical cluster for improving the process parameters. The goal of the investigation was to identify the least pushed power and most elevated MRR. Higher MRR was accomplished at 1250 rpm speed, 0.02mm/fire up feed, and a drill distance across of 8 mm and lower push power was acquired at speed of 800 rpm, feed at 0.02 mm/fire up, and drill measurement of 6mm. The experiments on AI 2014 combination utilizing HSS turned the drill at dry conditions [14, 15]. To upgrade the process parameters of drilling activity, the Taguchi technique and L-18 Orthogonal cluster were utilized. Shaft speed, feed rate, and point and helix angles were under consideration. A mix of a feed pace of 0.36 mm/fire up, point angle and Helix angle of 900/150, and a shaft speed of 200 rpm was ideal, which created the ideal consequences of opening width.

By using the Taguchi technique alongside L9 Orthogonal exhibit, Díaz-Álvarez et al. and Miller et al. [16, 17] did an improvement concentrate to advance parameters of drilling activity to acquire least surface roughness. The material utilized was solid metal, and the boring apparatus was HSS. Results showed that 4 mm measurement,

0.1 mm/fire up feed, and 80 rpm speed was the ideal blend [18–22]. Yahya et al. [6] did an examination on process parameter impacts on drilling of Ti-6Al-4V compound. The primary reactions during the examination were chip arrangement, opening quality, surface roughness, and burr height. Experiments were performed on a vertical machining center (VMC) by taking feed (0.05–0.15 mm/fire up), cutting velocity (12.5–25 m/min), and curve drill (900–1400 point angles). Results demonstrated that with higher speed, a decrement was seen in surface roughness, with a higher point angle and a decrement in feed rate. Various researchers [23–26] completed an investigation on angle of drill point while drilling on carbon fiber strengthened plastic. Experiments were performed utilizing cutting velocity (21–513 m/min) and feed (0.05–0.40 mm/fire up). A solidified carbide drill having shifting point angles (155°, 175°, and 185°) was utilized in experiments. Results showed that an increment in point angle and feed improves the nature of drill opening at entrance yet the quality at exit declines. Researchers [27–29] led drilling experiments on AISI 304 tempered steel and utilized the Taguchi technique and fluffy rationale approach to enhance the process of drilling. Using this methodology enabled them to arrive at the ideal process condition.

To obtain pivotal power and surface roughness, the researchers used ANFIS-GA, which is a crossover technique of drilling glass-strengthened polymer composite [30–33]. For the created model, the information process parameters were feed rate, shaft speed, boring tool width, and workpiece thickness. It resulted in hub power and surface roughness and thus met the researchers' expectations [34–36]. Somasundaram et al. [9] completed drilling activity on aluminum network composite sheets by utilizing fast steel. The target of this investigation was to determine roundness mistakes of the bored gaps under differing feed rate, axle speed, workpiece thickness, and rate weight of support particles. A relapse model condition for roundness blunder was made when using reaction surface strategy. Results demonstrated that with the expansion in level of support in aluminum grid composites, roundness blunder was diminished. Also, roundness mistake expanded with increment in different process parameters [37, 38]. Sharma et al. [39] directed drilling experiments utilizing L27 symmetrical cluster on mellow steel. The Taguchi technique was utilized to advance the surface roughness. Sign-to-noise proportion and ANOVA strategy were used to examine the outcomes. Results showed that lower cutting rate and feed rate alongside medium point angle was the ideal mix to limit the surface roughness. To limit the delamination factor in drilling activity of carbon fiber reinforced polymers (CFRPs), Sharma et al. [39] exhibited the strategy of improvement of the Taguchi technique. By taking the process parameters speed, feed, and point angle, experiments were done utilizing L27 symmetrical exhibit. To find the ideal process parameters for the delamination factor, ANOVA was utilized. Results showed that for the least delamination factor, the point angle, feed, and speed were the prominent factors.

Iyyappan et al. [40] conducted the drilling activity on Al/SiC/Mica crossover composites; they performed experiments on a vertical drilling machine and examined the different drilling parameters, which are axle speed feed rate, and drilling tool measurement. Dubey et al. [41, 42] assess the presentation of HSS k20 strong carbide and TiN-covered SS HSS tool for this dry drilling done on AISI 304 austenitic hardened steel. Input parameters were taken: axle speed, feed rate, drill point angle. Utilizing a design of experiment (DOE) technique and L9 symmetrical cluster, the outcome showed that surface roughness were unfortunate and tool wear with change of drilling parameters. TiN-covered HSS contort drill gives the best outcome, while HSS tool gives the most noticeably unfavorable outcome. Tiwari et al. [43] performed an experiment on AISI SS317L. The input parameters were speed and feed, and the yield parameters were MRR and surface roughness. The techniques used were the Taguchi strategy and ANOVA. Experiments were conducted on CNC mill machine to evaluate the better surface roughness and for this, input parameters such as feed rate, spindle speed, and tool type were selected. Taking a static analysis of response variables and signal-to-noise ratio, an optimal combination was developed by taking cutting parameters so surface roughness can be optimized. It is found that feed rate is more responsible for surface quality. The Taguchi parameter design was able to produce better surface roughness in this drilling operation [44].

After a review of various research papers, it has been found that very few researchers have worked on counterbore drilling operation, so present studies is focused on counterbore drilling operation.

12.2 RESEARCH METHODOLOGY

12.2.1 WORKPIECE AND PRODUCT SELECTION

This research study selected counterbore drilling to determine the role of process parameters on selective workpieces made of AL-6061 material.

12.2.2 RESPONSE SELECTION

The response parameter selected for this research study is total cutting time (TCT). TCT is a combination of two different cutting times measured during the experiment on an Al-6061 work piece material. The first cutting time is measured while making the hole, and the second is measured while making the counterbore on the hole. Both times are measured by using a stopwatch and display available on the machine.

TABLE 12.1
Chemical and Mechanical Properties of AL-6061

Chemical Composition

Mg	Si	Fe	Cu	Ti	Pb	Zn	Mn	Sn	Ni	Al
1–1.5	10–12	1	1–1.5	0.2	0.1	0.5	0.5	0.1	1.5	BAL

Mechanical Properties

Density	Hardness	UTS	TYS	MOE	Shear Strength
g/cc	brinell	MPa	MPa	GPa	MPa
2.7	95	310	276	68.9	207

TABLE 12.2
Various Factors and Range for Experimentation

Factor		−1	0	1	Unit
A	Feed rate	16	18	20	mm/min
B	Cutting speed	1000	1100	1200	rpm
C	Depth of cut	1.0	1.5	2.0	mm

TABLE 12.3
Experiment Table Made by PB Design Method

A (Feed)	B (Speed)	C (DoC)	DT (Drill Time)	CBT (Counterbore Time)	TCT (Total Cutting Time)	SR (Surface Roughness)
1	−1	−1	47.27	28.8	76.07	6.2
1	1	−1	45.5	28.4	73.90	5.5
1	1	1	38.84	27.22	66.06	4.9
−1	−1	−1	54.03	28	82.03	7.9
−1	−1	1	48.27	28.2	76.47	6.3
1	−1	1	39.79	28.4	68.19	7.1
1	1	−1	44.51	27.8	72.31	5.1
0	0	0	44.89	28.8	73.69	6.3
−1	−1	−1	54.5	28.5	83.00	7.1
1	1	1	38.7	28.6	67.30	5.2
−1	1	1	47.3	28	75.30	6.2
−1	1	1	47.3	27.9	75.20	5.8
1	−1	1	38.9	27.8	66.70	7.5
0	0	0	42.7	27.4	70.10	6.4
1	−1	1	39.6	28.3	67.90	8.5
0	0	0	42.8	28.6	71.40	6.4
−1	1	−1	53.4	27.1	80.50	5.8
1	−−1	−1	44.1	27.6	71.70	6.3
−1	−1	−1	54.1	29.1	83.20	7.5
−1	1	−1	53.3	28.3	81.60	4.8
−1	1	1	47.2	28.6	75.80	5.1
−1	−1	1	47.4	27.6	75.00	6.8
0	0	0	42.7	28.9	71.60	6.5
1	1	−1	44.1	27.3	71.40	5.5

12.2.3 EXPERIMENT TABLE MADE BY PB DESIGN METHOD

12.2.3.1 Signal-to-Noise Ratio (S/N) Analysis for Counterbore Study

S/N ratio analysis is performed to find the rank among factors as selective response variable. This test is performed using MINITAB software.

12.2.3.2 Rank Identification for TCT

Total cutting time for this study is present in the table for all 24 experiments performed. The S/N ratio analysis for total cutting time is present in the table, for rank identification.

TABLE 12.4
S/N Ratio Analysis for Total Cutting Time

Run	A (Feed)	B (Speed)	C (DoC)	DT	CBT	TCT	S/N Ratio
1	1	−1	−1	47.27	28.8	76.07	−35.38
2	1	1	−1	45.5	28.4	73.90	−34.91
3	1	1	1	38.84	27.22	66.06	−33.12
4	−1	−1	−1	54.03	28	82.03	−36.63
5	−1	−1	1	48.27	28.2	76.47	−35.47
6	1	−1	1	39.79	28.4	68.19	−33.62
7	1	1	−1	44.51	27.8	72.31	−34.56
8	0	0	0	44.89	28.8	73.69	−34.87
9	−1	−1	−1	54.5	28.5	83.00	−36.82
10	1	1	1	38.7	28.6	67.30	−33.41
11	−1	1	1	47.3	28	75.30	−35.22
12	−1	1	1	47.3	27.9	75.20	−35.20
13	1	−1	1	38.9	27.8	66.70	−33.27
14	0	0	0	42.7	27.4	70.10	−34.06
15	1	−1	1	39.6	28.3	67.90	−33.55
16	0	0	0	42.8	28.6	71.40	−34.36
17	−1	1	−1	53.4	27.1	80.50	−36.32
18	1	−1	−1	44.1	27.6	71.70	−34.42
19	−1	−1	−1	54.1	29.1	83.20	−36.86
20	−1	1	−1	53.3	28.3	81.60	−36.54
21	−1	1	1	47.2	28.6	75.80	−35.33
22	−1	−1	1	47.4	27.6	75.00	−35.15
23	0	0	0	42.7	28.9	71.60	−34.40
24	1	1	−1	44.1	27.3	71.40	−34.36

TABLE 12.5
Rank Identification for TCT

Level	Feed Rate	Speed	DoC
1	−35.75	−35.11	−35.68
2	−34.42	−34.42	−34.42
3	−34.06	−34.89	−34.33
Delta	1.69	0.69	1.34
Rank	1	3	2

Main Effects Plot for Means
Data Means

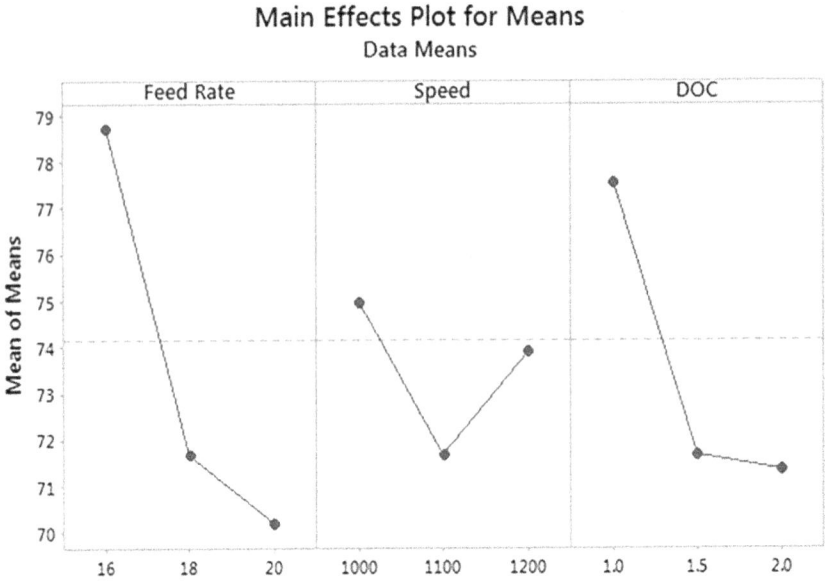

FIGURE 12.1   The graphs showing main effects plot for TCT response variable.

As seen in the table 12.5, the highest rank to maintain minimum TCT during the experiment is the feed rate provided to the drill machine, the second rank is set for depth of cut, and the last rank is set for spindle speed of machine. The same analysis is plotted in Figure 12.1, which helps to find the optimal solution for TCT minimum analysis.

As seen in the figure, the optimal solution for minimizing the TCT response variable is to select the levels in which the "Mean of Means" is on minimum value. So the optimal solution for TCT is the following: Feed rate must be at 20 mm/min, spindle speed is at 1100 RPM, and the last response variable is set at 2 mm.

12.2.3.3   Rank Identification for Surface Roughness (μm)

Surface roughness for this study is present in the table for all 24 experiments performed at Central Institute of Petrochemical Engineering and Technology, Jaipur. The S/N ratio data analysis for RA is also present in the table.

As seen in the table, the highest rank to maintain minimum SR during experiment is spindle speed provided to the drill machine, the second rank is set for DoC, and the last rank is set for the feed rate of the machine. The same analysis is plotted in the figure, which helps to find the optimal solution for SR minimum analysis.

As seen in the figure, the relation between the two factors feed rate and speed is shown for the TCT response variable. The plot shows that for the constant value of feed rate, if the speed value is going to change, then the TCT shows similar values, which means the effect of speed during the feed rate is not very high. But vice versa, if speed is constant and then the feed rate is going to change, then the TCT is also going to change and decrease for constant speed factor.

TABLE 12.6

S/N Ratio Analysis for Surface Roughness Response

Run	A (Feed)	B (Speed)	C (DoC)	Ra	S/N Ratio
1	1	−1	−1	6.2	−6.27
2	1	1	−1	5.5	−5.48
3	1	1	1	4.9	−4.76
4	−1	−1	−1	7.9	−8.05
5	−1	−1	1	6.3	−6.38
6	1	−1	1	7.1	−7.24
7	1	1	−1	5.1	−5.00
8	0	0	0	6.3	−6.38
9	−1	−1	−1	7.1	−7.24
10	1	1	1	5.2	−5.12
11	−1	1	1	6.2	−6.27
12	−1	1	1	5.8	−5.82
13	1	−1	1	7.5	−7.65
14	0	0	0	6.4	−6.49
15	1	−1	1	8.5	−8.63
16	0	0	0	6.4	−6.49
17	−1	1	−1	5.8	−5.82
18	1	−1	−1	6.3	−6.38
19	−1	−1	−1	7.5	−7.65
20	−1	1	−1	4.8	−4.64
21	−1	1	1	5.1	−5.00
22	−1	−1	1	6.8	−6.93
23	0	0	0	6.5	−6.60
24	1	1	−1	5.5	−5.48

TABLE 12.7

Rank Identification for Surface Roughness

Level	Feed Rate	Speed	DoC
1	−6.38	−7.24	−6.20
2	−6.49	−6.49	−6.49
3	−6.21	−6.11	−6.38
Delta	0.291	1.14	0.295
Rank	**3**	**1**	**2**

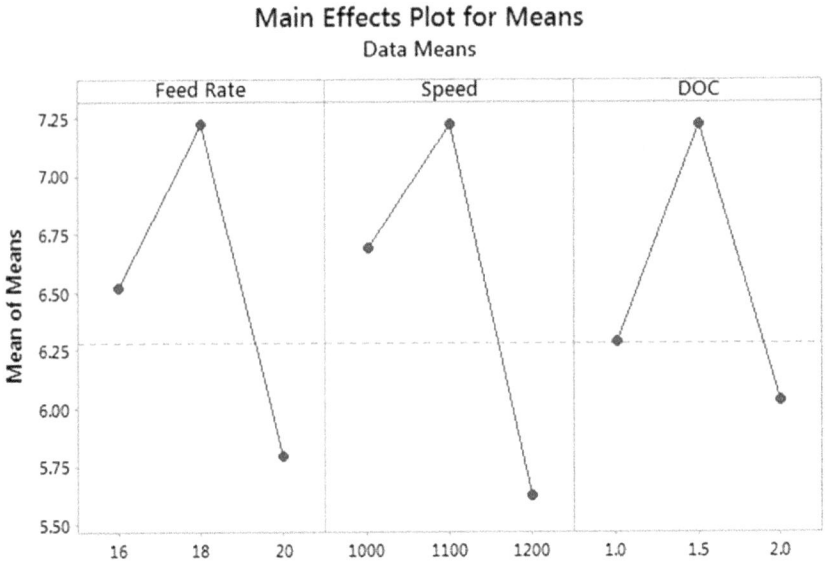

FIGURE 12.2   Mean effect plot for Ra response variable.

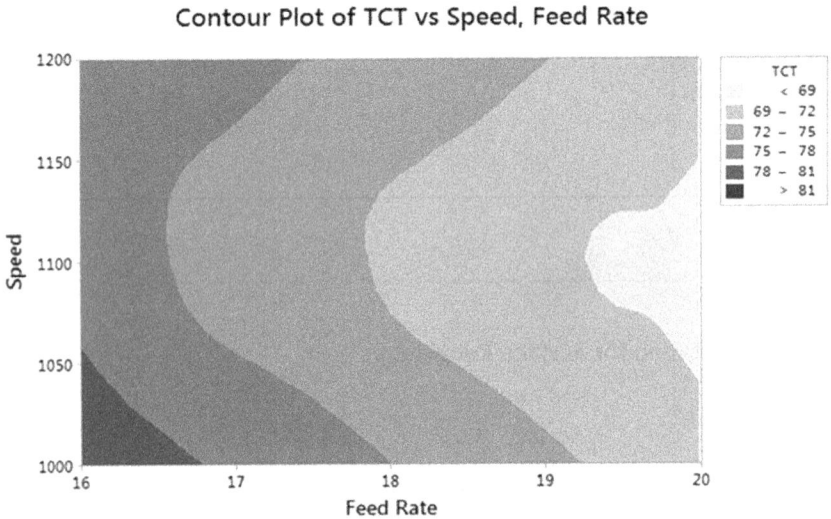

FIGURE 12.3   Contour plot analysis for TCT.

Contour Plot of TCT vs DOC, Feed Rate

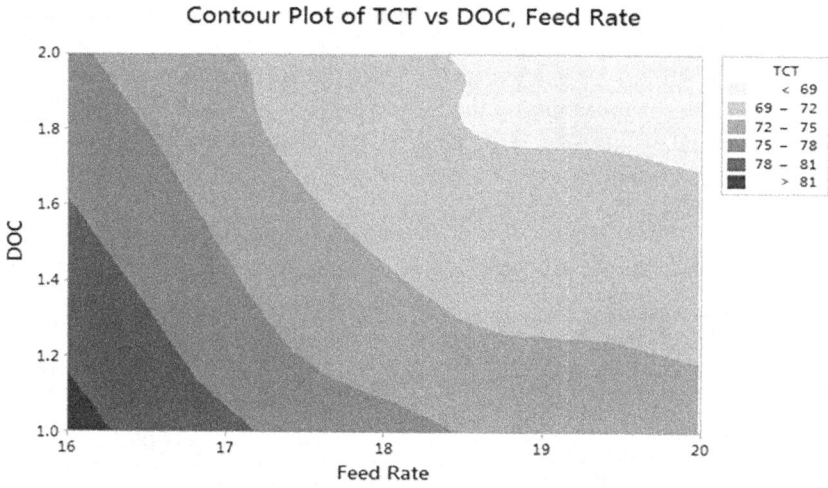

FIGURE 12.4 Contour plot of TCT vs DoC, feed rate.

Contour Plot of TCT vs DOC, Speed

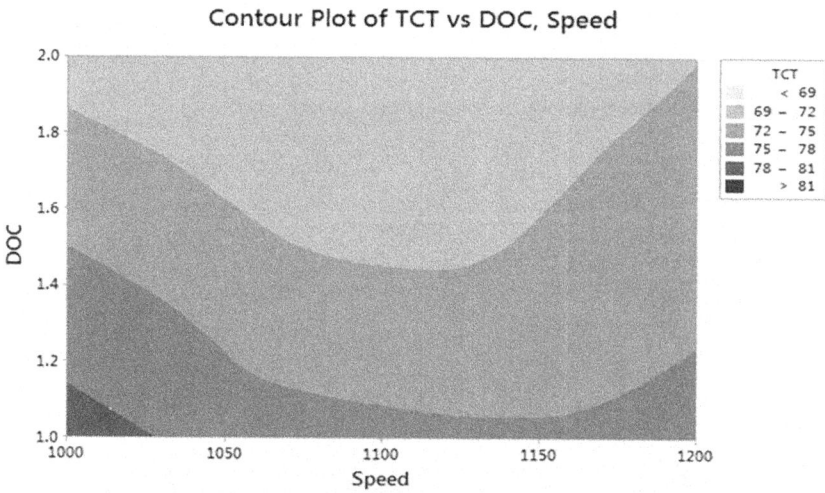

FIGURE 12.5 Contour plot of TCT vs DoC, speed.

As seen in the figure 12.6, the relation between the two factors feed rate and speed is shown for the SR response variable. The plot shows that for the constant value of feed rate, if speed value is going to change, then the SR shows similar values, which means the effect of speed during the feed rate is not very high. But vice versa, if speed is constant and then the feed rate is going to change, then the SR is also going to change and decrease for constant speed factor level. The two other plots are also shown in the figure.

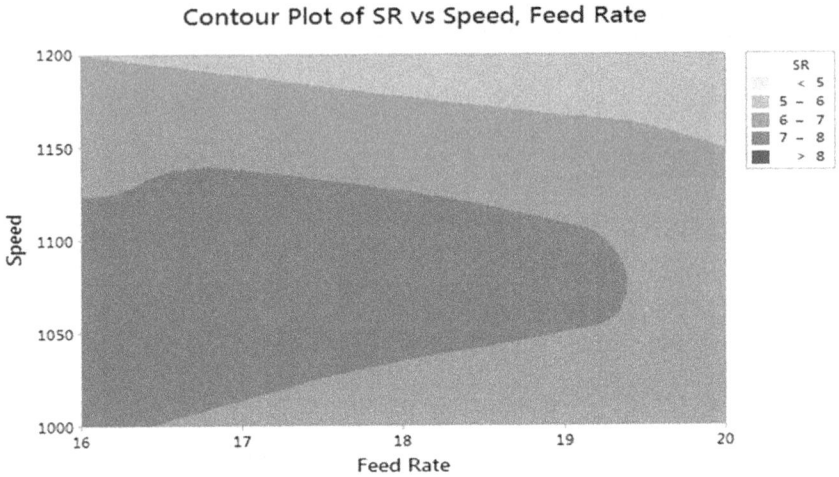

FIGURE 12.6   Contour plot of surface roughness vs speed, feed rate

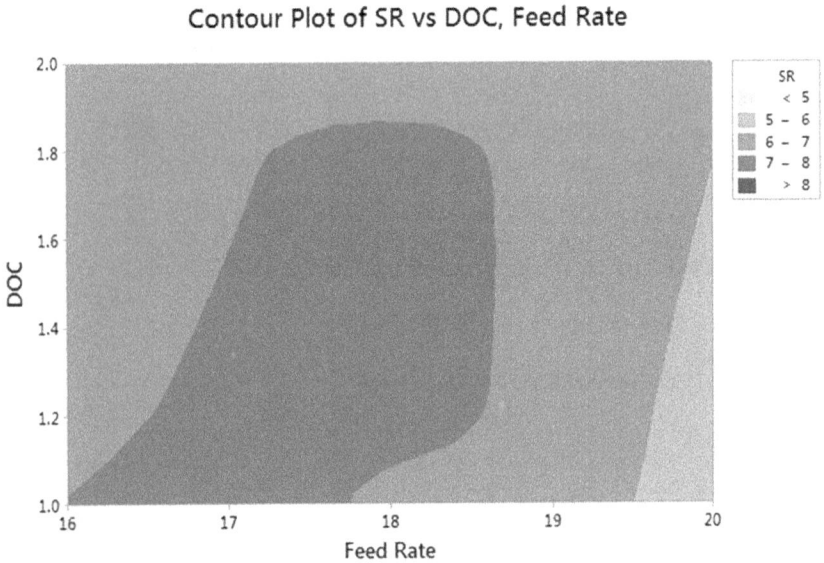

FIGURE 12.7   Contour plot of surface roughness vs DoC, feed rate.

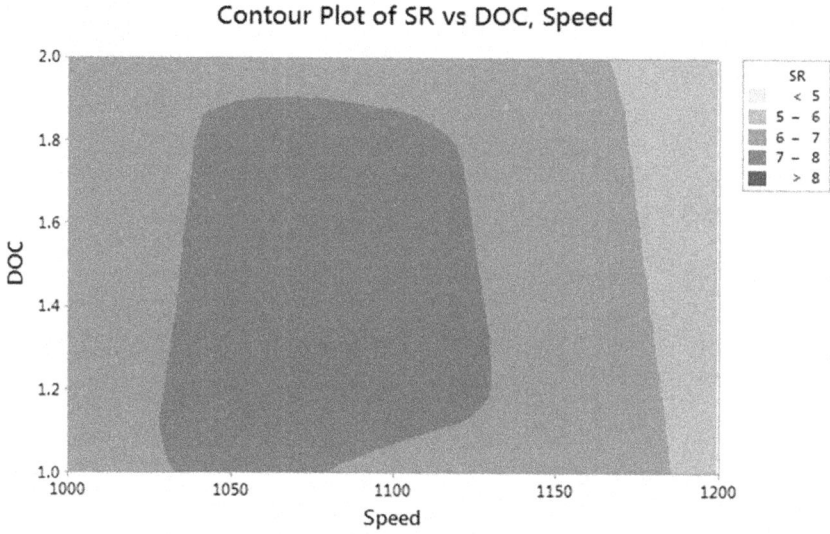

FIGURE 12.8   Contour plots of surface roughness vs DoC, speed.

12.2.4   ANOVA ANALYSIS FOR TCT (LINEAR MODEL)

TABLE 12.8
Table Showing ANOVA Analysis for Linear Model

Source	DF	Seq SS	Contribution	Adj SS	Adj MS	F-Value	P-Value
Model	3	571.548	91.77%	571.548	190.516	74.33	0
Linear	3	571.548	91.77%	571.548	190.516	74.33	0
Feed rate	1	374.718	60.17%	374.718	374.718	146.19	0
Speed	1	5.93	0.95%	5.93	5.93	2.31	0.144
DoC	1	190.9	30.65%	190.9	190.9	74.48	0
Error	20	51.263	8.23%	51.263	2.563		
Lack-of-fit	5	27.2	4.37%	27.2	5.44	3.39	0.03
Pure error	15	24.063	3.86%	24.063	1.604		
Total	23	622.811	100.00%				

Residual Plots for TCT

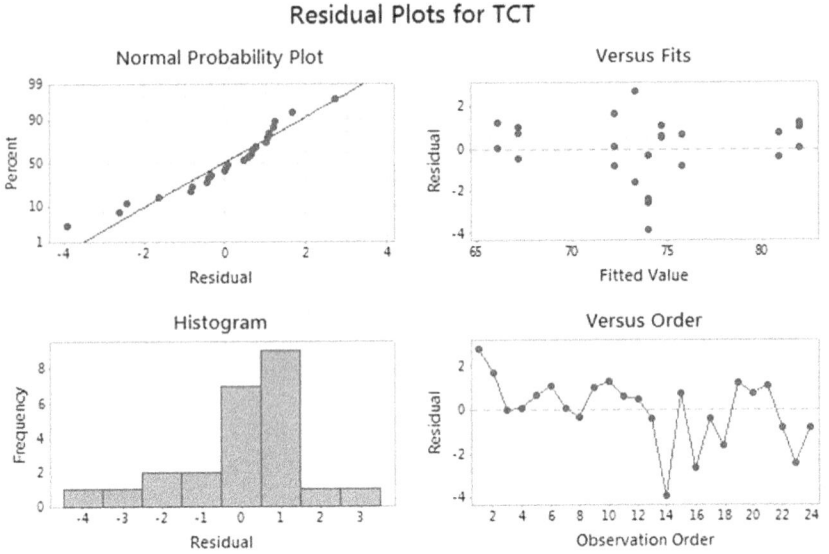

FIGURE 12.9　Residual plots for total cutting time.

TABLE 12.9
ANOVA Analysis for Total Cutting Time (Linear + Two-Way Model)

Source	DF	Seq SS	Contribution	Adj SS	Adj MS	F-Value	P-Value
Model	6	572.636	91.94%	572.636	95.439	32.34	0
Linear	3	571.548	91.77%	541.703	180.568	61.18	0
Feed rate	1	374.718	60.17%	352.193	352.193	119.33	0
Speed	1	5.93	0.95%	5.444	5.444	1.84	0.192
DoC	1	190.9	30.65%	184.066	184.066	62.36	0
Two-way interactions	3	1.088	0.17%	1.088	0.363	0.12	0.945
Feed rate*speed	1	0.022	0.00%	0.022	0.022	0.01	0.932
Feed rate*DoC	1	0.069	0.01%	0.069	0.069	0.02	0.88
Speed*DoC	1	0.997	0.16%	0.997	0.997	0.34	0.569
Error	17	50.175	8.06%	50.175	2.951		
Lack-of-fit	2	26.112	4.19%	26.112	13.056	8.14	0.004
Pure error	15	24.063	3.86%	24.063	1.604		
Total	23	622.811	100.00%				

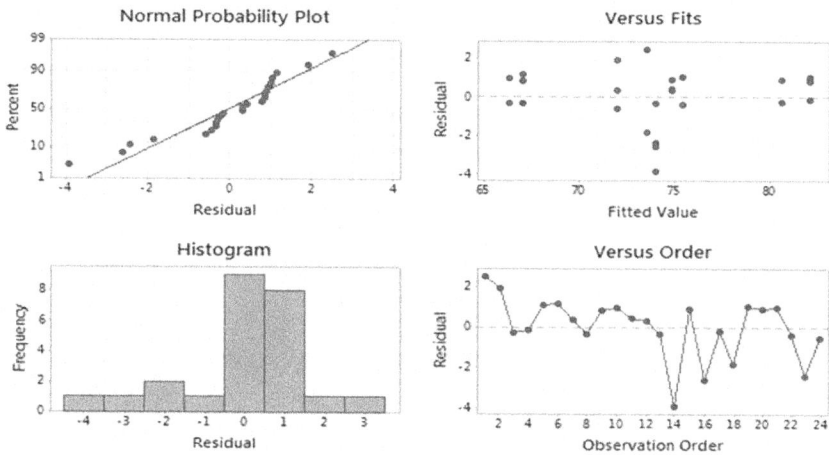

FIGURE 12.10   Graphs showing residual plots for total cutting time.

TABLE 12.10
ANOVA Analysis for TCT (Linear + Three-Way Model)

Source	DF	Seq SS	Contribution	Adj SS	Adj MS	F-Value	P-Value
Model	7	573.532	92.09%	573.532	81.933	26.6	0
Linear	3	571.548	91.77%	541.703	180.568	58.63	0
Feed rate	1	374.718	60.17%	352.193	352.193	114.35	0
Speed	1	5.93	0.95%	5.444	5.444	1.77	0.202
DoC	1	190.9	30.65%	184.066	184.066	59.76	0
Two-way	3	1.088	0.17%	1.088	0.363	0.12	0.948
Feed rate*speed	1	0.022	0.00%	0.022	0.022	0.01	0.934
Feed Rate*DoC	1	0.069	0.01%	0.069	0.069	0.02	0.883
Speed*DoC	1	0.997	0.16%	0.997	0.997	0.32	0.577
Three-way interactions	1	0.895	0.14%	0.895	0.895	0.29	0.597
Feed rate*speed*DoC	1	0.895	0.14%	0.895	0.895	0.29	0.597
Error	16	49.28	7.91%	49.28	3.08		
Lack-of-fit	1	25.217	4.05%	25.217	25.217	15.72	0.001
Pure error	15	24.063	3.86%	24.063	1.604		
Total	23	622.811	100.00%				

Pareto Chart of the Standardized Effects
(response is TCT, α = 0.05)

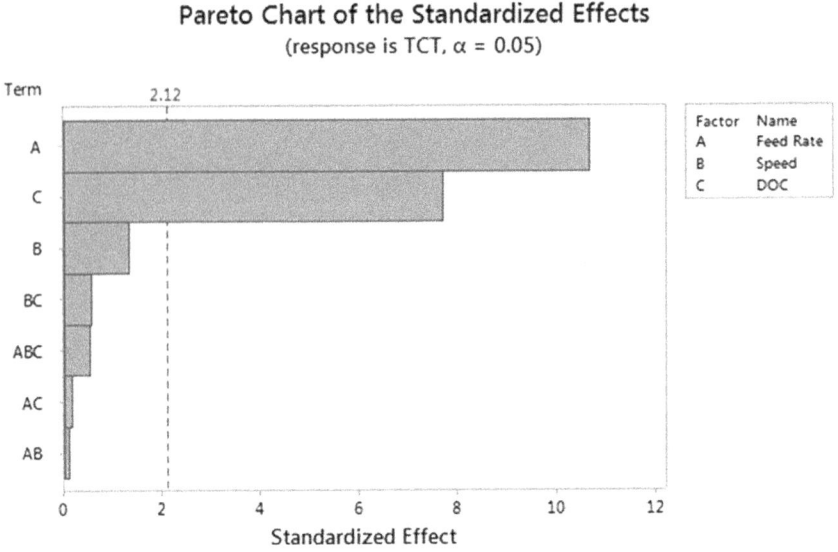

FIGURE 12.11   Pareto chart of the standardized effects.

Residual Plots for TCT

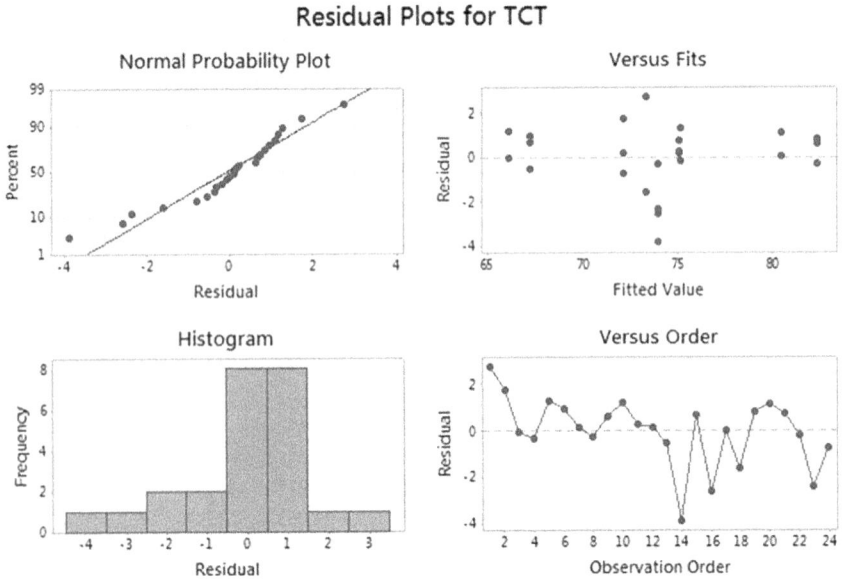

FIGURE 12.12   Residual plots for total cutting time.

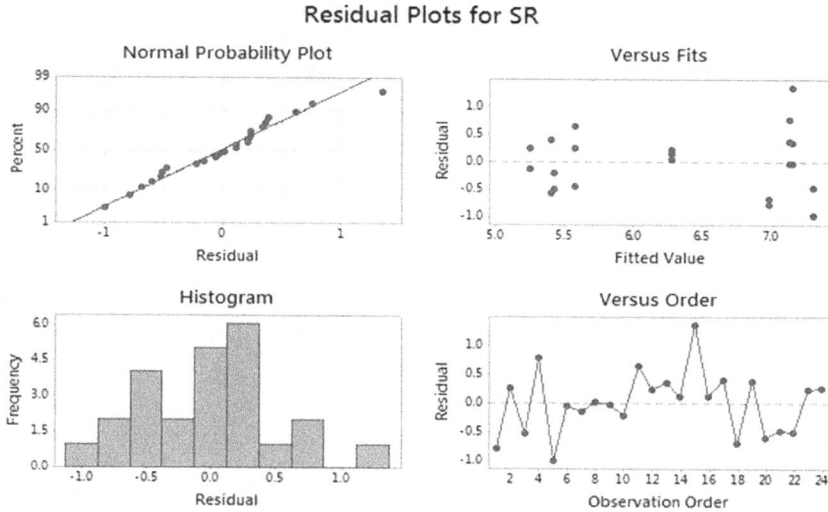

FIGURE 12.13 Residual plots for surface roughness.

TABLE 12.11
ANOVA Analysis for SR (Linear Model)

Source	DF	Seq SS	Contribution	Adj SS	Adj MS	F-Value	P-Value
Model	3	15.2215	69.32%	15.2215	5.0738	15.06	0
Linear	3	15.2215	69.32%	15.2215	5.0738	15.06	0
Feed rate	1	0.1125	0.51%	0.1125	0.1125	0.33	0.57
Speed	1	14.9645	68.15%	14.9645	14.9645	44.42	0
DoC	1	0.1445	0.66%	0.1445	0.1445	0.43	0.52
Error	20	6.7381	30.68%	6.7381	0.3369		
Lack-of-fit	5	3.9564	18.02%	3.9564	0.7913	4.27	0.013
Pure error	15	2.7817	12.67%	2.7817	0.1854		
Total	23	21.9596	100.00%				

TABLE 12.12
ANOVA Analysis for SR (Linear + Two-Way Model)

Source	DF	Seq SS	Contribution	Adj SS	Adj MS	F-Value	P-Value
Model	6	16.1937	73.74%	16.1937	2.699	7.96	0
Linear	3	15.2215	69.32%	13.2442	4.4147	13.02	0
Feed rate	1	0.1125	0.51%	0.1401	0.1401	0.41	0.529
Speed	1	14.9645	68.15%	13.0021	13.0021	38.34	0
DoC	1	0.1445	0.66%	0.1021	0.1021	0.3	0.59

(Continued)

TABLE 12.12 *(Continued)*
ANOVA Analysis for SR (Linear + Two-Way Model)

Source	DF	Seq SS	Contribution	Adj SS	Adj MS	F-Value	P-Value
Two-way interactions	3	0.9722	4.43%	0.9722	0.3241	0.96	0.436
Feed rate*speed	1	0.0701	0.32%	0.0701	0.0701	0.21	0.655
Feed rate*DoC	1	0.8501	3.87%	0.8501	0.8501	2.51	0.132
Speed*DoC	1	0.0521	0.24%	0.0521	0.0521	0.15	0.7
Error	17	5.7658	26.26%	5.7658	0.3392		
Lack-of-fit	2	2.9842	13.59%	2.9842	1.4921	8.05	0.004
Pure error	15	2.7817	12.67%	2.7817	0.1854		
Total	23	21.9596	100.00%				

TABLE 12.13
ANOVA Analysis for SR (Linear + Three-Way Model)

Source	DF	Seq SS	Contribution	Adj SS	Adj MS	F-Value	P-Value
Model	7	19.0134	86.58%	19.0134	2.7162	14.75	0
Linear	3	15.2215	69.32%	13.2443	4.4148	23.98	0
Feed rate	1	0.1125	0.51%	0.1401	0.1401	0.76	0.396
Speed	1	14.9645	68.15%	13.0021	13.0021	70.61	0
DoC	1	0.1445	0.66%	0.1021	0.1021	0.55	0.467
Two-way interactions	3	0.9722	4.43%	0.9722	0.3241	1.76	0.195
Feed rate*speed	1	0.0701	0.32%	0.0701	0.0701	0.38	0.546
Feed rate*DoC	1	0.8501	3.87%	0.8501	0.8501	4.62	0.047
Speed*DoC	1	0.0521	0.24%	0.0521	0.0521	0.28	0.602
Three-way interactions	1	2.8197	12.84%	2.8197	2.8197	15.31	0.001
Feed rate*speed*DoC	1	2.8197	12.84%	2.8197	2.8197	15.31	0.001
Error	16	2.9462	13.42%	2.9462	0.1841		
Lack-of-fit	1	0.1645	0.75%	0.1645	0.1645	0.89	0.361
Pure error	15	2.7817	12.67%	2.7817	0.1854		
Total	23	21.9596	100.00%				

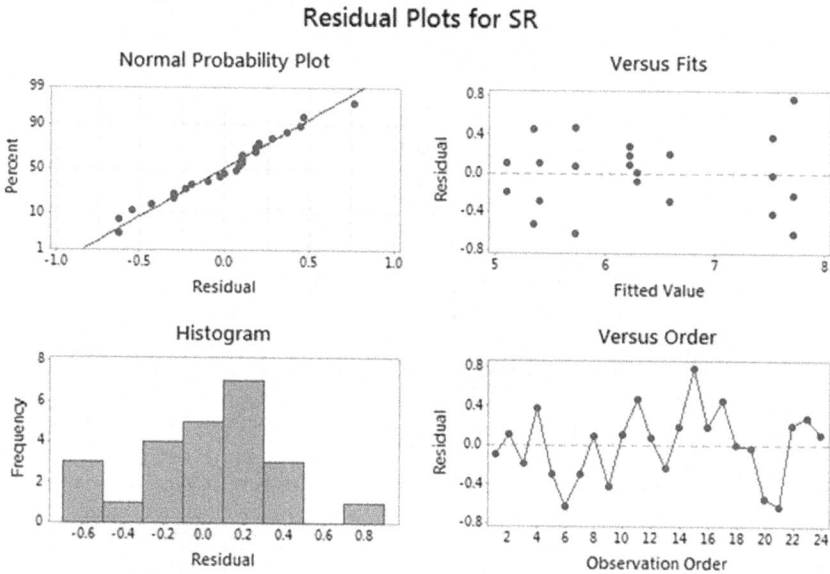

FIGURE 12.14   Residual plot for Ra (linear +three-way model).

TABLE 12.14
Different Levels for All Factors Used in This Research Work

		−1	0	1	Unit
A	Feed rate	16	18	20	mm/min
B	Cutting speed	1000	1100	1200	rpm
C	Depth of cut	1	1.5	2.0	mm

12.3   CONCLUSION AND FUTURE SCOPE

Three factors were selected for different levels combination to develop an experimental table, which are present in Table 12.14. The DOE technique used for this research work is the response surface method, and a total of 17 experiments were conducted.

The level selection was conducted via a local industrial survey and a literature survey, and the final range of factor levels was selected based on pilot experiments. It helped us to find the proper response parameters that were used in this research. All three responses were also selected by conducting pilot experiments.

REFERENCES

[1] L. Ozler and N. Dogru, "An experimental investigation of hole geometry in friction drilling," *Materials and Manufacturing Processes*, vol. 28, no. 4, pp. 470–475, 2013.

[2] M. B. Bilgin, K. Gök, and A. Gök, "Three-dimensional finite element model of friction drilling process in hot forming processes," *Proceedings of the Institution of Mechanical Engineers, Part E: Journal of Process Mechanical Engineering*, vol. 231, no. 3, pp. 548–554, 2017.

[3] H. M. Chow, S. M. Lee, and L. D. Yang, "Machining characteristic study of friction drilling on AISI 304 stainless steel," *Journal of Materials Processing Technology*, vol. 207, no. 1–3, pp. 180–186, 2008.

[4] A. Navanth and T. K. Sharma, "A study of Taguchi method based optimization of drilling parameter in dry drilling of Al 2014 alloy at low speeds," *International Journal of Emerging Technology and Advanced Engineering*, vol. 6, no. 1, pp. 65–75, 2013.

[5] A. S. Sumesh and M. E. Shibu, "Optimization of grinding parameters for minimum surface roughness using Taguchi method," *Journal of Mechanical and Civil Engineering*, pp. 12–20, 2014.

[6] Y. Hisman, "Investigating the effects of cutting parameters on the hole quality in drilling the Ti-6Al-4V alloy," *Materials Technology*, vol. 48, no. 5, pp. 653–659, 2013.

[7] U. Heisel and T. Pfeifroth, "Influence of point angle on drill hole quality and machining forces when drilling CFRP," *Procedia CIRP*, vol. 1, no. 1, pp. 471–476, 2012.

[8] S. A. El-Bahloul, H. E. El-Shourbagy, and T. T. El-Midany, "Optimization of thermal friction drilling process based on Taguchi method and fuzzy logic technique," *International Journal of Engineering Science,* vol. 4, no. 2, pp. 54–59, 2015.

[9] G. Somasundaram, S. Rajendra Boopathy, and K. Palanikumar, "Modeling and analysis of roundness error in friction drilling of aluminum silicon carbide metal matrix composite," *Journal of Composite Materials*, vol. 46, no. 2, pp. 169–181, 2012.

[10] T. Yogendra, C. Vedansh, and V. Jyoti, "Parametric optimization of drilling machining process using Taguchi design and ANOVA approach," *International Journal of Emerging Technology and Advanced Engineering*, vol. 2, no. 7, pp. 339–347, 2012.

[11] K. Sharma and A. Jatav, "Optimization of machining parameters in drilling of stainless steel", *International Journal of Scientific Research Engineering & Technology*, vol. 4, no. 8, 2015.

[12] J. Patel, A. Intwala, D. Patel, D. Gandhi, N. Patel, and M. Patel, "A review article on effect of cutting parameter on drilling operation for perpendicularity," *IOSR Journal of Mechanical and Civil Engineering*, vol. 11, no. 6, pp. 11–18,2015.

[13] M. Sundeep, M. Sudhahar, T. T. M. Kannan, P. V. Kumar, and N. Parthipan, "Optimization of drilling parameters on austenitic stainless steel (AISI316) using Taguchi's methodology," *International Journal of Mechanical Engineering and Robotics*, vol. 3, no. 4, pp. 388–394, 2014.

[14] P. V. G. Krishna, K. Kishore, and V. V. Satyanarayana, "Some investigations in friction drilling AA6351 using high speed steel tools," *ARPN Journal of Engineering and Applied Sciences*, vol. 5, no. 11–15, pp. 346–354, 2010.

[15] P. K. Bajpai, K. Debnath, and I. Singh, "Hole making in natural fiber-reinforced polylactic acid laminates: An experimental investigation," *Journal of Thermoplastic Composite Materials*, vol. 30, no. 1, pp. 30–46, 2017.

[16] A. Díaz-Álvarez, Á. Rubio-López, C. Santiuste, and M. H. Miguélez, "Experimental analysis of drilling induced damage in biocomposites," *Textile Research Journal,* vol. 88, no. 22, pp. 2544–2558, 2018.

[17] S. F. Miller, J. Tao, and A. J. Shih, "Friction drilling of cast metals," *International Journal of Machine Tools and Manufacture*, vol. 46, no. 12–13, pp. 1526–1535, 2006.

[18] P. D. Pantawane and B. B. Ahuja, "Parametric analysis and modelling of friction drilling process on AISI 1015," *International Journal of Mechatronics and Manufacturing Systems,* vol. 7, no. 1, p. 60, 2014.

[19] K. Abhishek, B. N. Panda, S. Datta, and S. S. Mahapatra, "Comparing predictability of genetic programming and ANFIS on drilling performance modeling for GFRP composites," *Procedia Materials Science,* vol. 6, no. ICMPC, pp. 544–550, 2014.

[20] W. L. Ku, C. L. Hung, S. M. Lee, and H. M. Chow, "Optimization in thermal friction drilling for SUS 304 stainless steel," *The International Journal of Advanced Manufacturing Technology,* vol. 53, no. 9–12, pp. 935–944, 2011.

[21] K. Lipin and P. Govindan, "A review on multi objective optimization of drilling parameters using Taguchi methods," *AKGEC International Journal of Technology,* vol. 4, no. 2.

[22] V. N. Gaitonde, S. R. Karnik, J. C. Rubio, A. E. Correia, A. M. Abrão, and J. P. Davim, "A study aimed at minimizing delamination during drilling of CFRP composites," *Journal of Composite Materials,* vol. 45, no. 22, pp. 2359–2368, 2011.

[23] P. K. Bajpai and I. Singh, "Drilling behavior of sisal fiber-reinforced polypropylene composite laminates," *Journal of Reinforced Plastics and Composites,* vol. 32, pp. 1569–1576, 2013. https://doi.org/10.1177/0731684413492866.

[24] G. D. Babu, K. S. Babu, and B. Goud, "Drilling uni-directional fiber-reinforced plastics manufactured by hand lay-up influence of fibers," *American Journal of Materials Science and Technology,* 2012. https://doi.org/10.7726/ajmst.2012.1001.

[25] U. Caydas, A. Hascalik, O. Buytoz, and A. Meyveci, "Performance evaluation of different twist drill in dry drilling of AISI 304 austenitic stainless steel," *Materials and Manufacturing Processes,* vol. 26, pp. 951–960, 2011.

[26] P. Nadarajan, "Multi objective optimization for drilling process stresses in SS317 material," *International Journal of Mechanical and Production Engineering Research and Development,* vol. 8, pp. 781–790, 2018. https://doi.org/10.24247/ijmperdapr201889

[27] P. S. Pankaj and M. P. Singh, "Optimization of process parameter by using CNC Wire Electrical Discharge Machine through Taguchi method," *International Journal of Engineering and Advanced Technology (IJEAT),* vol 9, pp. 518–522, 2020.

[28] Renu, A. Gupta, S. R. Kumar, C. Goswami, and P. Sharma, "Enhanced physical and mechanical properties of Al6061 alloy nanocomposite reinforced with nanozirconia," *Materials Today,* vol. 48, pp. 1130–1133, 2021.

[29] G. Tiwari, P. Sharma, and N. Sharma, "Study of Material Removal rate (MRR) for DSS 2205 steel in WEDM," *International Journal of Scientific Research in Engineering and Management (IJSREM),* vol. 6, pp. 31–40, 2022.

[30] P. Sharma, Y. Dubey, and M. P. Singh, "Experimental investigation for GTAW optimization using genetic algorithm on S-1 tool steel," *Materials Today: Proceedings. Elsevier,* 2023.

[31] Y. Zhang, L. Xu, C. Wang, Z. Chen, S. Han, B. Chen, and J. Chen, "Mechanical and thermal damage in cortical bone drilling in vivo," *Proceedings of the Institution of Mechanical Engineers, Part H: Journal of Engineering in Medicine,* vol. 233, no. 6, 621–635, 2019. https://doi.org/10.1177/0954411919840194.

[32] L. Kumar and R. K. Sharma, "Smart manufacturing and industry 4.0: State-of-the-art review," *Handbook of Smart Manufacturing,* pp. 1–28. https://doi.org/10.1201/9781003333760-1.

[33] A. Kumar, H. Singh, P. Kumar, and A. Gour (Eds.), *Forecasting the Future of Industry 4.0. Handbook of Smart Manufacturing.* CRC Press (2023). https://doi.org/10.1201/9781003333760.

[34] R. K. Mittal, *Incremental Sheet Forming Technologies: Principles, Merits, Limitations, and Applications.* CRC Press (2020). https://doi.org/10.1201/9780429298905.

[35] Ajay Kumar, P. Kumar, A. K. Srivastava, and V. Goyal (Eds.), *Modeming, Characterization, and Processing of Smart Materials.* IGI Global (2023). https://doi.org/10.4018/978-1-6684-9224-6.

[36] A. Kumar, R. K. Mittal, & R. Goel (Eds.), *Waste Recovery and Management: An Approach Toward Sustainable Development Goals.* CRC Press. https://doi.org/10.1201/9781003359784.

[37] A. Kumar, R. Mittal, and A. Haleem (Eds.), *Advances in Additive Manufacturing: Artificial Intelligence, Nature-Inspired, and Biomanufacturing.* Elsevier (2022). https://doi.org/10.1016/C2020-0-03877-6.

[38] M. P. Singh, P. Sharma, Y. Dubey, G. V. R. S. Rao, Q. Mohammad, and S. Lakhanpal, "Energy-efficient manufacturing: Opportunities and challenges," *The International Conference on Applied Research and Engineering,* vol. 505, 2024.

[39] A. Kumar and V. Gulati, "Experimental investigation and optimization of surface roughness in negative incremental forming," *Measurement,* vol. 131, pp. 419–430, 2019.

[40] S. Ganapathi Iyyappan, R. Sudhakarapandian, and M. Sakthivel, "Influence of silicon carbide mixed used engine oil dielectric fluid on EDM characteristics of AA7075/SiCp/B4Cp hybrid composites," *Materials Research Express,* vol. 8, no. 8, p. 086514, 2021.

[41] Y. Dubey, P. Sharma, M. P. Singh, G. V. R. S. Rao, Q. Mohammad, and S. Lakhanpal, "Green machining: Environmental and economic impacts of cutting fluids," *The International Conference on Applied Research and Engineering,* vol. 505, 2023.

[42] Y. Dubey, P. Sharma, and M. P. Singh, "Effect of TIG welding on Al alloy, Mg alloy, and stainless steel—A review," *International Journal of Mechanical Engineering,* vol. 7, no. 3, pp. 808–821, 2022.

[43] G. Tiwari, P. Sharma, and N. Sharma, "Machining of steel on wire electric discharge Machining: A review," *The International Journal of Advance Manufacturing Technology,* vol. 7, pp. 780–798, 2022a.

[44] G. Tiwari, P. Sharma, and N. Sharma, "Study of material removal rate (MRR) for DSS 2205 steel in WEDM," *International Journal of Scientific Research in Engineering and Management (IJSREM),* vol. 6, 2022b. ISSN: 2582-3930.

13 Application of IoT and Artificial Intelligence in Smart Manufacturing
Towards Industry 4.0

R. K. Jain[1] and S. K. Samanta[1]
1 CSIR-Central Mechanical Engineering Research Institute, Durgapur, West Bengal, India

13.1 INTRODUCTION

Smart manufacturing and assembly systems involve providing promising modern solutions which increase production features, flexible resources, and sustainable methods in a mechanized environment. Smart manufacturing consists of information data management systems at a central point which can be utilized for re-use of data so that the maximum output flow can be utilized in the organization. Nowadays, the manufacturing industry has been developed a variety of components through automation process where it is very difficult to exchange data in one step-to-another step of process parameters and optimize in one platform and run the business by maintaining the information of standards (Wang *et al.* 2020; Phuyala *et al.* 2020; He and Bai 2021). Smart manufacturing can be described as entirely collaborative and integrated manufacturing systems that act in online/real-time conditions to fulfill requirements per demands/customer needs within supply chain management. Smart manufacturing also provides the scope of future manufacturing by enhancing capabilities of factory automation using IoT, artificial intelligence (AI), and smart digital technology. IoT and AI can be used for failure identification of machines, prioritization of equipment maintenance, and recommendations to change spare part levels and increase production efficiency (Zawadzki et al. 2017; Pundir 2018; Panda *et al.* 2018; Gerekli *et al.* 2021).

IoT can be connected to industrial applications (i.e. Industry 4.0) to enhance substantial production systems and processes using smart/digital technology, AI, machine learning (ML), and big data for developing robust eco-friendly systems for industry (Sadiku *et al.* 2019; Arinez *et al.* 2020; Cioffi *et al.* 2020). This focuses on e-manufacturing and supply chain management systems and also helps in developing automation systems. Industry 4.0 completely transforms the production environment, and the aim is to enhance productivity and realize the highest mass production using such technologies. Industry 4.0 was an initiative by the German government in 2011 to develop a smart manufacturing environment for achieving digital industrial technology (Bushev and Maheva 2020; Clancy *et al.* 2022), where machine-to-machine (M2M) communication plays a vital role in IIoT technology. By developing different network communication systems, on-site visualization, full remote access, tele-service, data collection and storage, and easy computation through the IoT cloud, easy identification of failure through big data analytics and AI/ML can be achieved.

This chapter is organized as follows: a past research survey review and research work related to IoT and AI in smart manufacturing for Industry 4.0 are reported in Section 13.2. A detailed description of smart manufacturing is given in Section 13.3. Further, the key elements of smart manufacturing are discussed in Section 13.4. The development of a smart foundry for the Industry 4.0 concept is illustrated in Section 13.5. The advantages and challenges of smart manufacturing are presented in Section 13.6. This chapter ends with a conclusion in Section 13.7.

13.2 LITERATURE SURVEY RELATED TO IoT AND AI IN SMART MANUFACTURING FOR INDUSTRY 4.0

In the last decade, several researchers have presented reviews and research work where associated papers with IoT and AI in smart manufacturing for Industry 4.0 are discussed. Jeon *et al.* (2016) proposed a cyber-physical system (CPS)–based model-driven approach for solving the challenges of a smart factory and also presented a case study of a smart factory using IoT, CPS, and big data. K.Wang (2016) focused on a predictive maintenance system for Industry 4.0. By using this system, a manufacturing system can reach up to zero-defect manufacturing. S. Wang *et al.* (2016a) focused on an intelligent cooperation machines for cooperating to the agent each other where the different strategies are designed to prevent deadlocks by increasing the decision of agents. S. Wang *et al.* (2016b) have attempted the concept of the smart factory where a multi-agent system in a self-organization manner is applied which is supported by a feedback system using big data analytics. By applying this method, intelligent negotiation mechanisms are analyzed for cooperating. S. Wang *et al.* (2016c) emphasized a control system for the smart factory of Industry 4.0 where industrial wireless networks, the IoT cloud, and smart artifacts are incorporated. By providing feedback, cloud computing is realized and big data analytics is also performed. Wang *et al.* (2016c) proposed a big data-based framework for a smart factory where big data facilitates intelligibility for the decision-making control system. This also synchronizes manufacturing resources to accomplish high efficiency and more flexibility in the system. Strandhagen *et al.* (2017) investigated the Industry 4.0 technologies that can be used in the logistics industry to enhance the production environment. Multiple case studies were reported. Anita *et al.* (2017) established IoT technologies and data analytics for competitiveness in the market where IoT helps in the transformation of the manufacturing environment for running the plant more efficiently and productively. Cimini *et al.* (2017) presented reviews on technological advancement for Industry 4.0 by using suitable strategies which are faced in a business and operational environment, especially from the supplier's and customer's perspectives. Thoben *et al.* (2017) presented a review of smart manufacturing along with the prospective of CPS for designing the product, production system, logistics system, and maintenance of the system in Industry 4.0 where future research issues are also discussed. Debauche *et al.* (2018) explored the usage of AI/ML algorithms using a group of images and IoT for attempting parallel and cloud computing. Koleva (2018) proved the functionality of the production system in a manufacturing environment where opportunities and challenges for establishing the production system for Industry 4.0 are discussed. Kumar (2018) discussed the application of a variety of materials in smart manufacturing where IoT, different flexible sensors, and 3D printing processes are used for soft-skin robotic fingers. Nagy *et al.* (2018) emphasized the role and potential of Industry 4.0 and IoT for developing a business strategy along with value chain management where the case of Hungary is discussed. Xu *et al.* (2018) presented a brief state-of-the-art on Industry 4.0 and its relation to industries for Make-in-China. Further, the issues for upgrading China's industry (especially the manufacturing sector) are addressed.Li *et al.* (2018) developed a standards framework for smart manufacturing that is analyzed and evaluated with reference to smart manufacturing standards. Further, Liu *et al.* (2018a) attempted cyber-physical systems-based machine tools that can provide a solution for the connectivity of various devices such as sensors, actuators, etc. This also allows different types of real-time manufacturing data to run the system effectively and efficiently in manufacturing systems. Further, Liu *et al.* (2018b) focused on system architecture which allows machining processes, machine tools, real-time machining data, and AI-based algorithms for the integration of different types of networks. Different methodologies are discussed for developing a machine tool cyber twin. Santos *et al.* (2018) presented an overview of Industry 4.0 where IoT, AI, and cloud computing–based advanced technologies are used for generating new opportunities to overcome the present industrial challenges. Singh (2018) presented a comprehensive review of smart manufacturing, IoT-enabled manufacturing, cloud computing, and similar topics, which are parts of intelligent manufacturing. Zheng *et al.* (2018) presented a framework for attaining a smart manufacturing systems environment where smart machining and control processes are discussed. Anand and Nagendra (2019) presented the demand for smart manufacturing in Indian defense by establishing advanced technologies such as IoT, 3D printing, 5G connectivity, robotics and automation, and cyber security, which can have a good impact on industry stakeholders. Bhattarai *et al.* (2019) presented an application of big data analytics for power grids where the utility and challenges are also identified. These can help in the integration of big data analytics into power systems for creating a decision framework for planning and operation of the plant. Frank *et al.* (2019) proposed a framework of front-end technologies by considering four verticals for smart manufacturing. Through smart manufacturing, a smart production system and smart supply chain management can be established. Mahmoud and Grace (2019) focused on a framework for increasing the productivity of Industry 4.0 and its adoption in the market along with manufacturers, designers, and decision-makers so that smart manufacturing system capabilities can be enhanced. This framework combined three steps: the iterative process of application modeling, evaluation for the best possible arrangement and acceptance, and final accomplishment. Patricia *et al.* (2019) focused on specific conditions for the advancement of Industry 4.0 where human capital technological capability can be improved and economies can be enhanced. Tantawi *et al.* (2019) discussed smart manufacturing in the USA and how it can be implemented for the standardization of industrial robotic systems using IoT and AI. Yao *et al.* (2019) gave a brief report on cyber-physical system–based manufacturing where manufacturing-related issues are addressed, such as reconfigurations, decentralized decision-making, interoperation, intelligence, and proactiveness of existing integrated manufacturing systems. Karpagavalli (2019) focused on the smart factory where a self-organizing, dynamic, and real-time data processing system connects the various smart devices.Koh *et al.* (2019) focused on the fourth industrial revolution where operation, maintenance, and supply chain management can solve the problems of lean and scheduling in smart factories. Nicolae *et al.* (2019) presented a brief review of Industry 4.0 where IoT and IIoT have been applied and development directions are taken up. This shows the continuous improvement in the industry's technological aspect. Nguyen *et al.* (2019) presented details about the different components of a smart factory and the association between suitable mechanisms where IIoT, AI, and big data can be applied. Different case studies are reported to facilitate a smart manufacturing environment. Sreenivasulu and Chalamalasetti (2019) discussed the present scenario of IoT and its applications, which can provide solutions per market demand and customer requirements. The utilization of IoT can offer rapid investment returns and allow digital transformations in several aspects. Lampropoulos *et al.* (2019) described smart/intelligent manufacturing, which highlights the major IoT and Industry 4.0 issues for establishing a smart/intelligent manufacturing environment. Ahmadi *et al.* (2020) identified the main key technologies for smart manufacturing based on the analysis, policy, capability, and roadmaps of the US, Germany, China, and others. The concepts are explored for the realization of the fourth industrial revolution. Bidnur (2020) presented a review of the advancements in smart manufacturing, robotics, and automation technologies to achieve Industry 4.0 where robotics and automation technologies assist an important driver in synchronization of Industry 4.0. Nguyen *et al.* (2020) reviewed big data analytics, which can be applied in the gas and oil industry, where technical and nontechnical factors are discussed. These can help in the implementation of big data analytics technologies, and this also includes development platforms, system architecture, data privacy, and cyber security. Dornhofer *et al.* (2020) attempted multi-agent systems in different production units for the smart factory model where AI/ML methods provide feedback to the different machines in production units to decide in a faster manner.Oztemel and Gursev (2020) presented a review on the development of technological infrastructure in the aspect of physical systems, management systems, and business strategies for Industry 4.0 so that life can be made easier for practitioners in a simple way. Tan *et al.* (2020) focused on a fusion architecture for shop-floor applications where dynamic cyber-physical interactions using a closed-loop method are carried out. A dynamic re-configuration condition is demonstrated for modular smart assembly by deploying a physical assembly processes method. Singh (2020) proposed a high-tech strategy for future manufacturing industries where Industry 4.0 is highlighted for developing a smart manufacturing environment. Sony and Aithal (2020) conceptualized a model of Industry 4.0 for heavy and light engineering in Indian engineering industries which shows that the Indian engineering industries need to transition to manufacturing smart products and services by applying the Industry 4.0 concept. Cataldo *et al.* (2021) focused on the 4th industrial revolution where AI/ML-based optimization methods are applied to improve the smart manufacturing process. The design of the products and production system are also discussed for developing Industry 4.0.Mourtzis *et al.* (2021) proposed a design of layers of tactile internet based on 5G which is designed using the master/slave concept along with networking of the different layers. This enables the key technologies for establishing Society 5.0 which is routed through cyber-physical systems. Sevic and Keller (2021) proposed a representation of a smart factory where an autonomous manufacturing system is developed by combining two or more machine units such as handling units, machines, and warehouses. Each machine unit has been communicated using IoT technique. Soebandrija *et al.* (2021) focused on the sustainability of Industry 4.0, which depends upon product design, industrial engineering, and business engineering. This creates the transformation in a faster manner within manufacturing and production environments. Wang *et al.* (2021) presented a review on smart and intelligent manufacturing where both emerged in Industry 4.0 for providing intelligence in modern manufacturing and cyber-physical systems. Wang *et al.* (2021) reviewed smart and intelligent manufacturing, which can be applied to the Industry 4.0 concept for enhancing modern manufacturing capabilities in terms of human-cyber-physical systems. Drakaki *et al.* (2021) showed the potential of the multi-agent system and deep learning as efficient tools for developing CPS-enabled systems for the predictive maintenance of induction motor life. Amin *et al.* (2021) presented a review on smart manufacturing systems which is focused on the theoretical aspect of technology development and management of smart manufacturing systems. Lepasepp and Hurst (2021) described an efficient harmonious interface between man and machine for establishing smart factories. Using robotics, AI/ML, and big data analysis, the end-to-end solution can be analyzed and optimized in several ways, and there are a lot of challenges for implementations of smart manufacturing in the fields of development of medical devices and their manufacturing. Hingu *et al.* (2021) focused on product development using mechatronics systems where AI techniques are applied for attaining Industry 4.0. Rupp *et al.* (2021) presented an all-inclusive definition of Industry 4.0 where a bibliometric analysis is envisaged for the existence of Industry 4.0. Sufian *et al.* (2021) attempted a six-stage model for establishing the smart factory where the different industrial case studies are described for adopting the Industry 4.0.Sindhu *et al.* (2021) created a business model for applying AI and IoT in Industry 4.0 which gives a new direction for establishing manufacturing technologies such as 3D printing and big data analysis. Liagkou *et al.* (2021) discussed the major features of Industry 4.0 where AI algorithms are applied in a manufacturing environment to develop the smart factory concept. Mateo and Redchuk (2021) focused on a framework for adopting AI/ML and its incorporation with IIoT for the Industry 4.0 concept where the smart manufacturing framework is discussed for industrial process flow optimization. Sahoo (2021) applied AI techniques for predictive maintenance strategies for predictions and recommendations of systems based on databases. Nica *et al.* (2021) developed an IoT-based network system along with AI algorithms for decision-making where big data analytics is applied to establish digital twin smart manufacturing for Industry 4.0.Warke *et al.* (2021) focused on the digital twin framework development of smart manufacturing by facilitating technologies such as AI/ML, data-driven systems, and IoT, where the concept and advancement of digital twins are discussed, and the benefits and challenges are also addressed. Le and Pham (2021) presented a review of two emerging technologies, big data analytics and AI/ML, for Industry 4.0, which has a high impact on smart manufacturing for optimizing and automating production systems on a large scale and efficient handling as compared to conventional methods. Rai *et al.* (2021) discussed different ML techniques for various manufacturing applications where ML along with an edge computing, and CPS provide a smart manufacturing environment. Caldana *et al.* (2021) presented a compressive review of IoT and AI for Industry 4.0 which can be applied for the predictive maintenance of machines accordingly.Radanliev *et al.* (2021) presented cyber security necessities by integrating IoT with AI in Industry 4.0. Grounded theory methodology, modeling, and analysis are carried out using edge computing where taxonomic classifications are presented per the requirements. Further, Radanliev *et al.* (2022) presented a summary of the advancement of AI along with IoT networks where they address how such technologies can be utilized in industrial systems for organizational structure improvement. Chudasama (2022) provided a review of Industry 4.0 technologies that gives an outlook for implementing Industry 4.0 for future business in the market. Diao and Sun (2022) explored IoT and communication technologies in smart factories for enhancing the capabilities of smart manufacturing where the elliptic curve encryption algorithm is applied. Hughes *et al.* (2022) presented the significance of Industry 4.0 in the latest group of manufacturing systems, as well as key barriers and challenges. Further, it is aligned with the framework of Industry 4.0 and the United Nations Sustainability Goals. Maroor *et al.* (2022) applied a hybrid strategy for prediction/finding defects and forecasting, where the ML technique is exploited to predict the model complexity of the surroundings to apply it. Noor-A-Rahim *et al.* (2022) explored the concept of intelligent reflecting surfaces in smart manufacturing for the emergence of Industry 4.0 and Industry 5.0 with smart manufacturing. The potential, challenges, and opportunities in modern smart manufacturing are also discussed. Nguyen *et al.* (2022) focused on smart manufacturing with AI and big data analysis where the perspectives of such technology are discussed. Jagatheesaperumal *et al.* (2022) presented a review of explainable AI and IoT for enhancing the trust methods of different machines which can be used for smart manufacturing. Explainable AI frameworks, their characteristics, and IoT support are also addressed. Ramakrishna (2022) addressed Industry 4.0 and its huge impact on the Indian market where these technologies have huge application in industrial automation. Rubel *et al.* (2022) attempted a continuous real-time monitoring system that can help to develop a smart manufacturing system where in-process metrology is applied for a feedback control system to process real-time data using smart sensors and actuators. Petrovica *et al.* (2022) discussed awareness and development of Industry 4.0 in Latvia, which can support economics and the industrial competitiveness of the country by introducing Industry 4.0. Parthasarathy *et al.* (2022) presented the significance of the industrial AI ecosystem, which can define sequential thinking strategies for developing transformative AI systems. Kamal *et al.* (2022) attempted factory automation for Industry 4.0 where technologies related to 5G and LoRA communication for smart manufacturing automation are discussed. Karadgi (2022) focused on a framework for implementing a smart manufacturing system where lifecycles are also considered. This framework allows three layers from the bottom to the top layer. Kunju *et al.* (2022) focused on production and maintenance in existing industries which can reduce machine downtime and maintenance in industries. Du (2022) summarized the merits of applying IoT and smart supply chain management in Industry 4.0 for the enhancement of productivity. Stadnicka *et al.* (2022) focused on identifying the demand for industrial systems where AI, IoT, and edge computing can be implemented to solve industry problems. Further, the taxonomy can be established accordingly.Alabadi *et al.* (2022) presented the taxonomy of the IIoT challenge in smart manufacturing which includes system architecture and use cases. The relationship between enabling the technology and technical requirements is addressed using AI and edge computing. Hamid *et al.* (2022) discussed the Industrial Revolution 4.0 from the Malaysian smart manufacturing perspective where IoT, AI/ML, cloud computing, interconnectivity, and cyber-physical learning can be utilized for automation and exchanging the data in the manufacturing industry.Stief *et al.* (2022) developed a framework for implementing smart manufacturing through link4Smart using IoT and AI where the horizontal integration for the factory shop floor and vertical integration for human–machine planning are established. This can update information on a daily production basis. Sujatha *et al.* (2022) focused on AI-based decision tree algorithms for developing smart decision-making systems. By using them, energy consumption in machines and appliances can be monitored and future behavior predicted for detecting abnormalities. Tsaramirsis *et al.* (2022) presented the various support technologies for Industry 4.0 where the implementation schemes of these technologies are discussed for the smart manufacturing aspects. Further, a framework model of Industry 4.0 is also developed using IoT, AI/ML, big data, robots, and 3D printing. Yildiz and Ugur (2022) exploited the use of AI in developing a smart production system for Industry 4.0 where the relationship between smart production and Industry 4.0 using AI is established, and the benefits of the use of AI are also discussed. Haricha *et al.* (2023) gave a comprehensive review of existing efforts on smart manufacturing where different kinds of systems and the priority of requirements are described using IoT and AI. Konstantinidis *et al.* (2023) focused on a multi-sensor–based CPS system for the demonstration of Industry 4.0 where a multi-functional technique is applied. Kumar *et al.* (2023) focused on an IoT-based framework for establishing the infrastructure for IIoT-related activities in smart manufacturing industries. This kind of framework reflects the relationships between IIoT and smart manufacturing. The different IIoT layers are discussed and analyzed. Rahardjo *et al.* (2023) optimized production processes for a smart and sustainable manufacturing system where a case study for the fabrication of vacuum degassing equipment is discussed. This demonstrates the enhancement in the effective utilization of the controlling system. Ryalat *et al.* (2023) presented review of digital technologies and a design of a smart CPS-based system for realizing a smart factory where industrial robots and information and communication technologies (ICT) play a vital role for the realization of Industry 4.0. Rahim *et al.* (2023) presented a compressive review of wireless machine-to-machine communication systems that can be applied in smart manufacturing environments. Associated case studies are also presented for providing the impact on future smart manufacturing.

13.3   DESCRIPTION AND IMPLEMENTATION PLAN OF SMART MANUFACTURING FOR INDUSTRY 4.0

For developing Industry 4.0, the manufacturing sector demands an extensive study for developing intelligent product design to achieve suitable applications where the design of CPS-based manufacturing system has been customized according to targeted individualized products. While developing products, machine health prognosis and predictive maintenance are also considered in Industry 4.0. The Machine Tools 4.0 concept will be introduced in machining sites (Zheng *et al.* 2018). For an energy-efficient environment, energy management will also be implemented to attain the requirements of Industry 4.0. This creates an energy-efficient decision management system that transforms energy monitoring systems into autonomous systems. This can be one of the achievements of a self-optimized energy system in Industry 4.0. Therefore, a modified framework of intelligent manufacturing systems is shown in Figure 13.1. Typically, the Industry 4.0 smart manufacturing concept includes smart design, intelligent machining processing, smart monitoring systems, smart control systems, and manufacturing scheduling to achieve industrial applications. Another major consideration in Industry 4.0 is that different intelligent sensors and actuators are to be deployed to acquire data and big data analysis (BDA) for decision-making (Zywicki et al. 2017). In the Industry 4.0 concept, big data analysis plays a vital role in the smartness of activities using the IoT cloud so that synchronized data can be stored for the manufacturing scheduling purpose.

At first, the design engineer will have to develop a smart design per the requirements. After that, according to the requirements, CPS-enabled Machine Tools 4.0 will have to be used to develop physical products. For proper communication, CPS will be developed so that connectivity between virtual and physical systems can establish in the wireless network systems in the world. Therefore, the different objects can communicate with each other using IoT techniques. In Industry 4.0, production

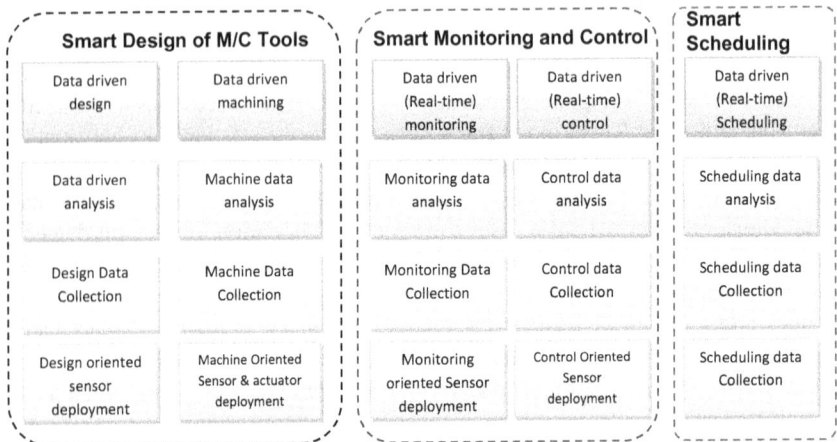

FIGURE 13.1   A framework for developing smart manufacturing systems for Industry 4.0.

systems develop into CPS, which consists of smart machines, production, and scheduling processes. These are digitally connected, which provides the feature of uninterrupted ICT-based integration (Kagermann *et al.* 2013). A typical smart machine tool system using ICT-based integration is designed as shown in Figure 13.2. RFID tags, Zigbee, and Wi-Fi modules are connected to significant parts, such as bearings, cutting tools, and spindles, so that the physical objects can be distinctively recognized. A variety of sensors and actuators such as dynamometers, accelerometers, cameras, data acquisition systems, and temperature sensors are also integrated in the machine tools arrangement for collecting real-time machining data. During the communication process, systems interact with an energy-efficient management system for collecting synchronized data from machining processes. For transmission of data, different wired/wireless communication technologies such as RS-232, Ethernet, Bluetooth, and a 4G network. Different exchange formats are also used for exchanging data from different sensors and actuators when the communication system has considerable challenges for data integration and its management. Standard data communication protocols can also be implemented. MT Connect is an open-source and royalty-free communication standard that can be used to attain faster response of devices and reduce the cost of data integration (MTConnect Standard 2009). MT Connect can transform data from various devices into the XML data format per ISO format (i.e. known as STEP format). Based on these standards, a proper communication service can be established to attain communication between the critical components and synchronized data for the particular conditions.

For smart visualization, this provides real-time data, and one of the advantages is that it can be integrated with any provider during operation through the internet. In this way, any essential part of the machine tools can be accessed visualized remotely using mobile phones, such as Android phones or tablets. The statistical data can be visualized, and reports can also be generated for establishing data management

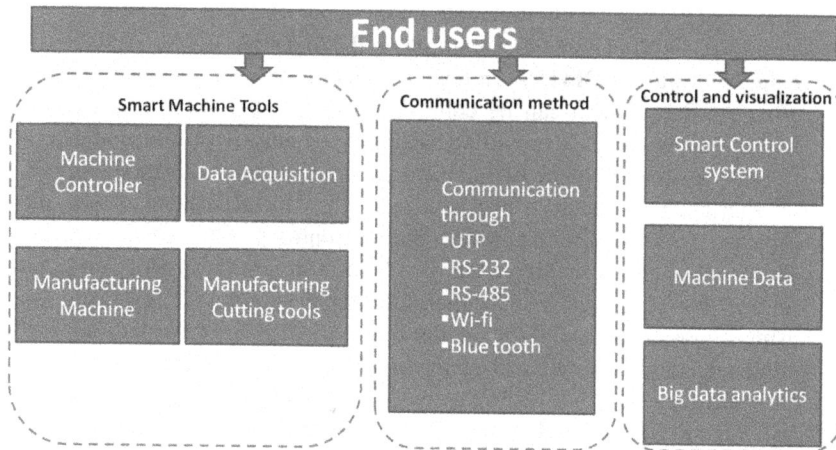

FIGURE 13.2 Typical ICT-based integration systems for smart machine tools in Industry 4.0.

systems, such as enterprise resource planning (ERP) modules. Using the IoT cloud, historical data can be stored and recorded in the cloud. Using AI techniques, different machine diagnostic management algorithms can be implemented to review/assessment of critical parts. Therefore, predictive maintenance can be done accordingly to minimize machine failure. Using augmented reality (AR), machining processes can be visualized, which can help in efficient interactions/communication between users/operators and machine tool devices.

13.4 KEY ELEMENTS OF SMART MANUFACTURING

Smart manufacturing must ensure that the following essential key components/elements for developing building blocks for Industry 4.0 (Zheng *et al.* 2018; Mittal *et al.* 2019a, 2019b).

1. **Smart Devices**: Smart devices can be used for developing intelligent functionality in existing industrial manufacturing systems, and automation provides solutions for the minimization of human intervention. This intelligent functionality also takes into account the suitable communication protocol and control functions, computation process, and networking with different smart devices and edge computing platforms so that legacy capital equipment will be maintained in different work cells, production lines, plants, and factories in the industry.

2. **Smart Interfaces**: The computation interface like tablets, phablets, and smart phones are required for human–machine interfaces to enhance production operations and maintenance actions. Such interface devices also provide solutions for remote monitoring and controlling processes in wired/wireless network systems. Using mixed reality technologies, automated digital instructions and diagnostic systems can be utilized. Using wired and wireless technologies, end-to-end devices can be connected through smart hubs/gateways to set up new information technology (IT)/operation technology (OT) interfaces between the two systems with help of the IIoT infrastructure, which plays an essential role in the industry.

3. **Edge Computing Devices**: In edge computing devices, the sensors visualize the data, which can be collected through edge computing devices within the network. By providing such devices, edge intelligence and control can be attained in the hierarchical manufacturing system, and it allows an interface with the edge networks instantly. This also interacts with the centralized IT management system and OT control system and its software. This kind of computing device is also helpful in taking action at the right time at the particular site location/machine location/fault detection. By using it, productivity and the quality of the product can be enhanced, and unplanned downtime reduced during handling of different devices in different industries.

4. **Software Platforms and Apps**: Using software platforms and app technologies, intelligent functionality can be visualized and controlled in each step at each hierarchical edge boundary within the manufacturing environment.

This consists of software platforms and apps that can support smart devices and smart interfaces through machine tools via OT solutions. This is also supported by third-party software platforms, and the apps can be customized accordingly, which provides retrofit intelligent functionality in the smart manufacturing environment. Open-source free software is available for developing mobile apps. Therefore, micro, small and medium enterprises, and startups can be easily utilized in industry.

5. **Data Management Systems**: Data management systems help manufacturers with data collection, analysis, and securing data, which boosts efficiency and reduces costs. The major challenge of smart systems is collection of massive amounts of data and storage of the data for analysis purposes. This kind of data management system helps to restore the data of micro, small, and medium enterprises (M SME) and startups (Dagadu and Gaikwad 2013–2014).

6. **Big Data Analytics**: In the big data analytics step, sophisticated analytical methods such as AI/ML statistics tools, health management, and condition-based maintenance systems provide information for predictive maintenance and health conditions of the system using different data sets. This will be helpful for preventive maintenance of the machine, life expectancy, and re-use of machine tools in industry.

7. **Safety and Security**: The main function of the safety and security step is to protect smart manufacturing systems proactively against any threat and cyber security vulnerabilities from the operator end or human–machine interactions. This also connects OT and IT systems using CPS. Safety and security are also demanded with configurations with IIoT infrastructure, which provides additional protection of layers during implementation over the system life cycle.

The smart IIoT systems provide a systematic, rapid, and sustainable impact on society in the various industrial sectors, like smart healthcare, automobiles, energy management systems, and transportation, etc., where different key enabling technologies (such as data technology (DT), analytics technology (AT), operation technology, and platform technology (PT)) can be utilized as shown in Figure 13.1. The machine-to-machine communication, cyber security, and data quality management along with an expert system are facing challenges in combining them for Industry 4.0, as shown in Figure 13.2.

To overcome these challenges, system methodologies for analyzing the smart manufacturing environment such as deep learning, simulation-based learning, relationship-based learning, and peer-to-peer learning can be applied with the help of developing software tools such as C, C++, R, Python, and MATLAB to achieve the desired features, which can interface with commercial IoT platforms such as Siemens, IBM, and Thing speaks, as shown in Figure 13.3. Using these IoT platforms, data can also be stored for data analytics purposes for prediction of the product life cycle. This kind of system can be operated through smart mobile phones from anywhere, which can save money, human power, and time in the industry.

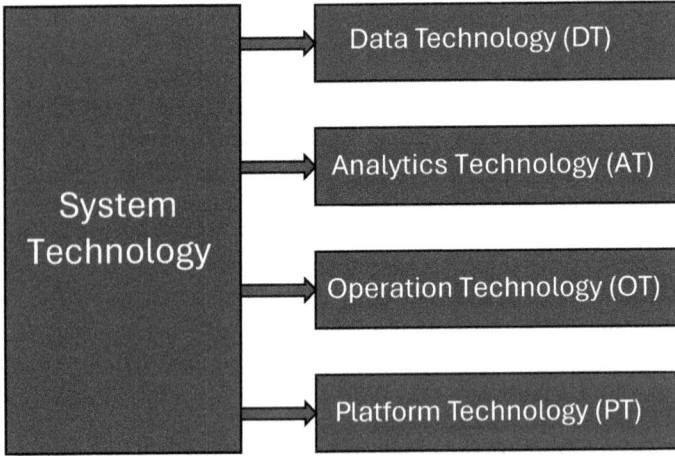

FIGURE 13.3 Different key enabling technologies for Industry 4.0.

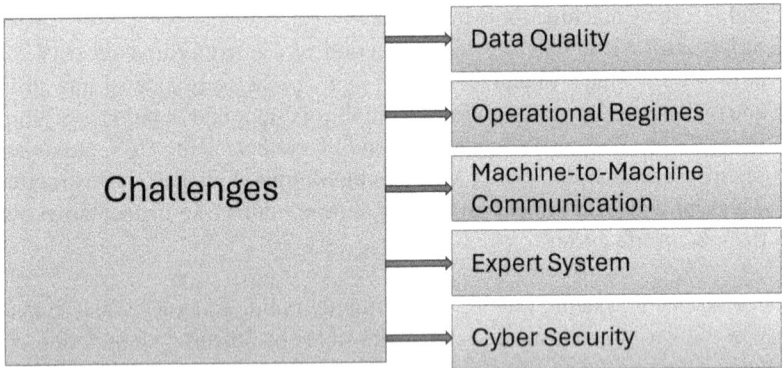

FIGURE 13.4 Different key challenges for Industry 4.0.

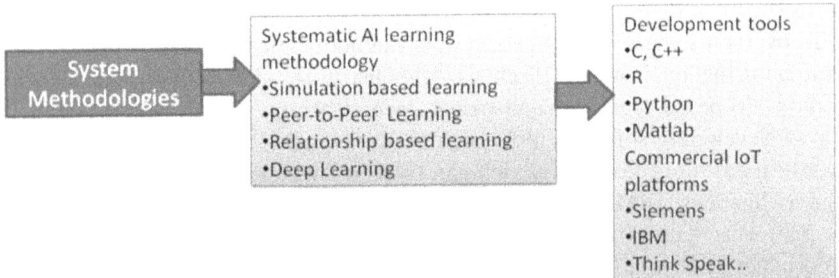

FIGURE 13.5 System methodologies for analyzing the smart manufacturing environment.

13.5 FEATURES OF SMART MANUFACTURING FOR INDUSTRY 4.0

To identify the smart features in smart manufacturing, several countries like Germany and the United States have initiated Industry 4.0, smart factories, CPS production, and intelligent manufacturing, which advanced manufacturing often uses (Zheng *et al.* 2018). Smart manufacturing can be accomplished after the combination of ICT, IoT, and AI/ML in the manufacturing industry. Using virtual/wireless network systems, the physical world can be connected. To establish a smart manufacturing environment, the following knowledge of different domain technologies is required. The framework of Industry 4.0 with domain technologies and manufacturing is shown in Figure 13.1. The domain technologies are robotics/automation, IoT, AI, cloud computing, big data analytics, and CPS, whereas the application areas in manufacturing are 3D printing, virtual and augmented reality, smart factories, smart logistics, and connection of smart devices.

Therefore, an IoT/AI-enabled system can help in decision-making with less human intervention where IoT devices interact using the internet, and AI builds the devices to learn from the data and experience using expert systems. Big data analytics plays a role in storing the data and processes the data for analytic purposes so that intelligent devices can be achieved, as shown in Figure 13.2.

To develop Industry 4.0, a functional model for product development in smart manufacturing is described in Figure 13.3. Industry 4.0 concept is developed using different promising technologies like physical model development, cyber-physical model, data analytics, etc. In the physical model development phase, engineers have to be focused on feature identification of the product per customer requirements. After feature identification, a CAD model will have to be developed with the help of

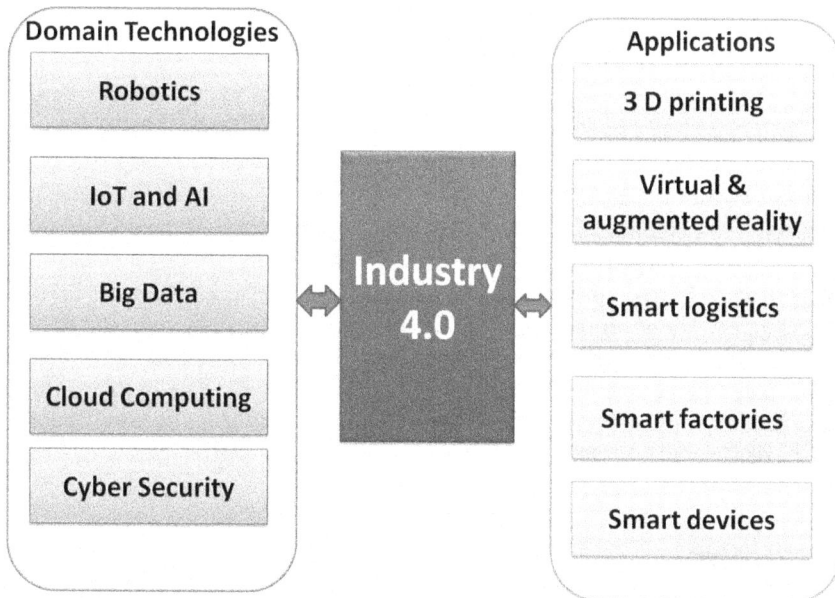

FIGURE 13.6 Digital technologies for smart manufacturing.

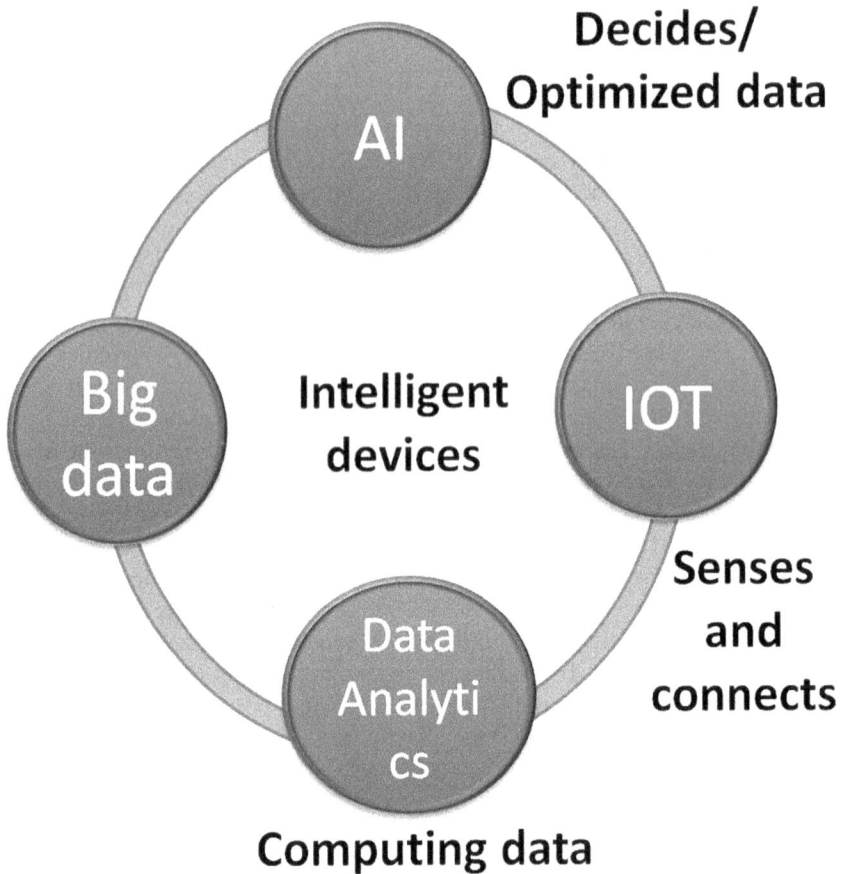

FIGURE 13.7   Inter-relationship of digital technologies in smart manufacturing.

CAD software in the cyber physical phase. Subsequently, using design tools, manufacturing the CAD model will be carried out for realization of the product design process and its life cycle, and geometric design parameters can be optimized using CAD software. Further, the product/prototype will be developed using the 3D printing technique. The machine-to-machine communication system will interact using IoT techniques for minimization of the process time, and integration of different sensors plays a vital role in acquiring the data and storing the data in IoT platforms. Using AI, analytics will have to be developed, and validation can be done in context with the design and its usage. Accordingly, a market plan and strategies will be established for future demands in the market.

13.6   DEVELOPMENT OF SMART FOUNDRY FOR INDUSTRY 4.0 CONCEPT

CSIR-CMERI is also closely working on the implementation of this technology in different machines. As a part of this initiative, CSIR-CMERI is working on a project

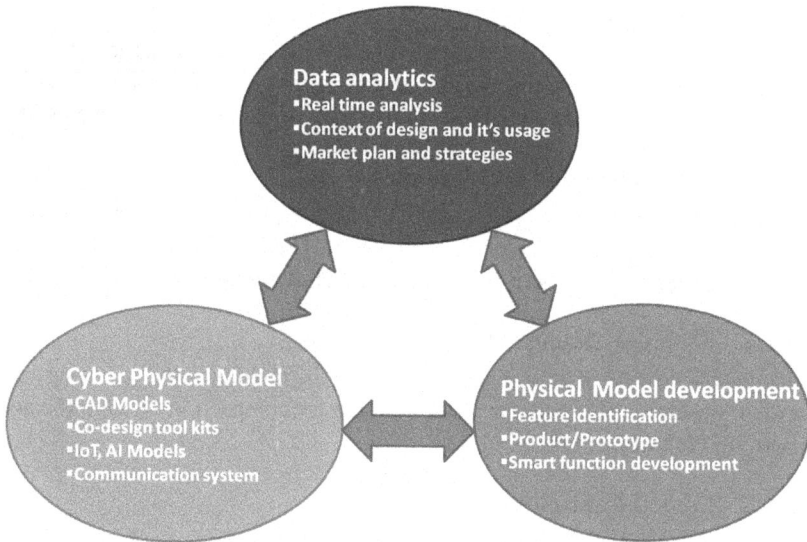

FIGURE 13.8 Functional model for product development in smart manufacturing.

named Smart Foundry. In this project, a complete advanced foundry is being developed for the casting of small aluminum components of up to 2 kg. The overall system footprint will fit inside a 12 × 12-ft room. The system is capable of running with minimum human interaction with one operator. The system is composed of six major system components.

13.6.1 SIMULATION SOFTWARE

Two different softwares are used to develop the products (the first for design methodology and the second for CFD-based multi-physics solver for casting process simulation). After optimization, the final design of product is forwarded for 3D printing.

13.6.2 3-D PRINTER

This is used for the 3D printing of patterns. From a 3D CAD model of the pattern with all the requisite allowances, the pattern is printed using ABS thermoplastics. After layer-by-layer printing, the pattern can also be treated by vapor polishing operation to achieve a superior surface finish per requirements. The required runner and risers are also printed with the pattern in this unit.

13.6.3 AUTOMATIC MOLDING MACHINE

This sub-system is used for the generation of a sand mold using the pattern printed by the 3D printer. This sub-system consists of five components.

13.6.3.1 Sand Container with Automated Valve

The appropriate amount of sand, resin, and harder are fed through a control system maintaining ingredients for sand molding process.

13.6.3.2 Sand Mixer

This is equipped with a motorized blade for mixing sand, resin, and hardener. The mixing speed and time can be controlled per process requirements. After completion of mixing, there is an actuator controlled gate to discharge the molding mass into the mold box.

13.6.3.3 Automatic Mold Box Assembly

The actuator-controlled automatic mold box assembly enables opening and closing of the mold box for ease of extraction of the mold. There is a provision in the match plate for providing a jolting action to evenly spread the molding mass.

13.6.3.4 Ramming Station

On receiving sensor input after pouring sand into the mold box, the ramming station comes into position and exerts the required force for compaction through the actuator-controlled ramming plate.

13.6.3.5 Central Controller

Equipped with a GUI for feeding process parameters, the central controller controls the sub-system operation where the GUI is designed in VC++ software.

13.6.4 Melting and Pouring Unit

After preparation of the mold, it is placed in this unit. It is equipped with a bottom pouring furnace that melts and pours the aluminum into the mold for casting of the component. There is a provision in the system for attachment units like a vacuum/inert gas unit and metal matrix composite.

13.6.5 IoT-Enabled Smart Features in Smart Foundry

A system architecture for an IoT-enabled device is designed for the controlling the process parameters of molding process. An automatic molding machine is developed in CSIR-CMERI where the control system is integrated with an automatic molding machine (Jain *et al.* 2020). A dashboard is also developed for the realization of IoT technology in the Smart Foundry concept, as shown in Figure 13.1. It is designed and developed using three layers:

- Hardware Layer: Sensor and actuators, motor, microcontroller
- Server/Internet Layer: Communication devices like gateway/router, internet server, cloud platform, configuration of IP in server
- End-User/Application Layer: Display interface devices like PC, mobile

13.6.6 Operation Steps

During operation, the sand is fed manually into the container. A butterfly-type valve is applied to supply sand from the container, which adjusts the supply of sand into the mixer. After that, the sand is discharged into the mixer. The hardener also blended in the stage. Afterwards, the pump releases the hardener into the mixer. The sand releases into the gate using pneumatic actuators to discharge the sand into the mold box. Then,

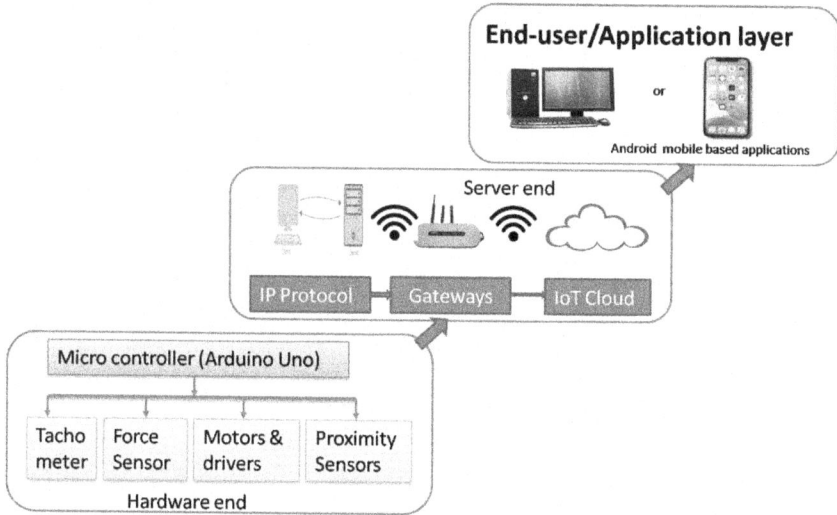

FIGURE 13.9 IoT-enabled system architecture for Smart Foundry.

the ramming station travels. After the ramming action, the mold box gate is automatically opened, and the two stages proceed in parallel. The mold is taken out with the base plate attached to it. The operation steps are shown in Figure 13.10. The aluminum brake shoe component generated using the Smart Foundry system is shown in Figure 13.11.

13.6.7 DATA SENSING AND ANALYTICS

In this part, a visual dashboard is developed for real-time monitoring of the entire operation through a computer or smart phone. All process parameters and other relevant data are collected from all physical sub-systems, streamed to the cloud, and displayed on the dashboard. Data analysis software is also connected with the dashboard for analysis of process data using IoT.

By developing this foundry system, IoT-enabled technology provides smart features in the smart foundry (Industry 4.0 concept). This provides easy machine-to-machine communication. It also helps create a self decision-making process that reduces human intervention. The advantages of IoT-enabled technology are given as follows;

- Easy process from design to product
- Self-decision in process
- Less time consumed
- Less manual intervention
- High precision and accuracy
- Availability of data storage
- Improvement in quality without increasing cost

Therefore, the smart manufacturing concept (Industry 4.0) can increase the overall operational efficiency in an effective manner in the industry.

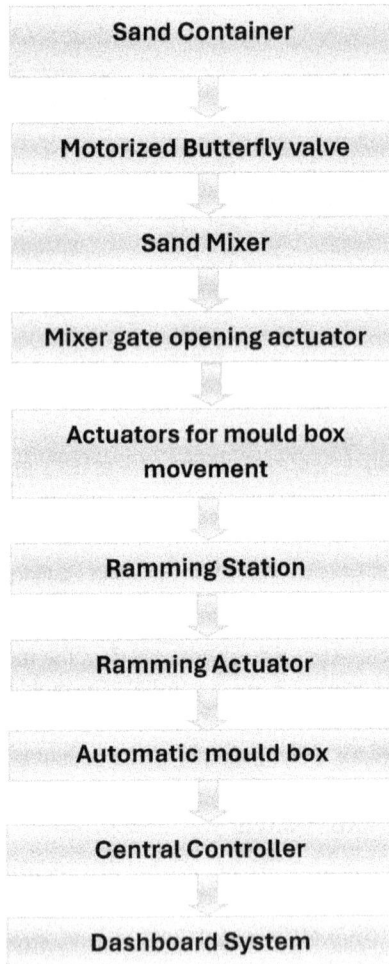

FIGURE 13.10　IoT-enabled smart architecture design for Smart Foundry.

FIGURE 13.11　Aluminum brake shoe component generated using Smart Foundry.

13.7 CONCLUSION

During the development of Industry 4.0, the Industrial IoT and AI are considered the main trends in smart manufacturing systems. To achieve an Industry 4.0 concept, a smart foundry is developed using IoT and AI techniques. By applying the IIoT, the smart manufacturing segment of the foundry can easily draw the required quantities of raw materials to supply the cores and casting iron, which safely enhances the product with less human intervention. This is shown by demonstration of the Smart Foundry for Industry 4.0 concept. With the implementation of IIoT and AI, smart manufacturing makes powerful tools for Industry 4.0 for easy access from anywhere in the world and provides continuous improvement of the product life cycle. IoT and AI are integral parts of sustainability technology that will enhance the global business environment in the manufacturing sector.

13.8 ACKNOWLEDGMENTS

The authors are grateful to the director, CSIR-CMERI, and Durgapur India for providing the permission to publish this paper. This work is carried out under the project "Smart Foundry 2020" (Project No. GAP-213212) and financially supported by the Department of Science and Technology, New Delhi, India.

REFERENCES

Ahmadi, A., C. Cherifi, V. Cheutet and Y. Ouzrout. 2020. Recent Advancements in smart manufacturing technology for modern industrial revolution: A survey. *Journal of Engineering and Information Science Studies.* hal-03054284. www.researchgate.net/publication/340066129_Recent_Advancements_in_Smart_Manufacturing_Technology_for_Modern_Industrial_Revolution_A_Survey

Alabadi, M., A. Habbal and X. Wei. 2022. Industrial internet of things: Requirements, architecture, challenges, and future research directions. *IEEE Access.* 10:66374–66400. https://doi.org/10.1109/ACCESS.2022.3185049.

Amin, M.A., M.T. Hossain and M.J. Islam. 2021. The technology development and management of smart manufacturing system: A review on theoretical and technological perspectives. *European Scientific Journal.* 17(43):170–193.

Anand, P., and A. Nagendra. 2019. Industry 4.0: India's defence industry needs smart manufacturing. *International Journal of Innovative Technology and Exploring Engineering (IJITEE).* 8(11S):476–485. https://doi.org/10.35940/ijitee.K1081.09811S19.

Anita, R., and B. Abhinav. 2017. Internet of Things (IoT)– Its Impact on manufacturing process. *International Journal of Engineering Technology Science and Research (IJETSR).* 4(12):889–895.

Arinez, J.F., Q. Chang, R.X. Gao, C. Xu and J. Zhang. 2020. Artificial intelligence in advanced manufacturing: Current status and future outlook. *Journal of Manufacturing Science and Engineering.* 142:111003–1–16.

Bhattarai, B.P., et al. 2019. Big data analytics in smart grids: State-of-the-art, challenges, opportunities, and future directions. *IET Research Journals.* 2(2):141–154.

Bidnur, A.V. 2020. A study on Industry 4.0 concept. *International Journal of Engineering Research & Technology (IJERT).* 9(4):613–618.

Bushev, S., and A. Maheva. 2020. Circular economy of foundry INDUSTRY 4.0. *International Scientific Journal "INDUSTRY 4.0".* V(1):45–47.

Caldana, V.M., F.D.G. Dasilva, R.A. Deoliveira and J.F. Borin. 2021. Internet of Things and artificial intelligence applied to predictive maintenance in Industry 4.0: A systematic literature review. *Proceedings of the International Conference on Industrial Engineering and Operations Management.* Sao Paulo, Brazil, April 5–8, 1387–1398.

Cataldo S.D., S. Lee, E. Macii and B.V. Heuser. 2021. Leading information and communication technologies for smart manufacturing: Facing the new challenges and opportunities of the 4th industrial revolution. *Proceedings of the IEEE.* 109(4):320–325. https://doi.org/10.1109/JPROC.2021.3064103.

Chudasama, N. 2022. Review on industry 4.0 technologies. *International Journal of Advanced Research in Science, Communication and Technology (IJARSCT).* 2(2):713–715. https://doi.org/10.48175/IJARSCT-2936.

Cimini, C., R. Pinto, G. Pezzotta and P. Gaiardelli. 2017. The transition towards industry 4.0: Business opportunities and expected impacts for suppliers and manufacturers. In: H. Lodding, R. Riedel, K.D. Thoben, G.V. Cieminski, D. Kiritsis (eds) *Advances in Production Management Systems. The Path to Intelligent, Collaborative and Sustainable Manufacturing, APMS-2017.* IFIP Advances in Information and Communication Technology. 513. https://doi.org/10.1007/978-3-319-66923-6_14.

Cio, R., M. Travaglioni, G. Piscitelli, A. Petrillo and F.D. Felice. 2020. Artificial Intelligence and machine learning applications in smart production: Progress, trends, and directions. *Sustainability.* 12:492 (26 pages). https://doi.org/10.3390/su12020492

Clancy, R., K. Bruton, D. O'Sullivan and D. Keogh. 2022. Industry 4.0 driven statistical analysis of investment casting process demonstrates the value of digitalisation. *Procedia Computer Science.* 200:284–297.

Dagadu, J.M., and S. Gaikwad. 2013–2014. Small scale industry and its present scenario in Indian industrialization. *Excel Journal of Engineering Technology and Management Science (An International Multidisciplinary Journal).* I(5):1–4.

Debauche, O., S.A. Mahmoudi, S. Mahmoudi and P. Manneback. 2018. Cloud platform using big data and HPC technologies for distributed and parallels treatments. *9th International Conference on Emerging Ubiquitous Systems and Pervasive Networks (EUSPN 2018). Procedia Computer Science.* 141:112–118.

Diao, Z., and F. Sun. 2022. Application of Internet-of-Things in smart factories under the background of industry 4.0 and 5g communication technology. *Mathematical Problems in Engineering.* Article ID 4417620:8 pages. https://doi.org/https://doi.org/10.1155/2022/4417620.

Dornhofer, M., S. Sack, J. Zenkert and M. Fathi. 2020. Simulation of smart factory processes applying multi-agent-systems-A knowledge management perspective. *Journal of Manufacturing Materials Processing.* 4:89. https://doi.org/10.3390/jmmp4030089.

Drakaki, M., Y.L. Karnavas, P. Tzionas and I.D. Chasiotis. 2021. Recent developments towards Industry 4.0 oriented predictive maintenance in induction motors. *Procedia Computer Science.* 180:943–949.

Du, J. 2022. The impact and challenges of the Internet of Things (IoT) on supply chain management. *Proceedings of 2nd International Conference on Enterprise Management and Economic Development (ICEMED 2022). Advances in Economics, Business and Management Research.* 656:1244–1249.

Frank, A.G., L.S. Dalenogare and N.F. Ayala. 2019. Industry 4.0 technologies: Implementation patterns in manufacturing companies. *International Journal of Production Economics.* 210(3). https://doi.org/10.1016/j.ijpe.2019.01.004.

Gerekli, I., T.Z. Celik and I. Bozkurt. 2021. Industry 4.0 and smart production. *TEM Journal.* 10(2):799–805. https://doi.org/10.18421/TEM102-37.

Hamid, M.S.R.A., N.R. Masrom and N.A.B. Mazlan. 2022. The key factors of the industrial revolution 4.0 in the Malaysian smart manufacturing context. *International Journal of Asian Business and Information Management.* 13(2):1–19. https://doi.org/10.4018/IJABIM.20220701.oa6.

Haricha, K., A. Khiat, Y. Issaoui, A. Bahnasse and H. Ouajji. 2023. Recent technological progress to empower smart manufacturing: Review and potential guidelines. *IEEE Access.* 11:77929–77951. https://doi.org/10.1109/ACCESS.2023.3246029.

He, B., and K.J. Bai. 2021. Digital twin-based sustainable intelligent manufacturing: A review. *Advances in Manufacturing.* 9:1–21. https://doi.org/10.1007/s40436-020.

Hingu, P.R., and D.N. Panchal. 2021. Industry 4.0-Advance to the mechatronic and management with approach of artificial intelligence. *The International Journal of Engineering and Science (IJES).* 10(4):32–36.

Hughes, L., Y.K. Dwivedi, N.P. Rana, M.D. Williams and V. Raghavan. 2022. Perspectives on the future of manufacturing within the Industry 4.0 era. *Production Planning and Control.* 33(2–3):138–158. https://doi.org/10.1080/09537287.2020.1810762.

Jagatheesaperumal, S.K., Q.V. Pham, R. Ruby, Z. Yang, C. Xu and Z. Zhang. 2022. Explainable AI over the Internet of Things (IoT): Overview, state-of-the-art and future directions. *IEEE Open Journal of the Communications Society.* 3:2106–2136. https://doi.org/10.1109/OJCOMS.2022.3215676.

Jain, R.K., P. Banerjee, D. Baksi and S.K. Samant. 2020. IOT Based Interface Device for Automatic Molding Machine towards "Smart Foundry-2020". *10th International Conference on Computing, Communication and Networking Technologies (ICCCNT), IEEE-45670,* IIT Kanpur, India, July 6–8. https://doi.org/10.1109/ICCCNT45670.2019.8944549.

Jeon, J., S. Kang and I. Chun. 2016. CPS-based model-driven approach to smart manufacturing systems. *INTELLI 2016: The Fifth International Conference on Intelligent Systems and Applications (includes InManEnt 2016),* 133–135.

Kagermann, H., W. Wahlster and J. Helbig, (2013). Securing the future of German manufacturing industry: Recommendations for implementing the strategic initiative INDUSTRIE 4.0. *Final Report of the Industrie.* 40:1–84.

Kamal, M.A., M.M. Alam, A.A.B. Sajak and M.M. Suud. 2022. Impact of LoRA and 5G on smart manufacturing from automation perspective. *Journal of Mobile Multimedia.* 18(5):1355–1378. https://doi.org/10.13052/jmm1550-4646.1852.

Karadgi, S. 2022. A framework towards realization of smart manufacturing systems. *IOP Conference Series: Materials Science and Engineering.* 1258:012018. https://doi.org/10.1088/1757-899X/1258/1/012018.

Karpagavalli, R. 2019. Smart factory of Industry 4.0. *International Journal of Research and Analytical Reviews.* 6(2):1–8.

Koh, L., G. Orzes and F. Jia. 2019. The fourth industrial revolution (Industry 4.0): Technologies disruption on operations and supply chain management. *International Journal of Operations and Production Management.* 39(6/7/8):817–828.

Koleva, N. 2018. Industry 4.0's Opportunities and challenges for production engineering and management. *International Scientific Journal "INNOVATIONS".* VI(1):17–18.

Konstantinidis, F.K., S. Sifnaios, G. Tsimiklis, S.G. Mouroutsos, A. Amditis and A. Gasteratos. 2023, Multi-sensor cyber-physical sorting system (CPSS) based on Industry 4.0 principles: A multi-functional approach. *4th International Conference on Industry 4.0 and Smart Manufacturing.* 217:227–237. https://doi.org/10.1016/j.procs.2022.12.218.

Kumar, A. 2018. Methods and materials for smart manufacturing: Additive manufacturing, internet of things flexible sensors and soft robotics. *Manufacturing Letters.* 15(Part B):122–125.

Kumar, A., H. Singh, P. Kumar and B. AlMangour (eds). 2023. *Handbook of Smart Manufacturing: Forecasting the Future of Industry 4.0.* CRC Press. https://doi.org/10.1201/9781003333760

Kunju, F.K.F., N. Naveed, M.N. Anwar and M.I.U. Haq. 2022. Production and maintenance in industries: Impact of industry 4.0. *Industrial Robot.* 49(3):461–475. https://doi.org/10.1108/IR-09-2021-211.

Lampropoulos, G., G. Lampropoulos and T. Anastasiadis. 2019. Internet of Things in the context of Industry 4.0: An overview. *International Journal of Entrepreneurial Knowledge*. 7(1):4–19.

Le, N.T.T., and M.L. Pham. 2021. Big data analytics and machine learning for Industry 4.0: An overview, Chapter 1. In: G. Rajesh, X.M. Raajini and H. Dang (eds) *Industry 4.0 Interoperability, Analytics, Security, and Case Studies*. CRC Press, Taylor & Francis Group. 1–12.

Lepasepp, T.K., and W. Hurst. 2021. A systematic literature review of Industry 4.0 technologies within medical device manufacturing. *Future Internet*. 13:264 (19 pages). https://doi.org/10.3390/fi13100264.

Li, Q., Q. Tang, I. Chan, H. Wei, Y. Pu, H. Jiang, J. Li and J. Zhou. 2018. Smart manufacturing standardization: Architectures, reference models and standards framework. *Computers in Industry, Computers in Industry*. 101:91–106.

Liagkou, V., C. Stylios, L. Pappa and A. Petunin. 2021. Challenges and opportunities in Industry 4.0 for mechatronics, artificial intelligence and cybernetics. *Electronics*. 10:2001. https://doi.org/10.3390/electronics10162001.

Liu, C., H. Vengayil, R.Y. Zhong and X. Xu. 2018a. A systematic development method for cyber-physical machine tools. *Journal of Manufacturing Systems*. 48(Part C):13–24. https://doi.org/10.1016/j.jmsy.2018.02.001.

Liu, C., X. Xu, Q. Peng and Z. Zhou. 2018b. MT Connect-based cyber-physical machine tool: A case study. *51st CIRP Conference on Manufacturing Systems. Procedia CIRP*. 72:492–497. https://doi.org/10.1016/j.procir.2018.03.059.

Mahmoud, M.A., and J. Grace. 2019. Towards the adoption of smart manufacturing systems: A development framework. *International Journal of Advanced Computer Science and Applications*. 10(7):29–35.

Maroor, J.P., N.L. Kumar, S. Khandelwal, B. Khan, P.J. Chate, E. Fantin and I. Raj. 2022. Novel management trends using IoT in Indian automotive spares manufacturing industries. *Journal of Pharmaceutical Negative Results*. 13(9):4887–4899.

Mateo, F.W., and A. Redchuk. 2021. IIoT/IoT and Artificial intelligence/machine learning as a process optimization driver under Industry 4.0 model. *Journal of Computer Science & Technology*. 21(2):170–176.

Mittal, S., M.A. Khan, D. Romero and T. Wuest. 2019a. Building blocks for adopting smart manufacturing. *Procedia Manufacturing*. 34:978–985. https://doi.org/10.1016/j.promfg.2019.06.098.

Mittal, S., M.A. Khan, D. Romero and T. Wuest. 2019b. Smart manufacturing: Characteristics, technologies and enabling factors. *Proceedings of the IMechE Part B: J Engineering Manufacture*. 233(5):1342–1361. https://doi.org/10.1177/0954405417736547.

Mourtzis, D., J. Angelopoulos and N. Panopoulos. 2021. Smart manufacturing and tactile internet based on 5g in industry 4.0: Challenges, applications and new trends. *Electronics*. 10:3175. https://doi.org/10.3390/electronics10243175.

MTConnect Institute. MTConnect Standard. Part 1—Overview and protocol. Version 1.0.1. 2009. https://static1.squarespace.com/static/54011775e4b0bc1fe0fb8494/t/55800405e-4b057e97372fe59/1434452997276/MTC_Part_1_Overview_v1.0.1R10_0 2_09.pdf

Nagy, J., J. Olah, E. Erdei, D. Mate and J. Popp. 2018. The Role and impact of Industry 4.0 and the Internet of Things on the business strategy of the value chain-The case of Hungary. *Sustainability*. 10:3491 (25 pages). https://doi.org/10.3390/su10103491.

Nguyen, H.D., K.P. Tran, P. Castagliola and F.M. Megahed. 2022. Enabling smart manufacturing with artificial intelligence and big data: A survey and perspective. In: C.I. Pruncu and J. Zbitou (eds) *Advanced Manufacturing Methods Smart Processes and Modeling for Optimization*. CRC Press.1–22.

Nguyen, H.D., K.P. Tran, X. Zeng, L. Koehl, P. Castagliola, et al. 2019. Industrial Internet of Things, big data, and artificial Intelligence in the smart Factory: A survey and perspective. ISSAT International Conference on Data Science in Business, Finance and Industry, DaNang, Vietnam, July, 72–76. https://hal.science/hal-02268119.

Nguyen, T., R.G. Gosine and P. Warrian. 2020. Systematic reviews of big data analytics for the oil and gas industry 4.0. *IEEE Access*. 8:61183–61201. https://doi.org/10.1109/ACCESS.2020.2979678

Nica, E., G.H. Popescu and G. Lazaroiu. 2021. Sustainable Industry 4.0 wireless networks, AI/ML algorithms, and internet of things-based real-time production logistics in digital twin driven smart manufacturing. *SHS Web of Conferences*. 129:04003. https://doi.org/10.1051/shsconf/202112904003.

Nicolae, A., A. Korodi and L. Silea. 2019. An Overview of Industry 4.0 development directions in the Industrial Internet of Things context. *Romanian Journal of Information Science and Technology*. 22(3–4):183–201.

Noor-A-Rahim, M., et al. 2022. Toward Industry 5.0: Intelligent reflecting surface in smart manufacturing. *IEEE Communications Magazine*. 60(10):72–78. https://doi.org/10.1109/MCOM.001.2200016.

Oztemel, E., and S. Gursev. 2020. Literature review of Industry 4.0 and related technologies. *Journal of Intelligent Manufacturing*. 31:127–182. https://doi.org/10.1007/s10845-018-1433-8.

Panda, S. 2018. Impact of AI in manufacturing industries. *International Research Journal of Engineering and Technology (IRJET)*. 5(11):1765–1767.

Parthasarathy, R.M., P. Ayyappan, S.S. Loong, N. Hussin and S.S. Riaz Ahamed. 2022. An Industry 4.0 vision with an artificial intelligence techniques and methods. *International Journal of Mechanical Engineering*. 7(1):1314–1322.

Patricia, A.C., M.G.C. Gerardo, R.V.J. Luis, R.T. Bernabe and C.A. Norma. 2019. Conditions for the development of industry 4.0 from the human capital technological competences perspective. *Revista de Ciencias Tecnológicas (RECIT)*. 2(4):159–165.

Petrovica, S., M. Strautmane and A.A. Naumeca. 2022. Awareness and Development of Industry 4.0: Case of Latvia. *CEUR Workshop Proceedings*. July 03–06, Riga, Latvia.

Phuyala, S., D. Bista and R. Bista. 2020. Challenges, opportunities and future directions of smart manufacturing: A state of art review. *Sustainable Futures*. 2:100023.

Pundir, K. 2018. The role of Industry 4.0 of small and medium enterprises in Uttar Pradesh. *ABS International Journal of Management*. 6(2):68–73.

Radanliev, P., D.D. Roure, R. Nicolescu, M. Huth and O. Santos. 2021. Artificial intelligence and the internet of things in industry 4.0. *CCF Transactions on Pervasive Computing and Interaction*. 3:329–338. https://doi.org/10.1007/s42486-021-00057-3.

Radanliev, P., D.D. Roure, R. Nicolescu, M. Huth and O. Santos. 2022. Digital twins: Artificial intelligence and the IoT cyber-physical systems in Industry 4.0. *International Journal of Intelligent Robotics and Applications*. 6:171–185. https://doi.org/10.1007/s41315-021-00180-5.

Rahardjo, B., F.K. Wang, R.H. Yeh and Y.P. Chen. 2023. Lean manufacturing in Industry 4.0: A smart and sustainable manufacturing system. *Machines*. 11:72. https://doi.org/10.3390/machines11010072.

Rahim, M.N.A., J. John, F. Firyaguna, D. Zorbas, H.H.R. Sherazi, S. Kushch, E. Connell, D. Pesch, B.O. Flynn, M. Hayes and E. Armstrong. 2023. Wireless communications for smart manufacturing and industrial IoT: Existing technologies, 5G, and beyond. *Sensors*. 23(1):73; https://doi.org/10.3390/s23010073.

Rai, R., M.K. Tiwari, D. Ivanov and A. Dolgui. 2021. Machine learning in manufacturing and industry 4.0 applications. *Internal Journal of Production Research*. 59(16):4773–4778. https://doi.org/10.1080/00207543.2021.195667.

Ramakrishna, G.N. 2022. Industry 4.0 and its impact on Indian economy. *EPRA International Journal of Research and Development (IJRD)*. 7(8):51–54.

Rubel, R.I., M.H. Ali and M.W. Akram. 2022. Role of in-process metrology in Industry 4.0 smart manufacturing. *Academic Journal of Manufacturing Engineering*. 20(2):12–18.

Rupp, M., M. Schneckenburger, M. Merkel, R. Borret and D.K. Harrison. 2021. Industry 4.0: A technological-oriented definition based on bibliometric analysis and literature review.

Journal of Open Innovation Technology Market Complex. 7:68 (20 pages). https://doi. org/10.3390/joitmc7010068.

Ryalat, M., H. ElMoaqet and M. AlFaouri. 2023. Design of a smart factory based on cyber-physical systems and Internet of Things towards Industry 4.0. *Applied Science.* 13:2156. https://doi.org/10.3390/app13042156.

Sadiku, M.N.O., O.D. Olaleye and S.M. Musa. 2019. Smart manufacturing: A primer, *International Journal of Trend in Research and Development.* 6(6):9–12.

Sahoo, P.R. 2021 Industry 4.0: AI enabled predictive maintenance strategies. *Journal of Multidisciplinary Engineering Science and Technology (JMEST).* 8(3):13683–13686.

Santos, B.P., F. Charrua-Santos and T.M. Lima. 2018. Industry 4.0: An overview. P*roceedings of the World Congress on Engineering,* London, U.K. July 4–6.

Sevic, M., and P. Keller. 2021. Model of smart factory using the principles of Industry 4.0. MM *Science Journal.* 4238–4243. https://doi.org/0.17973/MMSJ.2021_03_2020067.

Sindhu, V., G. Anitha and R. Geetha. 2021. Industry 4.0-A breakthrough in artificial intelligence the internet of things and big data towards the next digital revolution for high business outcome and delivery. *Journal of Physics: Conference Series.* 1937:012030. https://doi.org/10.1088/1742-6596/1937/1/012030.

Singh, A.K. 2018. Review on manufacturing context in intelligent industry. J*ournal of Emerging Technologies and Innovative Research (JETIR).* 5(7):14–17.

Singh, M. 2020. Industry 4.0: High-tech strategy for future manufacturing industries. *AKGEC International Journal of Technology.* 11(2):13–21.

Soebandrija, K.E.N., A.K. Samara, A.W. Wiratama, C.D. Sasabone, O. Gunanjar and Z. Toar. 2021. Industry X.0 and Sustainability through lens of industrial engineering, product design engineering and business engineering perspectives. *IOP Conference Series: Materials Science and Engineering.* 1115:012034. https://doi.org/10.1088/1757-899X/1115/1/012034.

Sony, M., and P.S. Aithal. 2020. Transforming Indian engineering industries through Industry 4.0: An integrative conceptual analysis. *International Journal of Applied Engineering and Management Letters (IJAEML).* 4(2):111–123.

Sreenivasulu, R., and S.R. Chalamalasetti. 2019. Applicability of Industrial Internet of Things in lean manufacturing: A brief study. *AKGEC International Journal of Technology.* 10(2):22–26.

Stadnicka, D., J. Sep, R. Amadio, D. Mazzei, M. Tyrovolas, C. Stylios, A.C. Coch, J.A. Merino, T. Zabinski and J. Navarro. 2022. Industrial needs in the fields of artificial intelligence, internet of things and edge computing. *Sensors.* 22:4501. https://doi.org/10.3390/ s22124501.

Stief, P., J.Y. Dantan, A. Etienne and A. Siadat. 2022. A new methodology to analyze the functional and physical architecture of existing products for an assembly oriented product family identification. *CIRP Design Conference,* May 2018, Nantes, France. Procedia CIRP. 107:1594–1599.

Strandhagen, J.W., E. Alfnes, J.O. Strandhagen and L.R. Vallandingham. 2017. The fit of Industry 4.0 applications in manufacturing logistics—A multiple case study. *Advanced Manufacturing.* 5:344–358. https://doi.org/10.1007/s40436-017-0200-y.

Sufian, A.T., B.M. Abdullah, M. Ateeq, R. Wah and D. Clements. 2021. Six-gear roadmap towards the smart factory. *Applied Science.* 11:3568 (35 pages). https://doi.org/10.3390/ app11083568.

Sujatha, M., N. Priya, A. Beno, T.B. Sheeba, M. Manikandan, I.M. Tresa, P.S. Jose, V.K. Peroumal and S.P. Thimothy. 2022. IoT and machine learning-based smart automation system for Industry 4.0 using robotics and sensors. *Journal of Nanomaterials.* Article ID 6807585:6 pages. https://doi.org/10.1155/2022/6807585.

Tan Q., Y. Tong, S. Wu and D. Li. 2020. Towards a next-generation production system for industrial robots: A CPS-based hybrid architecture for smart assembly shop floors with closed-loop dynamic cyber physical interactions. *Frontiers of Mechanical Engineering,* 15(1):1–11. https://doi.org/10.1007/s11465-019-0563-9.

Tantawi, K.H., I. Fidan and A. Tantawy. 2019. Status of smart manufacturing in the United States. *9th Annual Computing and Communication Workshop and Conference (CCWC)*, Las Vegas, NV, USA. 0281–028. https://doi.org/10.1109/CCWC.2019.8666589.

Thoben, K.D., S. Wiesner and T. Wuest. 2017. "Industrie 4.0" and smart manufacturing—A review of research issues and application examples. *International Journal of Automation Technology*. 11(1):4–16. https://doi.org/10.20965/ijat.2017.p0004.

Tsaramirsis, G., et al. 2022. A modern approach towards an Industry 4.0 model: From driving technologies to management. *Journal of Sensors*. Article ID 5023011:18 pages. https://doi.org/10.1155/2022/5023011.

Wang, B., F. Tao, X. Fange, C. Liu, Y. Liu and T. Freiheit. 2021. Smart manufacturing and intelligent manufacturing: A comparative review. *Engineering*. 7:738–757. https://doi.org/10.1016/j.eng.2020.07.017.

Wang, B., F. Tao and X. Fang, et al. 2020. Smart manufacturing and intelligent manufacturing: A comparative review. *Engineering*. 7(6):738–757. https://doi.org/10.1016/j.eng.2020.07.017.

Wang, K. 2016. Intelligent predictive maintenance (IPdM) system-Industry 4.0 scenario. *WIT Transactions on Engineering Sciences*. 113:259–268. https://doi.org/10.2495/IWAMA150301.

Wang, S., J. Wan, D. Li and C. Zhang. 2016a. Implementing smart factory of Industrie 4.0: An outlook. *International Journal of Distributed Sensor Networks*. Article ID 3159805:10 pages. http://dx.doi.org/10.1155/2016/3159805.

Wang, S., J. Wan, D. Zhang, D. Li and C. Zhang. 2016b. towards smart factory for Industry 4.0: A self-organized multi-agent system with big data based feedback and coordination. *Computer Networks*. 101(4):158–168. https://doi.org/10.1016/j.comnet.2015.12.017.

Wang, S., C. Zhang and D. Li. 2016c. A big data centric integrated framework and typical system configurations for smart factory. In: J. Wan, I. Humar and D. Zhang (eds) *Industrial IoT Technologies and Applications. Industrial IoT 2016*. Lecture Notes of the Institute for Computer Sciences. Social Informatics and Telecommunications Engineering. 173:12–23. https://doi.org/10.1007/978-3-319-44350-8_2.

Warke, V., S. Kumar, A. Bongale and K. Kotecha. 2021.Sustainable Development of smart manufacturing driven by the digital twin framework: A statistical analysis. *Sustainability*. 13:10139. https://doi.org/10.3390/su131810139.

Xu, L.D., E.L. Xu and L. Li, Industry 4.0: State of the art and future trends. *International Journal of Production Research*. 56(8):2941–2962. https://doi.org/10.1080/00207543.2018.1444806.

Yao, X., J. Zhou, Y. Lin, et al. 2019. Smart manufacturing based on cyber-physical systems and beyond. *Journal of Intelligent Manufacturing*. 30:805–2817. https://doi.org/10.1007/s10845-017-1384-5.

Yildiz, A., and L. Ugur. 2022. Use of artificial intelligence in smart production in the industrial 4.0 era. *International Journal of Pioneering Technology and Engineering*. 1(1):24–27. https://doi.org/10.56158/jpte.2022.19.1.01.

Zheng, P., H. Wang, Z. Sang, R.Y. Zhong, Y. Liu, C. Liu, K. Mubarok, S. Yu and X. Xu. 2018. Smart manufacturing systems for Industry 4.0: Conceptual framework, scenarios, and future perspectives. *Frontiers in Mechanical Engineering*. 13(2):137–150. https://doi.org/10.1007/s11465-018-0499-5.

Zywicki, K., P. Zawadzki and A. Hamrol. 2017. Preparation and production control in smart factory model. *Advances in Intelligent Systems and Computing*. 571: 519–527. https://doi.org/10.1007/978-3-319-56541-5_53.

14 Intelligent Manufacturing in the Financial Sector
Applications in Asset Management and Trading

Rosa Adamo[1], Domenica Federico[2],
Antonella Notte[2], and Maria Anastasia Arcuri[3]

1 Department of Business and Law, University of
Calabria, Arcavacata of Rende (CS), Italy

2 Department of Economics, eCampus
University, Novedrate (CO), Italy

3 University of Salerno, Salerno, Italy

14.1 INTRODUCTION

Over the last few decades, intelligent manufacturing has become increasingly important. It is a modern production system capable, on the one hand, of integrating the capabilities of people, machines and processes to obtain the best possible production result, and on the other, of obtaining the optimal use of production resources, minimizing waste and adding value to the company. The use of artificial intelligence in production systems means bringing flexibility into production processes, analyzing existing processes and their shortcomings, gathering information about them and using the information to formulate better processes.

All economic sectors, including the financial sector, benefit from the profound change underway. Indeed, intelligent manufacturing is increasingly being adopted by financial firms seeking to take advantage of the abundance of available data and the increased affordability of computing power that allow them to improve the quality of services and products offered to financial clients and increase the personalization and diversification of the products offered. Implementing intelligent manufacturing in finance can generate efficiencies by reducing friction costs (e.g. commissions and fees related to transaction execution) and improving productivity levels, which in turn leads to higher profitability. In particular, the use of automation and cost reductions enabled by technology allows for the reallocation of capacity, cost efficiency and greater transparency in decision making.

DOI: 10.1201/9781003405870-14

Such improvements are particularly evident in wealth management and investments. Indeed, a sizable portion of asset management firms are now using intelligent manufacturing to operate trading and investment platforms. In asset management, artificial intelligence concerns robo advice, management of portfolio strategies and risk management, while in trading it concerns artificial intelligence–driven algorithmic trading, automated execution, process optimization and back-office. Several studies have highlighted how artificial intelligence has a lot of applications for portfolio management and trading that allow the financial sector to be more efficient and compliant with investors' requests.

In the light of these considerations, this chapter describes the use of intelligent manufacturing applied to the asset management and trading sectors, combining the technical issues with the ethical and legal implications for data protection and management in order to generate greater trust and prevent the risk of unreliability.

The chapter starts with an overview of trends in intelligent manufacturing and the most common artificial intelligence techniques used in asset management and trading. Subsequently, the chapter deals with the importance of ethics in the main applications of advanced technology, defining its assumptions and indicating some solutions. Finally, the chapter illustrates the regulatory framework on artificial intelligence in Europe, highlighting the need for regulatory activity attentive to all phases of design and development of the most advanced procedures and forms of artificial intelligence.

14.2 INTELLIGENT MANUFACTURING IN FINANCE

14.2.1 Main Purposes

Intelligent production revolutionizes the times and methods of production, introducing companies' technologies and strategies that fall within the scope of Industry 4.0 [1, 2].

Indeed, new data collection and processing capabilities, machine learning, artificial intelligence and cloud and edge technologies enable a shift from reactive problem solving to increasingly proactive management of products, processes and human capital.

The purpose of intelligent manufacturing is to optimize processes, products and people.

Change adopts digital to work on innovative processes, define new products and services and develop new levels of integration and relationships, keeping the customer at the center of development [3]. These evolutions enable improvements in accuracy, efficiency, cost savings and better customer service [4].

Therefore, intelligent manufacturing constitutes a key element of the competitive advantage of economic sectors in developed countries. Indeed, intelligent manufacturing is revolutionizing all sectors globally and, among these, an important role is certainly played in the financial sector.

Today, financial firms are evaluating their internal systems and considering minimizing human intervention through cyber-physical systems and artificial intelligence networks. The production of financial products and services has therefore also been involved in this evolutionary process of intelligent manufacturing.

However, the financial sector is heavily regulated, the implementation of innovative systems is very complex, and often it is nipped in the bud by management risk considerations. Lack of compliance could pose an even greater danger than risk mitigation.

In general, it is known that in traditional production, abstractly described as a human-physical system (HPS), there are human beings and physical systems. In the traditional production of financial products and services, employees carry out all the main activities such as information collection, both in screening and monitoring, analysis activities and control and decision-making activities, as well as learning activities. All of this requires not only high skills and knowledge from employees but also high labor intensity, which can affect work efficiency, quality and ability to perform complex work.

It came naturally to move from traditional HPS production systems to human-cyber-physical systems (HCPSs), where a computer system can replace humans to complete some of the brain work and replace more manual work. This step has also been implemented in financial companies.

With the new intelligent generation of production systems, cognitive and learning functions are added into information systems [5]. In this way, computer systems acquire powerful sensing, analysis and control capabilities, as well as the ability to enhance learning and generate knowledge. Next-generation technology enables quality changes to HCPS. In financial firms, employees can transfer some of their cognitive and learning capabilities to the computer system, allowing the computer system to "learn and process".

Today, several studies show that intelligent automation is mainly used in the production and trading of financial instruments and in the provision of financial services [6, 7]. The latter is the correct combination of software robots to perform repetitive tasks previously performed by humans, artificial intelligence and business process optimization applied in a coherent way to achieve business objectives.

Until recently, the use of intelligent automation in the financial sector has allowed for cost reductions and creation of efficiency. Morgan Stanley [8] showed how the implementation of software robots allowed financial firms to increase cost reductions by 10–25%. This percentage can reach 30–50% with the use of the same technology enhanced with intelligent automation. Today, the possibility of offering automation directly to their customers, that is, the possibility of using this technology for their own front-line services, has also allowed financial firms to increase their revenues. From the Capgemini Report [9], it emerges that thanks to automation, approximately 35% of financial services companies boosted their growth by 2–5%, thanks to faster time-to-market and better cross-selling synergies.

Along with the benefits offered by automation in terms of costs and revenues, the impetus for financial firms to explore this technology has often come from the growing threat from non-traditional players, for example, bigtech, such as Amazon and Alphabet, which are real competitors, and financial technology (Fintech) companies which try to acquire new market shares with innovative transactional proposals [10]. Traditional institutions, having sufficient resources to address opportunities and challenges and review the process of offering financial products and services, are trying to adjust to competition from these digital-first organizations.

The outcomes are present in Table 14.1.

TABLE 14.1

Definitions, Effects and Connections

Focus	Description	References
Intelligent manufacturing and Industry 4.0	— Real-time data analysis — Artificial intelligence applications — Machine learning In the past, already, traditional centralized manufacturing systems were not able to meet the rapidly changing customer requirements. This has forced major changes in the production styles and configuration of manufacturing enterprises.	"manufacturing innovation and manufacturing intelligence technologies are developed to empower manufacturing excellence via soft computing, decision technologies, and evolutionary algorithms" [1, p. 32] "the internet has become the worldwide information platform for the sharing of information and data. Information processing is an important challenge in an internet-based manufacturing environment, and must facilitate distribution, heterogeneity, autonomy and cooperation" [2, p. 1]
Effects of intelligent manufacturing	— Production efficiency and competitiveness — Processing flexibility — Allocate resources reasonably — Zero-defect manufacturing (reduction of costs, energy consumption and scrap; faster lead times; increased planning and production ability; better problems managements; confidence in availability and output quality)	"The modeling of discrete intelligent manufacturing cells in intelligent factories can help intelligent enterprises improve production efficiency, allocate resources reasonably and improve comprehensive competitiveness" [3, p. 13] "We consider that intelligent manufacturing can reduce the cost of the new product by improving production efficiency, which creates a manufacturing benefit" [4, p. 1] "Such a zero-defect manufacturing approach allows for earlier identification of problems or issues, which will or already negatively affect the output" [5, p. 880]
Connections between intelligent manufacturing and finance	— The efficiency in the automating the workflow of back-office operations (personal financial assistance, collecting personal data and providing sound financial advice) — The potential of artificial intelligence in financial services (e.g., prescriptive decision-making algorithms) — Computing power evolution in finance and financial markets — The increase of the competition from bigtech and Fintech	"Data-driven solutions may support financial service firms from purely descriptive models and methods with strong predictive power, towards prescriptive decision-making algorithms" [6, p. 1] "The surveyed publications have given hope in harvesting improved efficiency, new data, information, advisory and management services, risk mitigation" [7, p. 189] "the term FinTech is not confined to specific sectors (e.g. financing) or business models (e.g. peer-to-peer (P2P) lending), but instead covers the entire scope of services and products traditionally provided by the financial services industry" [10, p. 5]

14.2.2 PROCESSES AND FORM

Intelligent manufacturing represents a concrete opportunity to improve the production processes of financial firms. It allows real-time data analysis, artificial intelligence applications and the use of machine learning in the production process of financial products and services.

Regarding data analysis, it is important to highlight how the development of new technologies has generated the creation of data from non-traditional sources that have taken on an independent value. They are big data and alternative data. Big data refers to a large volume of structured, semi-structured and unstructured data that require the use of computational capabilities and innovative algorithms to identify, store and analyze them [11]. Their use allows data science to extract information content that can be innovative. For example, very high frequency data relating to securities trading are big data in the dimensional sense because they offer new information content thanks to advances in available computational capacity, even if they are not particularly innovative or of recent availability. Alternative data refers to innovative, unstructured and heterogeneous data, which are not large enough to require new analysis technologies. In this they differ from big data [12]. The most requested data alternatives in finance concern the popularity and digital reputation of an activity (reviews, comments, social network) and geolocation [13]. For example, the social life of a top manager of a listed company, which can be extrapolated from social media, can be considered innovative and valuable information because it can have a predictive power for returns and stock volatility [14].

Regarding artificial intelligence, it allows one to delegate to computers tasks that previously could only be performed by humans. In particular, artificial intelligence allows not only the application of machine learning of large volumes of data but also the provision of forecasts that allow automated decision-making processes even in areas that require complex choices, based on multiple factors and non-predefined criteria [15].

In the machine learning application, artificial intelligence is able to make predictions by often processing personal data (through algorithmic models) and allowing for faster and potentially more accurate automated decisions, resulting in innovative finance.

Artificial intelligence in finance has been a research area of great interest for many decades, and it is encompassing all spheres of activity, including capital markets, trading, banking, insurance, lending, investing, asset management and assets, risk management, marketing, compliance and regulation, payments, contracting, auditing, financial infrastructure, blockchain, financial operations, financial services, financial security and ethical finance [16].

Applications in finance have been primarily driven by advances in next-generation artificial intelligence data science (AIDS) [17], which are innovating, transforming and synthesizing financial assets. With the use of artificial intelligence in finance, the results—probably—will be more accurate because with the use of machines, it is possible to reduce psychological biases and increase the accuracy of financial action [18]. Through the use of artificial intelligence, financial firms obtain more relevant information on customer satisfaction and can personalize the experience of the customers themselves. For example, in assessing a customer's creditworthiness, financial firms can use artificial intelligence–provided solutions by also taking into account other data relating to the customer's financial history (e.g., its repayment habits, payment process debts, etc.). Through this information it is possible to customize the interest rate of the individual customer. Artificial intelligence also allows clients to manage their portfolios more efficiently. This can be done through robo advisors and digital wealth management.

Finally, machine learning, which is considered a subset of artificial intelligence, represents the tool through which it is possible to extract valuable information content from the increasingly large set of available data [19]. Machine learning techniques are based on statistical-mathematical models that make it possible to process large amounts of data and different types of data (unstructured data such as text, images, videos). The use of hardware and statistical-mathematical optimization techniques allow high computational efficiency. For example, machine learning can help prevent fraud like credit card cloning and digital data theft. By recording the customer's habitual behavior, the algorithms can report any anomalies (if the customer's profile is connected via a social profile, the customer is immediately informed of the anomaly).

Machine learning techniques can be classified into three main categories: supervised learning, unsupervised learning and reinforced learning.

In supervised learning, the algorithm is programmed to learn the relationship between a set of input observations and a target output variable so that it can reproduce the output variable for new input sets. In this method, two approaches can be distinguished: the regression approach, which predicts the output on the basis of the inputs, and the classification approach, which classifies the output into categories.

Unsupervised learning allows you to directly identify the data structure. In this method, only the input observations are available and not the previous output target values. In this way, it is possible to group the examples into internally cohesive classes, whose elements are similar to each other based on the optimization of a predefined criterion, but without having the labels expected in supervised learning. From a statistical point of view, factor analysis allows one to identify the main drivers underlying the data, while cluster analysis allows one to identify subgroups of homogeneous data within them.

In the latest machine learning technique, reinforced learning, the algorithm does not receive instructions on what to do but must discover, based on trial and error, which actions give greater reward in the long term. Basically, the system learns to choose autonomously to achieve certain objectives through interaction with the environment. The environment provides reinforcing signals in terms of "reward", but it does not explicitly indicate the correct choice.

A subset of machine learning is deep learning. It is a system based on neural network architectures, according to a sophisticated approach that aims to replicate the functioning of the human nervous system. Learning data characteristics happens on its own, without explicit instructions, and requires a large volume of training data. Deep learning has applications in image recognition, understanding and translating a language and automating complex tasks. An example of application is customer assistance by the use of chatbots which are able to guide the client towards the desired information, orienting themselves on the basis of keywords contained in the customer's request [20].

The outcomes are presented in Table 14.2.

TABLE 14.2

Applications of Intelligent Manufacturing in the Production Processes of Financial Firms

Focus	Description	References
Real-time data analysis	— Big data — Alternative data — Financial application — Influence the business models of financial operators	"After reviewing the literature, this study found some financial areas directly linked to big data, such as financial markets, internet credit service-companies and internet finance, financial management, analysis, and applications, credit banking risk analysis, risk management, and so forth" [11, p. 9] "Alternative data provide new views for managers and investors to understand more about the enterprise activities, and then to benefit business management and investment in finance markets" [12, p. 13] "The availability of new technologies together with increasing data and computing power could significantly influence the business models of operators with specific regard to activities such as financial advice, risk management, client identification and monitoring, trading, and portfolio management" [13, p. 9]
Artificial intelligence	— Application of machine learning of large volumes of data — Automated decision-making processes — Innovative finance — AIDS	"Computing technologies play an important role in the transformation of modern financial services" [17, p. 65] "With the latest advancement in technology, blockchain technology has emerged. It has many advantages and applications in various areas such as cryptocurrencies, banking and financial services, many internet things, managing risk etc." [18, p. 1]
Machine learning	— Supervised learning — Unsupervised learning — Reinforced learning — Deep learning	"The study considers six financial domains: stock markets, portfolio management, cryptocurrency, forex markets, financial crisis, bankruptcy and insolvency" [19, p. 1] "All previous studies prove the efficiency of the machine and deep learning techniques in achieving the best results and the highest accuracy rates when used in the process of building chatbots" [20, p. 1230]

14.3 USE OF INTELLIGENT MANUFACTURING IN INVESTMENT DECISIONS

14.3.1 APPLICATIONS IN PORTFOLIO ASSET MANAGEMENT

The implications of the use of intelligent manufacturing on investors' financial choices are increasingly relevant [21, 22]. In theory, the composition of the managed portfolio is a very complex process which evolves gradually and which aims to create investment opportunities capable of producing value. First of all, the asset allocation alternatives which envisage the composition of a financial portfolio made up of securities of various kinds (shares, bonds, money market instruments, etc.) are studied. Subsequently, the sector/security selection hypotheses which concern the subdivision of the managed portfolio among securities belonging to different economic sectors or geographical areas for which a better performance is expected are analyzed. Finally, stock picking decisions are made that pertain to the selection of individual stocks that can be placed in a managed portfolio [23, 24].

To carry out this process of composition of the managed portfolio, therefore, financial firms have to request and obtain from the client detailed information that is necessary to verify the knowledge and experience in the investment sector, the financial situation and the investment objectives. It is therefore clear that financial firms should be able to acquire both market data and information and specific information on individual issuers (prospectuses, financial statements, reports and accounts, etc.), the analysis and evaluation of which imply the existence of complete and correct data as well as adequate financial expertise. In investment choices, therefore, the reliability and completeness of the information available influence the financial decisions that can lead to a higher return. Furthermore, once the information has been acquired, the financial firm evaluates that the portfolio management service, as well as the operations recommended as part of the advisory service, correspond to the client's investment objectives.

Intelligent manufacturing allows the use of real-time data analytics and, therefore, allows the use of data in a better (in terms of reliability) and greater (in terms of completeness) way [25, 26].

As for asset management, the use of data to make financial decisions is even more important. Traditional data are now considered a commodity accessible to all, and they are gradually joining big data and alternative data. Financial firms are increasingly seeking to derive competitive advantages from information with the aim of creating new investment strategies with high active returns on investment [27]. In fact, new artificial intelligence technologies make it possible to automatically identify factors with the highest predictive power of stock returns.

Also in portfolio management, artificial intelligence and machine learning are used to extract information and improve the decision-making process. These applications can have positive effects on risk management, operational workflow efficiency, performance and customer relationships [28].

The applications of new technologies can concern four main phases of the portfolio management value chain: data collection, investment process, relationship with customers and cross-chain activities (Table 14.3).

TABLE 14.3
The Portfolio Management Value Chain

Phase	Activities	Benefits	Technology Impacts
1. Data collection	Collection, management and analysis of data	— Better customer segmentation — Greater availability of big data and alternative data	— Artificial intelligence models (using historical and prospective data) — Machine learning techniques (extracting relevant information content from heterogeneous data)
2. Investment process	Creation and optimization of the portfolio of financial instruments	Identify the weights of the asset class that compose the portfolio to satisfy the client's objectives of active return on investment and manager performance	— Artificial intelligence makes it possible to estimate more accurate risk and return and offer alternative portfolio optimization approaches — Genetic algorithms can solve the mean-variance problem
3. Relationship with customers	Innovation of activities of marketing, business operations, back-office operations and investment administration	— Better customer segmentation and intelligent design of distribution models — More opportunities of cross-selling multiple products to the same customer	— Artificial intelligence systems can enable deeper knowledge of customer decision journeys, identifying an opportunity for action or change — Big data allows collection of information about potential and existing customers
4. Transversal activities	Risk management and prevention	Constant monitoring of the entire production process	— Artificial intelligence improves the modeling of market risk through the use of qualitative information — Operational risk can be controlled through intelligent manufacturing techniques

In the first phase, data collection, activities of collection, management and analysis of data are carried out. Through this phase, it is possible to benefit from better customer segmentation thanks to the large set of information on individual preferences and choices (collected, for example, from social media). Artificial intelligence models can be trained on the basis of historical but also prospective data, such as buy or sell recommendations from analysts. Furthermore, machine learning techniques can be used to extract relevant information content from heterogeneous sources (company financial statements, specialized press articles, information available on social media, etc.).

In the second phase, the investment process, the results of the data analysis are used in the process of creating and optimizing the portfolio of financial instruments to identify the weights of the assets that compose it according to the clients' objectives in terms of active return on an investment and manager performance (e.g., tracking a benchmark or maximizing the Sharpe ratio).

In portfolio management, artificial intelligence systems make it possible to obtain more accurate risk and return estimates than those produced by traditional methods and offer alternative portfolio optimization approaches to those generated by traditional linear techniques. In this way it is possible to overcome some problems related to the application of the Markowitz mean-variance model. The main problem, in fact, lies in the fact that the expected return appears substantially difficult to determine, and a lot of data are needed to elaborate the variance-covariance matrix, in addition to the hypothesis that the correlation between the assets is stable. A very popular method, used as a tool to generate a portfolio in which the number of assets is limited by solving the mean-variance problem, is that of genetic algorithms [29].

In the third phase, relationship with customers, big data and artificial intelligence systems are very useful. In fact, the technological innovations that are transforming manufacturing also affect marketing and business operations, as well as back-office operations and investment administration. At this stage of the chain, it is possible to benefit from better customer segmentation and an intelligent design of distribution models in line with the characteristics of each segment. In this way, cross-selling opportunities for multiple products to the same customer can be identified.

The fourth phase of the chain involves a series of cross-chain activities such as risk management and prevention. In fact, risk management and prevention can derive considerable benefits from the application of intelligent production systems thanks to the constant monitoring of the entire production process. With reference to risks in finance, market risk or operational risk can obtain benefits. With regard to market risk, artificial intelligence can improve the modeling of market risk through the use of qualitative information, which makes it possible to estimate financial or economic variables at the aggregate and company level in a more accurate way than traditional data. Operational risk, due to its pure risk nature, can be largely controlled through intelligent manufacturing techniques. In fact, operational risk is the risk of losses originating from failures or errors of internal processes, human resources and technological systems or deriving from external events.

In the study of the value chain, it is evident that the value generated by the changes introduced by intelligent manufacturing seems destined to grow and to compensate for the development and use costs that such complex systems entail.

In this context, an important decision concerns the methods of acquiring new technologies, that is, the degree of outsourcing of activities. Internalization allows one to collect, store and analyze raw data in a way that respond to specific needs, but supporting high investments in research and in the acquisition of adequate technologies and human resources. Partial outsourcing allows one to achieve a compromise, being able to combine, for example, the purchase of data (raw, processed or semi-processed) with the internal analysis of the same data. Total outsourcing, on the other hand, has the advantage of not requiring investments in infrastructure and skills and the disadvantage of the non-exclusivity of the analysis and projects acquired.

For example, as part of the first phase of the chain (data collection), the manager can choose to operate in-house or to turn to external subjects. In particular, in the collection activity, data may be provided by specialized data providers (e.g., bigtech);

in the data archiving activity (storage), the manager can have a computer platform or use the platform of an external provider (cloud computing). For data reliability (so-called data-cleaning), an external database as a model can be used.

On the positive impact about the perception of value in the portfolio management market, artificial intelligence–powered funds register significant performance compared to traditional funds, confirming the expectations of managers in relation to new technologies [29].

14.3.2 Applications in Trading

In the area of investment services, trading activity, that is, the purchase and sale of financial instruments (such as shares, options, currencies, cryptocurrencies, futures), is also evolving with intelligent applications for operating in the financial markets [29].

In theory, the process by which the trading of financial instruments takes place can be divided into two main phases in the market value chain: trading and post-trading (Table 14.4). The trading phase consists in the match between supply and demand of securities. The post-trading phase is represented by a set of procedures (clearing, settlement and custody) which guarantee the successful completion of trading operations on the markets.

For some time now, the development of technology has revolutionized trading activity by spreading contracting methods no longer based on the physical meeting of the subjects admitted to trading but on the activation of increasingly efficient telematic circuits.

With the application of intelligent manufacturing mechanisms, it is possible to revolutionize the trading of investments. The dissemination of the mechanisms underlying intelligent manufacturing, that is, real-time data analysis, artificial intelligence and machine learning, involved the two main phases of investment strategies.

In the first phase, the collection of market data (data/input) and the evaluation of user preferences to then execute purchase and sales orders (task/output) in the market that meet the needs of customers are realized. Collecting and analyzing data is very complex. Indeed, the dynamic nature of markets prevents reliance on models that have static rules, as they quickly become outdated.

In this sense, machine learning can be very useful: thanks to massive data analysis, it recognizes "weak signals", thus becoming a valuable support in decisions or automating some processes that require great timeliness.

The algorithm created for trading is, therefore, able to make the analysis much faster and reduces the impact of emotion that often leads to rash decisions.

Specifically, rules are defined which indicate from time to time whether to place orders to buy or sell financial instruments. The rules that make up a trading system can be originated in different ways, and potentially the investor could decide on any element as a source of investment signals, from technical analysis indicators to daily weather forecasts.

These rules allow one to give a rational logic to the investment process and thus to be able to control human emotion. For this reason, in addition to the definition of indicators that provide buy or sell signals, automatic functions are introduced to ensure profits or limit losses, according to pre-set levels. These are stop-loss and

TABLE 14.4
The Trading Value Chain

Phase	Activities	Benefits	Technology Impacts
1. Trading	— Collection of market data (data/ input); — Evaluation of user preferences; — Execution of purchase and sales orders (task/output) in the market	— Definition of indicators that provide buy or sell signals — Automatic functions to ensure profits or limit losses, according to pre-set levels — Accuracy, efficiency and speed of trading decisions — Reduction of the impact of emotion	— Real-time data analysis — Artificial intelligence identifies market trends, formulates trading strategies, analyzes financial news, determines the impact of market events and evaluates the investment risk — Machine learning recognizes some signals; thus, it can support decisions and automate some processes that require timeliness — Reinforcement learning (analytical and calculating capacity of a machine can be more suitable than the human mind for interacting with the financial market)
2. Post-trading	— Clearing: positions of all participants in the securities market are ascertained — Settlement: the contract is fulfilled — Custody: The securities are physically owned and managed	— More efficient and accurate anomaly detection — Data quality verification and checks — Automated extraction of information from unstructured sources	— In clearing and custody: artificial intelligence and machine learning develop calculation algorithms capable of processing a growing amount of input data and detecting anomalies using a predictive analysis model — In settlement: artificial intelligence can allow for greater efficiency (e.g., optimizing the allocation of liquidity)

take-profit orders. Stop-losses are lower limits, and upon reaching them, the position is closed to avoid excessive loss; take-profits, on the other hand, represent upper limits, after which the position is closed to prevent the accumulated profit from subsequently decreasing.

After this, the trading phase can begin. This process can occur in two ways: manually or automatically. In the first mode, the client evaluates the signals received and decides whether to convert them into market orders. In the second mode, the investment systems are managed directly by a computer which, through the algorithms established by the trader, automatically sends the market orders. This second method defines artificial intelligence trading, that is, the use of computer algorithms and artificial intelligence to operate on the financial markets. Some applications of artificial intelligence trading in financial markets can help identify market trends and formulate trading strategies based on those trends or analyze financial news and determine the impact of specific market events or evaluate the investment risk. The use of artificial intelligence in trading has greatly improved the accuracy, efficiency and speed of trading decisions in the financial sector.

Increasingly accurate learning methodologies such as reinforcement learning are also applied in this phase to improve trading strategies. In fact, the analytical and calculating capacity of a machine can prove more suitable than the human mind for interacting with the financial market.

Several studies [30, 31] highlight that the theoretical application of machine learning can be a useful approach to predict stock prices. In fact, forecasting stock returns is considered one of the most challenging tasks because the stock market is dynamic [32, 33].

The second phase, that is, the post-trading phase, includes clearing, settlement and custody.

Clearing is the stage in which the positions of all market participants are ascertained, that is, who is to deliver, which securities and to whom are determined. Clearing makes it possible to reduce the movement of securities and cash.

Settlement is the stage in which the contract is fulfilled; the seller delivers the securities, and the buyer delivers the money. There is material certainty that the initial order has been executed, and the exchange is completed.

Custody refers to the material possession of the securities and includes management of the securities on behalf of the holders (e.g. accounting, coupon collection, securities lending).

In clearing and custody, for example, central counterparties (which have the task of absorbing counterparty risk) and central securities depositories (which have the task of safekeeping financial securities in the form of paper or computerized certificates) begin to use artificial intelligence and machine learning to develop calculation algorithms capable of processing a growing amount of input data and detect anomalies using a predictive analysis model.

Also in settlement, the use of artificial intelligence can allow for greater efficiency, for example, by optimizing the allocation of liquidity to predict the probability that a given contract will not be settled due to an inefficient distribution of these amounts.

In essence, in post-trading, some data reporting and trade repository service providers have started to develop artificial intelligence and machine learning solutions for more efficient and accurate anomaly detection, data quality verification and checks and data quality control–automated extraction of information from unstructured sources, although currently they mostly turn to cloud services offered by third-party providers.

14.4 ETHICS IN THE USE OF ARTIFICIAL INTELLIGENCE

14.4.1 CONDITIONS

The attention to intelligent manufacturing for productive finance cannot disregard the close link existing with the use of data and information in technological networks.

Access to technological networks becomes fundamental for the growth of productivity. It is necessary to acquire data and information that improve the ways of doing business and end up affecting the criteria that regulate competitiveness and open up new operational opportunities.

The availability of data, together with the inclusion of these in ad hoc algorithms, assumes importance for the design and increase of the various phases of the economic processes that put intelligent manufacturing in production processes.

The consequence is an economic process in which the core of the professional activity in practice does not change, while the technical methods of the action change and improve, allowing the achievement of useful and rapid results.

A framework emerges in which real-time data analysis, artificial intelligence and machine learning are aimed at solving numerous production choices, so that their use is rapidly expanding in the most varied production systems, as well as in finance.

The impact in ethical terms must concentrate on identifying the risks of the growing use of digital tools, governed by algorithms and more or less sophisticated forms of artificial intelligence.

The need to implement guarantee measures for the transparency of the processes based on these technologies appears undoubted, also due to the pressing difficulty in controlling the data due to the opacity of the collection methods, of the conservation places, of the selection criteria and of analysis.

Furthermore, it is necessary to take charge of the evaluation of the discriminatory potential that derives from their use, as a function of increasingly precise and analytical profiling, to the point of canceling the uniqueness of the person, their value and their exceptional nature.

The European data protection authorities have repeatedly underlined the need to accompany these phenomena with a more rigorous ethical and general responsibility approach [34].

The legislative resolutions adopted in October 2020 by the European Parliament [35–37] are aimed in this direction. To this end, recently, on February 16, 2023, an independent European body responsible for ethics issues was established [38].

14.4.2 SOME SOLUTIONS

In production processes, financial conduct inspired by profiles of transparency and fairness should be pursued with reference to each application of real-time data analysis, artificial intelligence and machine learning, in a sort of "by design" approach.

Whenever an intelligent manufacturing business case is proposed, if there is no compliance with the standards of ethics—transparency and correctness—the application should be refused.

The transparency of financial processes has the objective of making customers aware of the essential elements of the contractual relationship, as well as their variations, in compliance with the autonomy of negotiations.

The condition of transparency makes it possible to create application models of intelligent manufacturing mechanisms that can be directly interpreted and understood by humans. It follows the achievement of a state of awareness capable of generating trust and preventing the risks of unreliability.

Another important behavior to follow in a reliable intelligent manufacturing model is its correctness of installation. The principle of fairness requires that the model guarantee, for example, data protection to prevent discrimination against individuals or groups of them. To operate correctly, the model should trace the collection of data to explain how the results and decisions were reached, demonstrating respect for consumer rights and their interests.

In Europe, it is necessary to comply with Regulation (EU) 2016/679 of the European Parliament and of the Council [39].

14.5 ARTIFICIAL INTELLIGENCE STANDARDS AND REGULATION IN EUROPE

14.5.1 THE NEED TO REGULATE

The use of data cannot be separated from the maintenance of a state of trust.

With the Resolution of the European Parliament of March 14, 2017, on the implications of big data for fundamental rights [40], the foundations of a rule of law are being laid so that compliance with data protection legislation, together with robust scientific and moral assumptions, is the basis for building trust in big data processing.

Big data, according to the aforementioned Resolution of the European Parliament, concern the collection, analysis and processing of large volumes of data, including personal data, deriving from a slew of distinct sources. It is a question of referring to the use of them in an automated process, which provides for the use of computer algorithms and advanced data processing techniques. The processing involves both stored and streamed information, with the aim of tracing specific correlations, trends and patterns (so-called big data analyses).

The European Union, over the years, has been trying to outline a rigorous legal framework, whose objectives are, among others, to guarantee the protection of personal data together with the privacy of citizens. The main objective is to develop an adequate and potentially suitable body of standards for a number of emerging artificial intelligence applications [41].

EU directives related to data protection and management are numerous [42–49].

Certainly, the proliferation of directives on the protection of personal data and the privacy of citizens denotes the centrality of the subject in allowing the development of European countries in harmony with the conditions enshrined in the EU treaties and the Charter of Fundamental Rights of the EU, bound by respect for human dignity, in which humans benefit from a unique and inalienable moral status.

There is no doubt that there is a need to give EU countries a uniform and certain framework of rules that is accompanied, however, by the need for mechanisms for updating the discipline: big data is a field of application of artificial intelligence which constitutes a difficult object to regulate.

Artificial intelligence, even more than other innovative technologies, is characterized by its great development strength, which quickly makes obsolete any intervention aimed at regulating it.

On the other hand, artificial intelligence, in its technical expressions such as machine learning and deep learning, is characterized by its own autonomous and equally unpredictable functioning, which also often entails the impossibility of explaining its internal processes (the so-called black box phenomenon), causing a potential source of risks, which cannot be calculated ex ante [50] and, therefore, are in need of timely and specific regulatory prescriptions.

Furthermore, artificial intelligence applications, from time to time, affect several sectors: justice, health care, transport, energy, public administration and finance [51], even affecting the risks of protection, safety, consumer rights and fundamental rights differently according to the scope of application and, consequently, requiring regulatory activity which should periodically list the sectors concerned. Also, the impact for interested persons, which can be more or less significant in terms of risks on legal effects, on the rights of a person [52] or of a company, on material damage or immaterial, should be taken into consideration.

A regulatory framework functional to the various applications of artificial intelligence would be able to guarantee legal certainty.

In particular, for the purposes of this chapter, this need is paramount in finance, where, at present, most of the activity is carried out in an automated form, in compliance with rules imposed in algorithmic matrix programs [53]. For example, artificial intelligence applications occur to:

- Automate institution–client, institution–institution and institution–regulator relationships;
- Analyze the trend of markets, instruments and securities;
- Develop forecasts on future trends;
- Develop new business models;
- Improve data protection or cybersecurity.

The applications of artificial intelligence in relation to customers create critical issues both for the protection of personal data, especially in the case of adoption of automated customer assessment systems for the offer of "personalized" investments related to risk appetite, and for the security, truthfulness and objectivity of the information (big data) processed in the algorithms of artificial intelligence systems.

Financial decisions based on unreliable data could produce discriminatory phenomena. A control and audit activity is needed on the logic underlying AI applications and also on the dynamics that lead to big data.

14.5.2   FROM THE START-UP STANDARD TO THE REGULATORY FRAMEWORK: A CONSTRUCTION SITE STILL IN PROGRESS

14.5.2.1   Personal and Non-Personal Data Processing

The main European legislation on the protection of personal data is the General Data Protection Regulation (GDPR) 2016/679 [54]. The GDPR intends to establish the rules for the protection of personal data and privacy in European countries that have adopted specific regulations with regard to the use of artificial intelligence.

Nonetheless, in artificial intelligence applications, the use of big data gives rise, in the development of various algorithms both in quantitative and qualitative terms, to the interweaving of personal and non-personal data, increasingly showing the difficulty—downstream—in being able to distinguish the processing of personal versus non-personal data. Furthermore, only as a result of computer applications will it be possible to have the case that from personal data we arrive at anonymous data, and, vice versa, from anonymous data we arrive at personal data.

Certainly, the very nature of big data (of considerable variety, with rapidly growing volumes) together with the manifest criticality dictated by the difficulty of keeping personal data discrete during artificial intelligence applications determines the efforts of the principles set in vain by the GDPR, such as, for example, the achievement of the accountability of the data controller.

Specifically, the GDPR, with the principle of accountability (e.g. article 5, paragraph 2 of the GDPR), invokes the guarantee and contextual demonstration of compliance with the principles related to the processing of personal data. In essence, the intention is to implement the principle of transparency to demonstrate the choices that are made by the algorithms, and not only this. Together with the principle of transparency with accountability, the principles of accuracy, data minimization, purpose limitation, conservation limitation and the consent of the interested party are envisaged.

First of all, the different data and their volume make the accuracy verification process difficult to implement, also considering the speed of data processing. Then, the creation of meta-data prevents the initial definition of the purposes of data collection and the study of the information underlying it, contributing to the failure to achieve both the principle of data minimization and purpose limitation. It also follows the difficulty of pursuing the principle of limitation of conservation, considering the potentially inexhaustible availability of data for various purposes.

Finally, attention must be paid to the consent of the interested person because in big data, there is not always informed consent of the interested person for the further purposes pursued. The reason for this can be found in the structural and knowledge gap between the final user and the data processing manager. These are physiological information asymmetries to telematic procedures and more. It is neither easy nor even possible for the interested party to have all the information available to be able to express informed consent, bearing in mind that even if he or she did have it, he or she would not have sufficient know-how to understand it, since these procedures are based on algorithms and software, that is, "techniques" of a computer language for "professionals".

Consequently, the automated management of data collected via networks must be carried out in compliance with the GDPR not only to avoid incurring penalties but to strengthen users' trust in artificial intelligence.

In this regard, if it is possible to identify—at least a priori—the rule governing the processing of personal data in EU Regulation 2016/679, that relating to non-personal data is identified in EU Regulation 2018/1807 [55]. The latter aims to eliminate the barriers that deny the freedom of circulation, cross-border within the EU, of data not attributable to identified or identifiable natural persons.

Finally, attention is drawn to the Final Report of the 2020 fact-finding survey on big data. This report identifies the main challenges for the optimal use of big data. We start by considering the centrality of the data, to be considered also an economic asset and therefore worthy of protection as a fundamental right of the person.

14.5.2.2 Algorithmic Applications

The use of artificial intelligence involves the application of algorithms.

The security of the algorithms is of fundamental importance, as they support and replace human decisions.

Therefore, first of all, a problem of an ethical rather than a legal nature emerges, in which the personal freedom of a subject, their privacy, their personal data and all their fundamental rights collide with a decision-making system that is not human but "manipulated" by humans.

Attention must focus on the role played by algorithms within decision-making processes, pronunciations of sentences and drafting of laws.

The final objective of ethics of algorithms (so-called algorethics) is to ensure that machines, which are not endowed with conscience, are at the service of humans, par excellence endowed with conscience.

Machines are able to select the correct choice on the basis of data provided by humans; for this reason it is necessary to identify the rules to orient artificial intelligence to program and guide the algorithms.

On the ethical and legal implications of artificial intelligence applications, the EU is undergoing an increasingly careful regulatory process.

Starting from the Resolution of the European Parliament of February 16, 2017 [56], the European Data Protection Board has started to issue opinions and guidelines to support the fulfillment of the obligations under the GDPR such as the document "Guidelines on Automated Individual Decision-Making and Profiling" [57], as a guideline for those who have to apply artificial intelligence.

Subsequently, in 2019, the European Commission appointed a group of experts to initially draft the ethical guidelines and then prepare policy and investment recommendations to develop trustworthy artificial intelligence applications. All of this was documented by the Commission in the publication entitled "Creating Trust in Anthropocentric Artificial Intelligence" [58]. In the document, the Commission sets out a guideline for reliable artificial intelligence based on a principle of anthropocentric artificial intelligence, because it is able to guarantee the centrality of human values in various forms of development of artificial intelligence applications.

In 2020, the foundations for real European legislation were laid through the publication of the white paper on artificial intelligence [59].

The introduction of a white paper on artificial intelligence is a very important signal; it demonstrates the will to proactively propose actions in the field of artificial intelligence, pushing for a common European approach. In fact, the white papers, unlike the green ones which instead aim to undertake a consultation activity within the EU, propose a comparison with the whole community, with the subjects and the national and European institutions involved, in order to achieve unanimous political consensus. A shared approach on artificial intelligence applications can only aspire to broad and widespread consensus in order not to incur a fragmentation of operations in this field within the EU and risk undermining the rule of law. We need to maintain citizens' trust and contribute to a progressive and vital vision of the EU.

The strong will to define a common regulatory framework in the EU aims to ensure constant human surveillance of artificial intelligence applications, together with verification of the technical robustness and security of the algorithms, for data confidentiality and, as previously noted, compliance with the principle of accountability established by EU Regulation 2016/679.

14.5.2.3 Use in Finance

The main reference to regulation in the finance sector is by the European Commission [60].

The proposal identifies only one high-risk artificial intelligence application, that is, applications to be used "to assess the creditworthiness of natural persons or to establish their creditworthiness" [61]. However, the European Commission foresees the possibility of adapting, in the future, to the developments of artificial intelligence in finance. For high-risk artificial intelligence applications, there is an ex-ante compliance assessment based on the risk management system, oversight of governance practices, and database management and collection. This should be followed by an ex-post evaluation, through the automatic recording of events during the operation of the artificial intelligence system, which will allow the data responsible for the decision to be traced.

The proposal also provides for a system of transparency towards users, to understand artificial intelligence applications in terms of accuracy, historical performance and purpose.

Finally, the proposal requires a minimum of human oversight of artificial intelligence applications and introduces a set of rules to regulate and hold accountable those offering high-risk artificial intelligence applications.

With this proposal, the EU gives weight to the risks and dangers that artificial intelligence entails and also considers the mandatory measures to be adopted, moving away from a regulation inspired by self-regulation measures.

In fact, the proposal is based on a risk-based approach [62] which assesses the risk of artificial intelligence applications, imposing an absolute ban on applications that harm human rights (e.g., if they manipulate people's conduct or if they exploit the vulnerabilities of some subjects) or of social scoring systems and biometric identification systems for law enforcement purposes.

Then, it adds obligations for "high risk" applications regarding the use and quality of data, risk management (i.e. a process of identification and assessment of known and foreseeable risks before marketing or in post-market monitoring), as well as obligations for documentation of the characteristics and functioning of artificial intelligence applications, transparency towards users and human oversight.

Last, the proposal also pays attention to "low or minimal risk" artificial intelligence applications because they represent the largest part within the single European market.

14.6 CONCLUSIONS

The mechanisms underlying intelligent manufacturing, that is, real-time data analysis, artificial intelligence and machine learning in the production process of financial products and services, will increasingly develop advanced finance, capable of optimizing portfolio and investment strategies.

Artificial intelligence is the engine of this new finance, while big data and machine learning are its fuel. On the one hand, in portfolio decisions, these new mechanisms are able to predict and analyze future data trends and provide insights aimed at maximizing investment returns. On the other hand, in investment strategies, the use of big data on behavior in the financial sphere will make it possible to manage the emotional aspect of financial operators and customers.

Despite this, questions continue to arise regarding the challenges that real-time data analysis, artificial intelligence and machine learning pose to the maintenance of ethical and legal principles on the protection of natural persons with regard to the management of personal data. The production of studies on the subject is accompanied by standards of ethics, declarations of principles, technical regulations and resolutions from international and national institutions.

An approach to the subject that accompanies the development of intelligent manufacturing, in itself a harbinger of great economic and social benefits, must go hand in hand with the implementation of a regulation that is suitable for protecting users from the inevitable negative externalities deriving from use of big data.

The development and diffusion of the mechanisms underlying intelligent manufacturing should be guaranteed in a socio-technical framework, inclusive of standards, in which individual interests and the social good are preserved and valued.

REFERENCES

[1] Oztemel, Ercan, and Samet Gursev. 2020. "Literature review of Industry 4.0 and related technologies." *Journal of Intelligent Manufacturing* (31):127–182. DOI:10.1007/s10845-018-1433-8

[2] Tian, Gui Yun, Guofu Yin, and David Taylor. 2002. "Internet-based manufacturing: A review and a new infrastructure for distributed intelligent manufacturing." *Journal of Intelligent Manufacturing* (13):323–338. DOI:10.1023/A:1019907906158

[3] Lan, Xiaoyi, and Hua Chen. 2023. "Simulation analysis of production scheduling algorithm for intelligent manufacturing cell based on artificial intelligence technology." *Soft Computing* (27):6007–6017. DOI: 10.21203/rs.3.rs-2551439/v1

[4] Li, Kai, Limin Zhang, Hong Fu, and Bohai Liu. 2023. "The effect of intelligent manufacturing on remanufacturing decisions." *Computers & Industrial Engineering* (178):1–31. DOI: 10.1016/j.cie.2023.109114

[5] Lindström, John., Erik Lejon, Petter Kyösti, Massimo Mecella, Dominic Heutelbeck, Matthias Hemmje, Mikael Sjödahl, Wolfgang Birk, and Bengt Gunnarsson. 2019. "Towards intelligent and sustainable production systems with a zero-defect manufacturing approach in an Industry 4.0 context." *Procedia CIRP* (81):880–885. DOI: 10.1016/j.procir.2019.03.218

[6] Boute, Robert N., Joren Gijsbrechts, and Jan A. Van Mieghem. 2022. "Digital lean operations: Smart automation and artificial intelligence in financial services." *Springer Series in Supply Chain Management* (11):175–88. DOI: 10.2139/ssrn.3747173

[7] Milana, Carlo and Arvind Ashta. 2021. "Artificial intelligence techniques in finance and financial markets: A survey of the literature." *Strategic Change* (30):189–209. DOI: 10.1002/jsc.2403

[8] Morgan Stanley. 2017. *The Rise of the Machines*, Report, September, New York.

[9] Capgemini. 2018. *World Wealth Report 2018*, Paris.

[10] Arner, Douglas W., Barberis Janos Nathan, and Ross P. Buckley. 2015. "The evolution of fintech: A new post-crisis paradigm?" *University of Hong Kong Faculty of Law Research Paper* (47):1–45. DOI: 10.2139/ssrn.2676553

[11] Hasan, Md Morshadul, Jozsef Popp, and Judit Oláh. 2020. "Current landscape and influence of big data on finance." *Journal of Big Data* (7): 1–17. DOI: 10.1186/s40537-020-00291-z

[12] Sun, Yunchuan, Lu Liu, Ying Xu, Xiaoping Zeng, Yufeng Shi, and Ajith Abraham. 2022. *A Survey on Alternative Data in Finance and Business: Emerging Applications and Theory Analysis*, 28 June. Available at SSRN: https://ssrn.com/abstract=4148628 or http://dx.doi.org/10.2139/ssrn.4148628

[13] Caschera, Maria Chiara, Arianna D'Ulizia, Fernando Ferri, and Patrizia Grifoni. 2019. "MONDE: A method for predicting social network dynamics and evolution." *Evolving Systems* 10 (3):363–379. DOI: 10.1007/s12530-018-9242-z.

[14] Linciano, Nadia, Valeria Caivano, Daniela Costa, Paola Soccorso, Tommaso Nicola Poli, and Gianfranco Trovatore. 2022. "L'intelligenza artificiale nell'asset e nel wealth management." *Quaderni FinTech*, CONSOB (9):1–97.

[15] McCarthy, John, Marvin L. Minsky, Nathaniel Rochester, Claude E. Shannon. 2006. "A proposal for the Dartmouth summer research project on artificial intelligence, August 31, 1955." *AI Magazine* 27 (4):12–14. DOI: 10.1609/aimag.v27i4.1904

[16] Boscia Vittorio, Cristina Schena, and Valeria Stefanelli. 2020. *Digital banking e FinTech. L'intermediazione finanziaria tra cambiamenti tecnologici e sfide di mercato.* Rome: Bancaria Editrice.

[17] Qi, Yuan, and Jing Xiao. 2018. "Fintech: AI powers financial services to improve people's lives." *Communications of the ACM* 61 (11):65–69. DOI: 10.1145/3239550

[18] Singh, Gurinder, Vikas Garg, and Pooja Tiwari. 2020. Application of artificial intelligence on behavioral finance." In *Recent Advances in Intelligent Information Systems and Applied Mathematics*, 342–353. Cham: Springer. DOI: 10.1007/978-3-030-34152-7_26

[19] Nazareth, Noella, and Yeruva Venkata Ramana Reddy. 2023. "Financial applications of machine learning: A literature review." *Expert Systems with Applications* (219). DOI: 10.1016/j.eswa.2023.119640

[20] Alazzam, Bayan A., Manar Alkhatib, and Khaled Shaalan. 2023. "Artificial intelligence chatbots: A survey of classical versus deep machine learning techniques." *Information Sciences Letters* (12):1217–1233. DOI: 10.18576/isl/120437

[21] Xie, Minzhen. 2019. "Development of Artificial intelligence and effects on financial system." *Journal of Physics: Conference Series* (1187):1–6. DOI: 10.1088/1742-6596/1187/3/032084

[22] Guidolin, Massimo, Monia Magnani, and Paola Mazza. 2021. *Big data e sentiment analysis. Il futuro dell'asset management.* Milan: Egea.

[23] Sanyapong Petchrompo, and Ajith Kumar Parlikad. 2019. "A review of asset management literature on multi-asset systems." *Reliability Engineering & System Safety* (181):181–201. DOI: 10.1016/j.ress.2018.09.009

[24] Adamo Rosa, Domenica Federico, and Antonella Notte. 2022. "Equity mutual funds in healthcare sector: A comparison of performance and risk through an age-cohort analysis." *American Journal of Economics and Business Administration* 14 (1):44–54. DOI: 10.3844/ajebasp.2022.44.54.

[25] Abe, Masaya, and Hideki Nakayama. 2018. "Deep learning for forecasting stock returns in the cross-section." In *Advances in Knowledge Discovery and Data Mining*, 273–284. Cham: Springer. DOI:10.1007/978-3-319-93034-3_22

[26] Rasekhschaffe, Keywan, and Robert Jones. 2019. "Machine learning for stock selection." *Financial Analysts Journal* 75 (3):70–88. DOI: 10.1080/0015198X.2019.1596678

[27] Borghi, Riccardo, and Giuliano De Rossi. 2020. "The artificial intelligence approach to picking stocks." In *Machine Learning for Asset Management: New Developments and Financial Applications*, 115–166. New York: Wiley. DOI:10.1002/9781119751182.ch4

[28] IOSCO. 2021. "The use of artificial intelligence and machine learning by market intermediaries and asset managers." *IOSCO Final Report Review*, Madrid, 7 September.

[29] Bartram, Söhnke M., Jurgen Branke, and Mehrshad Motahari. 2020. "Artificial intelligence in asset management." CFA Institute Research Foundation Literature Reviews, August. DOI: 10.2139/ssrn.3510343

[30] Paiva, Felipe D., Rodrigo T. N. Cardoso, Gustavo P. Hanaoka, and Wendel M. Duarte. 2019. "Decision-making for financial trading: A fusion approach of machine learning and portfolio selection." *Expert Systems Applications* (115):635–655. DOI:10.1016/j.eswa.2018.08.003

[31] Patel, Jigar, Sahil Shah, Priyank Thakkar, and Ketan Kotecha. 2015. "Predicting stock and stock price index movement using Trend Deterministic Data Preparation and machine learning techniques." *Expert Systems with Applications* (42):259–268. DOI: 10.1016/j.eswa.2014.10.031

[32] Ballings, Michel, Dirk Van den Poel, Nathalie Hespeels, and Ruben Gryp. 2015. "Evaluating multiple classifiers for stock price direction prediction." *Expert Systems with Applications* (42):7046–7056. DOI:10.1016/j.eswa.2015.05.013

[33] Ampomah, Ernest A., Zhiguang Qin, and Gabriel Nyame. 2020. "Evaluation of tree-based ensemble machine learning models in predicting stock price direction of movement." *Information* 11 (6):332–352. DOI: 10.3390/info11060332

[34] European Commission. 2019. *Ethics Guidelines for Trustworthy AI, High-Level Expert Group on Artificial Intelligence*, Brussels, 8 April.

[35] European Parliament. 2020. *European Parliament resolution of 20 October 2020 on intellectual property rights for the development of artificial intelligence technologies (2020/2015(INI))*. Brussels, 20 October.

[36] European Parliament. 2020. *European Parliament resolution of 20 October 2020 with recommendations to the Commission on a civil liability regime for artificial intelligence (2020/2014(INL))*. Brussels, 20 October.

[37] European Parliament. 2020. Eur*opean Parliament resolution of 20 October 2020 with recommendations to the Commission on a framework of ethical aspects of artificial intelligence, robotics and related technologies (2020/2012(INL))*. Brussels, 20 October.

[38] European Parliament. 2023. *European Parliament resolution of 16 February 2023 on the establishment of an independent EU ethics body (2023/2555(RSP))*. Strasbourg, 16 February.

[39] European Parliament. 2016. *Council of the European Union: Regulation (EU) 2016/679 of the European Parliament and of the Council of 27 April 2016 on the protection of natural persons with regard to the processing of personal data and on the free movement of such data, and repealing Directive 95/46/EC (General Data Protection Regulation)*. Brussels, 4 May.

[40] European Parliament. 2018. *European Parliament resolution of 14 March 2017 on fundamental rights implications of big data: Privacy, data protection, non-discrimination, security and law-enforcement (2016/2225(INI))*. Brussels, 25 July.

[41] Taddei Elmi, G. and Sofia Marchiafava. 2022. "Sviluppi recenti in tema di Intelligenza Artificiale e diritto: una rassegna di legislazione, giurisprudenza e dottrina." *Rivista italiana di informatica e diritto* (4):123–139. DOI: 10.32091/RIID0106

[42] European Parliament. 1995. *Council of the European Union: Directive 95/46/EC of the European Parliament and of the Council of 24 October 1995 on the protection of individuals with regard to the processing of personal data and on the free movement of such data*. Brussels, 23 November.

[43] European Parliament. 1996. *Council of the European Union: Directive 96/9/EC of the European Parliament and of the Council of 11 March 1996 on the legal protection of databases*. Brussels, 27 March.

[44] European Parliament. 2002. *Council of the European Union: Directive 2002/58/EC of the European Parliament and of the Council of 12 July 2002 concerning the processing of personal data and the protection of privacy in the electronic communications sector (Directive on privacy and electronic communications).* Brussels, 31 July.

[45] European Parliament. 2016. *Council of the European Union: Directive (EU) 2016/680 of the European Parliament and of the Council of 27 April 2016 on the protection of natural persons with regard to the processing of personal data by competent authorities for the purposes of the prevention, investigation, detection or prosecution of criminal offences or the execution of criminal penalties, and on the free movement of such data, and repealing Council Framework Decision 2008/977/JHA.* Brussels, 4 May.

[46] European Parliament. 2016. *Council of the European Union: Directive (EU) 2016/681 of the European Parliament and of the Council of 27 April 2016 on the use of passenger name record (PNR) data for the prevention, detection, investigation and prosecution of terrorist offences and serious crime.* Brussel, 4 May.

[47] European Parliament. 2016. *Council of the European Union: Directive (EU) 2016/943 of the European Parliament and of the Council of 8 June 2016 on the protection of undisclosed know-how and business information (trade secrets) against their unlawful acquisition, use and disclosure.* Brussels, 15 June.

[48] European Parliament. 2016. *Council of the European Union: Directive (EU) 2016/1148 of the European Parliament and of the Council of 6 July 2016 concerning measures for a high common level of security of network and information systems across the Union.* Brussels, 19 July.

[49] European Parliament. 2019. *Council of the European Union: Directive (EU) 2019/1024 of the European Parliament and of the Council of 20 June 2019 on open data and the re-use of public sector information.* Brussels, 26 June.

[50] Scherer, Matthew U. 2016. "Regulating artificial intelligence systems: Risks, challenges, competencies and strategies." *Harvard Journal of Law & Technology* 29 (2):365–400. DOI: 10.2139/ssrn.2609777

[51] European Commission. 2021. *Communication from the Commission to the European Parliament, the Council, the European economic and social Committee and the Committee of the regions 2030. Digital Compass: The European way for the Digital Decade.* COM/2021/118 final, Brussel, 9 March.

[52] Buzzelli, Dario, and Massimo Palazzo. 2022. *Intelligenza artificiale e diritti alla persona.* Pisa: Pacini Giuridica.

[53] Berruti, Federico, Emily Ross, and Allen Weinberg. 2017. "The transformative power of automation in banking." McKinsey & Company, 3 November.

[54] European Parliament. 2020. *The Impact of the General Data Protection Regulation (GDPR) on Artificial Intelligence.* Brussels: European Union.

[55] European Parliament. 2018. *Council of the European Union: Regulation (EU) 2018/1807 of the European Parliament and of the Council of 14 November 2018 on a framework for the free flow of non-personal data in the European Union.* Brussels, 28 November.

[56] European Parliament. 2018. *European Parliament resolution of 16 February 2017 with recommendations to the Commission on Civil Law Rules on Robotics (2015/2103(INL)).* Brussels, 18 July.

[57] European Commission. 2018. *Guidelines on Automated individual decision-making and Profiling for the purposes of Regulation 2016/679.* Brussels, 22 August.

[58] European Commission. 2019. *Communication from the Commission to the European Parliament, the Council, The European Economic and Social Committee and the Committee of the Regions Empty. Building Trust in Human-Centric Artificial Intelligence.* COM/2019/168 final. Brussels, 8 April.

[59] European Commission. 2020. *White Paper on Artificial Intelligence—A European approach to excellence and trust,* Brussels, 2 February.

[60] European Commission. 2021. *Proposal for a regulation of the European Parliament and of the Council laying down harmonized rules on artificial intelligence (Artificial Intelligence Act) and amending certain union legislative acts. COM/2021/206 final.* Brussels, 21 April.

[61] European Parliament. 2020. *Regulation (EU) 2020/852 of the European Parliament and of the Council of 18 June 2020 on the establishment of a framework to facilitate sustainable investment, and amending Regulation (EU) 2019/2088.* Brussels, 22 June.

[62] Yeung, Karen. "Response to European Commission white paper on artificial intelligence." *SSRN Electronic Journal*, 13 June:1–24. DOI: 10.2139/ssrn.3626915

15 Machine Learning Applications in Industry 4.0
Opportunities and Challenges

Ezekiel Tijesunimi Ogidan[1], Oluwaseun Priscilla Olawale[2], and Kamil Dimililer[3]

1 Department of Computer Engineering, Near East University, Nicosia, North Cyprus, via Mersin 10, Turkey

2 Department of Software Engineering, Near East University, Nicosia, North Cyprus, via Mersin 10, Turkey

3 Department of Electrical and Electronic Engineering, Near East University, Nicosia, North Cyprus, via Mersin 10, Turkey Applied Artificial Intelligence Research Centre (AAIRC), Near East University, Nicosia, North Cyprus, via Mersin 10, Turkey

15.1 INTRODUCTION

Over the years, as the world has moved towards globalization and capitalism, the demand for high-quality goods and services made available the soonest at cheap prices has increased. There have also been a number of further requirements of manufacturing, such as environmental sustainability, responsible production, worker safety, and reduced energy consumption, even up to product acceptance at the retail level [1]. The intense competition brought on by capitalism and pressing demands caused by consumerism put enormous pressure on manufacturers to optimize their manufacturing processes. Parallel to this is the increase in the collection and availability of data [2]. This demand for large amounts of data in a timely fashion has even caused the creation of dedicated information enterprises and has led many organizations to establish data analysis departments. In line with these developments, in modern-day manufacturing, large amounts of data are collected using various data acquisition techniques from the production lines to distribution links and even up to consumers [3]. These large amounts of data create an opportunity for leveraging the capabilities of machine learning. This chapter aims to give some insight into the importance of leveraging machine learning methods for manufacturing, the methods

DOI: 10.1201/9781003405870-15

being used so far, and their level of effectiveness, as well as ways that machine learning can be integrated into manufacturing and further research to be done on the topic.

15.1.1 DEFINITION OF MACHINE LEARNING FOR MANUFACTURING

Machine learning is responsible for major advancements in many fields. The ability for computer systems to learn from data by analyzing patterns, mapping the patterns to corresponding consequences, and making deductions with almost no human help makes it a useful tool in many fields. The more specific machine learning tasks, such as classification and computer vision, have also been very instrumental in advancements in many fields.

In the case of manufacturing, it has the potential to be one of the most important tools in the advancement to Industry 4.0, which may also be referred to as intelligent industry [4]. One of the advantages of intelligent industry is cost reduction [5]. Some of the problems aimed to be solved with the move to Industry 4.0, such as quality control, defect detection, and predictive maintenance, could be solved with correct applications of machine learning algorithms. Table 15.1 shows a breakdown of manufacturing tasks and problems and what corresponding machine learning algorithms can be applied to them.

15.1.2 IMPORTANCE OF MACHINE LEARNING IN MANUFACTURING

Machine learning models are necessary to resolve critical issues in industries [23]. Combining manufacturing and machine learning has spurred the use of smart devices [24] to create factories that gather vast amounts of data on production. The

TABLE 15.1
Manufacturing Tasks and Corresponding Applicable Machine Learning Algorithms

Manufacturing Task	Machine Learning Algorithms	References
Predictive maintenance	Decision trees, support vector machines, artificial neural networks, multiple linear regression, fuzzy logic, logistic regression, reinforcement learning	[2, 6–9]
Quality control	Artificial neural networks, linear regression, support vector machines, decision trees	[10–16]
Production scheduling	Reinforcement learning, genetic algorithms, simulated annealing, linear regression	[17]
Supply chain management	Neural networks, recurrent neural networks, decision trees, support vector machines, reinforcement learning, linear regression	[18–20]
Process optimization	Bayesian optimization, artificial neural networks, support vector machines, decision trees	[14, 15, 21, 22]

field of machine learning has had an important role in this trend, as it provides the means to process and analyze this data, generating valuable insights that can improve manufacturing efficiency with minimal resource consumption. In particular, machine learning models excel at generating predictive insights that enable the identification of complex patterns in manufacturing data. Another way to manage enormous amounts of data is using data mining [25] to help machines make optimum decisions [26]. This capability is essential for the development of intelligent decision support systems that can assist with tasks like predictive maintenance, quality control, process optimization, supply chain management, and so on. Although machine learning has been applied to many manufacturing tasks, many challenges remain, including managing data, understanding data patterns, and enabling real-time processing. Other topics, such as edge computing and cybersecurity, are also relevant to the ongoing development of smart manufacturing [27]. These capabilities make machine learning undeniably important for the advancement of manufacturing to Industry 4.0.

15.1.3 BRIEF OVERVIEW OF THE CHAPTER

So far, we have discussed the meaning of machine learning for manufacturing and why machine learning is important for manufacturing and the evolution of Industry 4.0.

Section 15.2 gives an overview of machine learning techniques. Details of how they work and their distinguishing factors will be discussed.

In Section 15.3, some specific applications of machine learning to manufacturing will be highlighted, pointing out what algorithms can be used for specific aspects of manufacturing. This section discusses the subtopics of quality control and defect detection, predictive maintenance and asset management, inventory management and supply chain optimization, and productive optimization and process control. We also discuss the benefits of machine learning for each of these aspects of manufacturing.

Section 15.4 addresses the challenges faced in implementing machine learning in manufacturing. The challenges discussed in this chapter include acquisition and management of data for intended machine learning systems, data quality and cleaning, model selection and optimization, and integration of developed machine learning systems to existing manufacturing systems.

Section 15.5 features case studies on the areas of manufacturing discussed in Section 15.3. We discuss real-life cases where machine learning has been used for these aspects of manufacturing, the intended effects at the start of development of these systems, and the outcomes of the developments.

In Section 15.6, new directions and future work being done on machine learning systems for manufacturing are discussed. The intended effects of these new developments on manufacturing and how much work has already been done to make these developments a reality are discussed. The directions for research that are discussed include explainable AI, edge computing and IoT, human–robot collaboration, and digital twins.

In conclusion, Section 15.7 summarizes key points discussed in this chapter, the implications of incorporating machine learning to manufacturing, and research recommendations for the future.

15.2 MACHINE LEARNING TECHNIQUES FOR MANUFACTURING

Machine learning has led to major advancements in many fields, especially medicine, forestry [28], the textile industry [7], education, and image processing [29, 30], where it has been applied. The ability for computer systems to learn from data by analyzing patterns, mapping the patterns to corresponding consequences, and making deductions with almost no human help makes it a very useful tool in many fields. The more specific machine learning tasks, such as classification and computer vision, have also been very instrumental in advancements in many fields. Machine learning techniques allow computer systems to learn patterns from data and draw conclusions or give predictions on future outcomes based on this obtained knowledge. Typically, the machine learning model is trained on data, known as the training data, and then tested to measure its level of accuracy. This training process aims to establish a relationship between the data and a target or expected output by mapping the data to the output. As will be shown later in this section, this target output might sometimes be predefined and sometimes not.

Machine learning algorithms are often classified based on the learning method applied as supervised, unsupervised, or reinforcement learning. Another category called deep learning describes a class of machine learning techniques where a neural network is developed with multiple layers to emulate the human brain.

15.2.1 SUPERVISED LEARNING

Supervised learning is a technique of machine learning algorithms where the input data is known. In this example, the data to be used as input in training the algorithm is known and often labeled to denote a mapping to the expected output [31]. During the learning process, the model aims to reduce the variance between the obtained output of each iteration (epoch) and the expected output. This difference is obtained using the loss function. When the error gets to an acceptable level, the model is considered trained and can be used. Supervised learning is often used for regression and classification problems.

15.2.2 UNSUPERVISED LEARNING

Unsupervised learning is a technique of machine learning used with input data that has no known labels or related output. As such, the mapping between the input data and the obtained outcome is not known but is rather determined by the model. Unsupervised learning is often used to detect unknown trends in data, and as output, they categorize or cluster this data into corresponding groups based on the detected patterns. It is a very helpful technique for cases where patterns or relationships are not obvious and might even be misinterpreted as unstructured noise [32]. It also has the unique ability to detect new associations [33]. Unsupervised learning is often used for clustering and association problems.

15.2.3 REINFORCEMENT LEARNING

Reinforcement learning is a technique of machine learning that is based on trial and error. The model, known as the intelligent agent, is tasked with finding the best possible solution to a problem and is rewarded or punished based on its decisions. The intelligent agent then adjusts its behavior accordingly to get a reward and achieve the desired outcome. Reinforcement learning is best suited for cases that require adaptive decision making where the intelligent agent is able to adjust accordingly to changes in its immediate environment or entirely new environments [34]. Reinforcement learning can be applied to problems such as decision making, where a clear reward can be defined for decisions within the space of the environment that bring the intelligent agent closer to the desired outcome and corresponding punishments for decisions that get the agent further from the desired outcome.

15.2.4 DEEP LEARNING

Deep learning essentially features an advanced neural network. These consist of three groups of layers—input, hidden, and output layers—that are interconnected in a way to simulate the connection of neurons in the human brain. Deep learning models contain a vast number of nodes contained in multiple layers and interconnected. This allows the model to identify complex structures that could be hidden in very large amounts of data [35, 36]. Deep learning is often used for classification and detection problems.

15.3 APPLICATIONS OF MACHINE LEARNING IN MANUFACTURING

Machine learning has significant potential for reducing the workload and increasing the efficiency of manufacturing engineers, who are increasingly faced with a growing variety of products [37]. The following sections describe some specific tasks within the scope of manufacturing where machine learning can be applied.

15.3.1 QUALITY CONTROL AND DEFECT DETECTION

The ability to detect and correct defects before they impact the end product is critical to ensuring product quality, reducing production costs, and maintaining customer satisfaction. When mechanical items are manufactured, defects frequently appear as a result of factors like poor design, malfunctioning machine production equipment, and adverse working circumstances. These flaws put consumers' safety at risk by raising expenses for businesses, shortening product lifespans, and wasting resources. Defects must thus be found for businesses to increase product quality and lower costs without compromising production. Due to its great accuracy and efficiency in operating in unsuitable conditions, automatic defect-detection technology is superior to manual detection. The development of defect-detection technologies can lower production costs, boost productivity, and create the framework for a thoughtful transformation of the manufacturing sector. Douard et al. [38] demonstrated an application of machine learning with electron beam melting (EBM), a new technology in additive manufacturing [39]. Various algorithms were tested during the production of a component to determine which could best predict the risks of deformation versus the quality of the finished product. Specifically, the study focused on a few parameters of support structures, which were varied to assess the algorithms' effectiveness.

Machine learning models can be trained on large datasets of historical manufacturing data to detect trends and anomalies indicating the presence of defects. They could be used to analyze images of manufactured products to detect defects. For example, a machine learning algorithm could be trained on a large dataset of images of good and defective products to identify the characteristics that distinguish between the two. Once trained, the algorithm could be used to analyze new images of products in real time to identify defects and alert operators to take corrective action.

On the other hand, many manufacturing processes involve the use of sensors to monitor equipment performance and detect anomalies signaling defects. Machine learning models can be trained on data from these sensors to detect patterns and anomalies that indicate defects. After this, the algorithm could be used to analyze new sensor data in real-time to identify defects and alert operators to take corrective action. By automating the detection process, manufacturers can reduce manual inspections, which can be time consuming and are often inaccurate. This can result in significant cost savings and improve overall production efficiency.

According to He et al. [40], the processing of metal workpieces can lead to uncontrollable defects on their surfaces, which can impact the quality of the finished product. Detecting these surface defects is important for ensuring high-quality production and efficiency. Despite the availability of various detection methods, the volume and irregularity of these defects in metal workpieces often results in manual inspections. This approach can lead to missed or false inspections, highlighting the need for improved defect detection methods.

15.3.2 PREDICTIVE MAINTENANCE AND ASSET MANAGEMENT

According to Singh et al. [41], the goal of maintenance management is to ensure optimum machine availability and reliability for producing goods of desired quality in a certain time frame at an affordable cost. It also aims to optimize the entire asset life cycle. Any organization's cost and operating profits are significantly impacted by the maintenance decisions and strategies. The industry has embraced a number of strategies, such as preventive and predictive maintenance. Predictive maintenance entails the forecasting of equipment diagnostics and prognostics. Preventive maintenance, on the other hand, uses AI-enabled smart machines to make decisions about how to prevent problems. In order to perform preventive maintenance, sensors, converters, processors, and other intelligent devices are used to collect and process machine health indications. Machine learning techniques are also used for prognosis and defect categorization. Preventive maintenance has become more common with the introduction of Industry 4.0, providing real-time health monitoring as well as targeted early alerts of issues.

In order to forecast equipment breakdowns and lower the likelihood of downtime, machine learning algorithms can assess real-time sensor data to improve asset performance and lower maintenance costs. These algorithms may predict possible breakdowns well in advance by finding patterns and anomalies in sensor data, enabling operators to perform preventive maintenance. By determining the best time to execute maintenance tasks based on equipment usage, wear and tear, and other factors, machine learning can help to optimize maintenance schedules. In general, industrial firms are using machine learning more and more to improve asset performance and save maintenance costs.

15.3.3 Inventory Management and Supply Chain Optimization

The requirements for a product are frequently uncertain due to the impact of the rapidly changing marketing environment. Additionally, the uncertainty of the supply chain with respect to these requirements may result in an uncertain throughput or yield, thereby making the initial storage of the product uncertain as well. In cases where production and inventory systems are subject to multiple uncertainties, decision-making regarding whether or not to produce, how to produce, and the optimal production batch size becomes critical for managers. To ensure optimal outcomes, managers must carefully consider these issues [42, 43] carried out a study to improve stock management in the industry by predicting the estimated time of arrival. Storage and processing of vast amounts of data has become more manageable due to recent technological advancements. To conduct this research, the Hadoop infrastructure was utilized, which includes HDFS for data storage and Hive for data querying. Spark and machine learning modules were employed for data analysis and building data models, which will serve as the dataset for estimating the stock's arrival time. Effective inventory management is critical for organizations that depend on the acquisition and conversion of raw materials into useful products. Therefore, it is vital to ensure that all parties involved in the process have access to transparent data concerning stock availability, required storage space, and individual stock arrival dates. To optimize the stock management system, it is essential to develop an access point that stakeholders can use as a reference to better understand the current stock and inventory. This information will enable them to make informed decisions.

At present, businesses are placing a significant emphasis on creating sustainable manufacturing supply chains that minimize environmental impact in the production of goods and products. By utilizing machine learning-based supply chain methods, numerous sectors can enhance safety for workers, consumers, communities, and products alike. In order to manage the enormous volumes of data generated by industrial activities and to find a solution for sustainability issues, machine learning is an essential tool in supply chain management. By statistically analyzing published scientific papers, Yadav et al. [44] used a neutral and objective bibliometric review methodology to assess the state of the field and identify potential paths for further research on the application of machine learning in the development of sustainable supply chains. Their work shows how supply chain difficulties can be handled with machine learning techniques.

Machine learning has proven to be a valuable tool in the fields of inventory management and supply chain optimization. To make data-driven decisions regarding inventory management, demand forecasting, and supply chain optimization [45], businesses can use machine learning algorithms. By analyzing historical data, machine learning algorithms can determine appropriate inventory levels, enhance inventory accuracy, and forecast future demand trends, resulting in cost reductions and improved operational efficiency. Machine learning can assist companies in identifying and managing supply chain risks, such as interruptions and delays, by providing real-time insights into supplier performance and logistics. Machine learning applications in inventory management and supply chain optimization have become increasingly critical for organizations looking to maintain a competitive edge in the global market.

15.3.4 PRODUCTION OPTIMIZATION AND PROCESS CONTROL

In the modern industrial field, certain complex processes consist of multiple homogeneous subsystems with varying parameters and several control parameters. When these subsystems exhibit nonlinear characteristics, they become challenging to manually adjust. Wang et al. [46] developed a distributed immune algorithm based on the notion of human immune networks for managing these intricate industrial processes. The program can increase the control knowledge through immune evolution and learning and can learn control information from many operators online. The outcomes of experiments and simulations show that the algorithm outperforms manual control and is capable of distributed learning. With the help of this method, complicated industrial processes that use homo-structural variable-parameter systems can be controlled intelligently.

15.4 CHALLENGES IN IMPLEMENTING MACHINE LEARNING IN MANUFACTURING

As alluring as the prospects of leveraging machine learning tools for manufacturing might be, there are some challenges that are faced in the process. These challenges either affect the adoption of machine learning technologies and methodologies or the effectiveness of their application for manufacturing purposes. Some of these challenges discussed in this section include the acquisition and management of data to be used with the models, the quality of the data, the selection of the model, and its integration into the manufacturing system.

15.4.1 DATA ACQUISITION AND MANAGEMENT

It has already been established that a major component for developing effective machine learning models is data. For high quality systems to be developed, enormous amounts of data need to be acquired accurately, precisely, and quickly. As a result, automated data acquisition processes are often preferred. Resources such as sensors and actuators as well as automated identification systems such as barcodes and RFID are recommended [47]. A number of other components could also be used to form a network for the collection, transmission, and preprocessing and storage of the data acquired. The Industrial Internet of Things (IIoT) is an important component to this development.

Since most of the decisions to be made within the space of manufacturing could be affected by lots of factors from the managerial level of the organization to production and supply chains and even further on to retail, it is necessary that different types of data be acquired at these different levels and allowed to be used in an integrative way. Data acquisition and management processes have also seen their fair share of innovation, with dedicated information enterprises developing data acquisition and management systems to be used to gather data about products, raw materials, machinery, staff working conditions, and consumer feedback. The management systems are also sophisticated and can combine these varying types of data intricately to help the machine learning models arrive at inferences that are not ordinarily obvious. Management standards like MTConnect are used to aid communication between these devices, machinery, and enterprises and help process the data that is acquired.

15.4.2 DATA QUALITY AND CLEANING

Data quality and cleaning are crucial steps in preparing data for machine learning applications in manufacturing. The effectiveness of machine learning models largely depends on the quality of its input data. Bad data can lead to inaccurate predictions, biased models, and unreliable insights. Therefore, it is essential to ensure that the data is accurate, complete, consistent, and relevant to the problem at hand.

Raw manufacturing data can be noisy and incomplete and contain outliers, which could affect the accuracy of the developed model. For this reason, data cleaning is needed. This is the process of finding and fixing errors, inconsistencies, and missing values in the data. It entails eliminating duplicates, adding missing values, eliminating outliers, and adjusting data types. Data cleaning can be a time-consuming and resource-intensive procedure.

In manufacturing, data cleaning can be particularly challenging because of the large amounts and complexity of the data. The data often comes from a wide range of sources, such as sensors, machines, and humans, and may contain noise and errors. Therefore, it is important to have a systematic approach to data cleaning that involves domain experts and uses appropriate tools and techniques.

One approach to data cleaning is to use statistical methods to identify and remove outliers and errors. For example, a common method is to use mean and standard deviation to detect outliers and replace them with a suitable value. Another approach is to use machine learning algorithms, such as clustering and classification, to identify patterns and anomalies in the data.

15.4.3 MODEL SELECTION AND OPTIMIZATION

Selecting the appropriate machine learning model and optimizing its parameters can be challenging. The first step is to define the problem and identify the type of learning required, such as supervised, unsupervised, reinforcement, or deep learning. Then, several models can be evaluated and compared to select the one that best fits the problem. The selection process should consider several factors, including the complexity of the model, interpretability, scalability, and computational efficiency.

After selecting the appropriate machine learning model, the next step is to adjust its hyperparameters to achieve the best performance. Hyperparameters are parameters that are set before the training process, such as the learning rate or regularization strength. This process is known as model optimization. There are a number of techniques used for model optimization such as grid search, random search, and Bayesian optimization.

Grid search involves exhaustively testing all combinations of all feasible hyperparameters. Random search involves randomly sampling all feasible hyperparameters. Bayesian optimization uses a prospective model to predict the performance of different hyperparameter combinations and selects the best combination based on these predictions.

Feature selection is another important step in the development process. Feature selection involves selecting the most relevant features from the available data to considerably improve accuracy. There are various methods for feature selection, such as wrapper methods, filter methods, and embedded methods.

Wrapper methods evaluate the model's performance with different feature subsets and select the subset that produces the best results. Filter methods evaluate the relationship between features and the output and select the most relevant features based on these relationships. Embedded methods use regularization to select the most relevant features during the model training process.

15.4.4 INTEGRATION WITH EXISTING MANUFACTURING SYSTEMS

Implementing machine learning systems in existing manufacturing processes requires careful consideration of the challenges associated with integration. Some of the challenges and best practices for integrating machine learning systems with existing manufacturing systems are as follows.

Challenges:

1. Data integration: Machine learning models require high-quality data to be effective. Integrating machine learning with existing systems can be challenging if the data is scattered across different sources, in different formats, or with varying degrees of quality. Data cleansing and standardization are crucial steps in preparing data for machine learning models.
2. Compatibility issues: Machine learning models must be compatible with existing software and hardware systems. Integration can be complicated if the machine learning model requires specific software or hardware configurations that are not compatible with the existing system.
3. Security concerns: Manufacturing systems are often subject to cybersecurity threats, and the integration of machine learning models can increase the risk of security breaches. The models may require access to sensitive data, which must be protected from unauthorized access or misuse.

Best Practices:

1. Understand the existing system: Before integrating machine learning systems, it is essential to have a thorough understanding of the existing manufacturing system. This includes understanding the data sources, data quality, and processes involved.
2. Start with a small project: Starting with a small pilot project can help identify and address integration challenges early on. A small project also allows for testing and validation of the machine learning models before full-scale integration.

3. Collaborate with experts: Collaboration with experts in machine learning and manufacturing can help identify potential challenges and develop solutions. This includes involving domain experts in the data cleansing and standardization process.
4. Ensure compatibility: Machine learning models should be compatible with the existing software and hardware systems. This can be achieved by selecting machine learning tools that are compatible with existing systems or by upgrading existing systems.
5. Address security concerns: Security concerns should be addressed from the beginning of the integration process. This includes identifying potential threats and setting up security measures to protect sensitive data.

The integration of machine learning systems with existing manufacturing systems can be challenging but is essential for improving manufacturing efficiency, reducing downtime, and improving product quality. By understanding the challenges and implementing best practices, manufacturers can successfully integrate machine learning systems into their existing processes.

15.5 CASE STUDIES

15.5.1 QUALITY CONTROL AND DEFECT DETECTION

Douard et al. [38] demonstrated an application of machine learning with EBM, a new technology in additive manufacturing. Various algorithms were tested during the production of a component to determine which could best predict the risks of deformation versus the quality of the finished product. Specifically, the study focused on a few parameters of support structures, which were varied to assess the algorithms' effectiveness.

The manual inspection of printed circuit boards (PCBs) is a lengthy and error-prone process, which may result in some defects being missed. As a result, machine learning and deep learning algorithms have been implemented to enhance the quality of inspection. Recently, several new deep learning-based methods have been put forth for more effective inspection of PCBs. Aggarwal et al. [48] provided a comprehensive overview of machine learning and deep learning techniques used for automatic detection of defects in PCBs, discussing the latest models used to inspect single-layer and multi-layer PCBs.

In the textile industry, spotting fabric flaws is essential since they have a direct impact on the value of the finished product. Traditional detection techniques, like physical inspection, on the other hand, are ineffective and inaccurate, and they put the health of the workers at risk. Consequently, a system for automatically detecting fabric defects is needed. As a result, a new strategy called the "edge-cloud" collaborative fabric defect detection architecture was put forth by Zhao et al. [49]. The design offers closed-loop optimization for this task with sensing, analysis, decision-making, and execution by utilizing an edge layer, platform layer, and application layer. They contend that the suggested architecture is better suited to adapting detection environments and can increase their machines' environmental flexibility.

The detection of surface defects in steel manufacturing is crucial, but manual visual inspection is often used due to the complexities of automated visual inspection methods. However, automated methods could provide faster and more cost-effective quality control. An approach to tackle the expensive process of preparing annotated training data was proposed by Damacharla et al. [50] using the transfer learning-based U-Net (TLU-Net) framework. ResNet and DenseNet encoders were explored and compared using random initialization and pre-trained networks from the ImageNet dataset. The tests were conducted on Severstal data and showed that transfer learning outperformed random initialization in defect classification and segmentation, with a greater gain as the training data decreased. The convergence rate with transfer learning was also found to be better than random initialization.

A useful framework for advanced quality-based process control (AQPC) for linked manufacturing processes was put out by Schmitt et al. [51]. Based on the recorded process parameters, this framework used machine learning techniques to forecast the anticipated product quality. The best control decisions were then derived using the projected quality information. In order to achieve this, the authors created a mathematical optimization model that factors in possibilities like order reassignment and process parameter adaptation in order to identify the ideal set of control decisions. The optimization is gradually deconstructed into steps that enable integration into manufacturing processes that are application-specific.

15.5.2　Predictive Maintenance and Asset Management

Predictive maintenance is a crucial aspect of asset management in various industries, as it can help reduce downtime and increase revenue. Singh et al. [41] proposed a data-driven approach that utilizes a multi-classifier model to implement predictive maintenance in a manufacturing plant with 100 machines. The objective was to identify potential failures, their nature, and the causes that contribute to them to prevent machine downtime and idle time. To achieve this, the study employed the use of two machine learning algorithms: gradient boosting tree classifier and random forest classifier. The proposed approach provided a reliable means of early failure detection and preventive maintenance planning for optimal asset management in the manufacturing industry.

15.5.3　Inventory Management and Supply Chain Optimization

Having effective supply chain operations is essential for firms to deliver their finished goods to clients in today's cutthroat market. When retailers place orders, they must carefully balance inventory levels and account for future demand. However, many supply chains struggle to maintain inventory levels that can match consumer demand, which causes ordering processes to fluctuate more frequently [45].

In [52], the current status of research identifying potential areas for exploration in the use of machine learning for sustainable supply chain management was explored. It was shown that machine learning methods can indeed help the industrial sector overcome supply chain challenges.

Raghuram et al. [53] optimized inventory levels in the face of demand uncertainty and supply complexity for biomedical equipment, particularly knee implants, by employing inventory positioning and reliable machine learning-based forecasting. A discrete event simulation model that takes into account a two-level supply chain analyzes the product flow. Arena is used to model and simulate inventory replenishment. A number of parameters, including the reorder point, order quantity, supplier lead time, and inventory costs, are taken into consideration. Then, OptQuest is used to optimize these parameters in order to reduce total costs and backorders. To meet service level limitations, the model calculates the stock inventories needed at each echelons to stay safe. The underlying source of this issue is demand unpredictability, which can be decreased by using more accurate forecasts.

In a dynamic supply chain context, [9] presented a new strategy for optimizing multi-echelon inventory policy. The suggested method handles the broad range of planning variables and the intricate dynamics of the supply chain environment by combining deep reinforcement learning with local search strategies. The effectiveness of three approaches—scatter search, deep Q-learning, and the suggested combined strategy—was compared by the authors. The simulation results showed that, in terms of profitability, adaptability, and solution time efficiency, the combined strategy performs better than the other techniques.

15.5.4 PRODUCTION OPTIMIZATION AND PROCESS CONTROL

One powerful tool is Petri nets modeling for the performance analysis of industrial systems, and it was proposed by Kumar et al. [54]. The capacity to handle real working settings is the primary aspect that sets it apart from other widely used simulation techniques in terms of effectiveness. In order to depict the actual behavioral pattern of the various installed sub-systems in the plant, the Petri nets modeling method uses data gathered from such a production system, which is typically full of uncertainty. By operating the plant virtually, the Petri nets–based simulation method ensures the subsystems' availability for a lengthy period of time. The analysis's findings can be used to pinpoint the subsystems that have the biggest impact on the system's availability. By creating separate maintenance schedules for each subsystem, production loss caused by a subsystem's unavailability can be minimized.

The quality of polyester fiber in the direct spinning process is significantly affected by its performance during the polymerization process. Intrinsic viscosity, as a crucial process parameter, serves as an important indicator of the polyester's quality. As a result, it is essential to accurately predict the intrinsic viscosity of polyester to guarantee the quality of the final product. To optimize this process, Wu et al. [55] proposed a model that utilizes the K-nearest-neighbor algorithm, binary coding, and the atom search optimization algorithm to select features from industrial data. The weight, threshold, and activation function of an improved extreme learning machine was then optimized using the selected data set. A prediction model was established, which was used in the production of polyester fiber. The model had high accuracy levels, proving that it could be used in the industrial production process.

15.6 FUTURE DIRECTIONS IN MACHINE LEARNING FOR MANUFACTURING

While there are some real-life use cases of machine learning in manufacturing that are already being implemented, there are still a number of ways that these machine learning systems can be improved, augmented, or supported to make them more productive and efficient. These areas could be built up through research, allowing for even more innovation and development. Some important new directions for machine learning in manufacturing are as follows.

15.6.1 EXPLAINABLE AI

Explainable artificial intelligence (XAI) is an emerging field that aims to make artificial intelligence systems more transparent and interpretable. In the context of manufacturing, XAI can help engineers and operators understand the decisions made by AI systems and ensure that those decisions align with the company's goals and objectives.

One of the benefits of XAI in manufacturing is its ability to identify and diagnose issues in production quickly and accurately. By providing detailed explanations of how an AI system arrived at a decision, engineers and operators can identify potential sources of error or bias in the data used to train the model and make adjustments to improve the accuracy and reliability of the system.

XAI can also help to improve the overall efficiency of manufacturing operations. For example, an AI system that is designed to optimize the scheduling of production lines can provide detailed explanations of its decisions, such as why it chose a particular sequence of tasks or why it assigned certain workers to specific jobs. This can help managers to identify opportunities to streamline processes and reduce waste.

Another important application of XAI in manufacturing is in predictive maintenance. By analyzing data from sensors and other sources, AI systems can predict equipment failures and schedule maintenance before they occur. With XAI, engineers and operators can better understand the reasoning behind these predictions and take action to prevent failures before they happen.

15.6.2 EDGE COMPUTING AND IOT

Edge computing and the Internet of Things (IoT) are revolutionizing machine learning applications in manufacturing. By enabling data processing and analysis at the edge of the network, closer to data sources, manufacturers can leverage real-time insights and improve the accuracy of machine learning models. Here are some examples of how edge computing and IoT are used in machine learning applications in manufacturing:

1. Predictive Maintenance: Maintenance staff can schedule repairs in advance using machine learning models that are trained to anticipate when equipment is likely to fail. Edge computing and IoT provide real-time analysis of data from sensors and other sources, resulting in more precise predictions and lower downtime risks.

2. Quality Control: Machine learning models can identify defects in products, reducing waste and improving product quality. With edge computing and IoT, manufacturers can collect data from sensors and cameras in real time, allowing machine learning models to quickly identify defects and take corrective action.

3. Process Optimization: Machine learning can help optimize manufacturing processes, reducing energy usage and improving efficiency. With edge computing and IoT, data from sensors and other sources can be analyzed in real time, allowing machine learning models to identify opportunities for optimization and make adjustments on the fly.

4. Supply Chain Optimization: Machine learning models can help optimize supply chain operations, improving logistics and reducing costs. With edge computing and IoT, manufacturers can collect data on inventory levels, shipping times, and other factors, allowing machine learning models to optimize logistics and reduce delays.

As these technologies continue to evolve, they will become even more essential for manufacturers seeking to stay ahead in the fast-paced world of modern manufacturing.

15.6.3 Human–Robot Collaboration

Human–robot collaboration (HRC) holds promise for manufacturing. By combining the strengths of humans and robots, manufacturers can achieve greater efficiency, productivity, and flexibility. Here are some examples of how HRC is used in manufacturing:

1. Co-Bots: Collaborative robots, or "co-bots", are designed to work alongside humans, often performing repetitive or physically demanding tasks. Co-bots are smaller and more lightweight than conventional industrial robots, which makes them easier to integrate into existing production lines. By automating these tasks, manufacturers can reduce worker fatigue and improve safety while freeing up human workers to perform more complex tasks.

2. Augmented Reality: Augmented reality (AR) can be used to enhance human–robot collaboration by providing real-time information about the production process. For example, AR can be used to display instructions or schematics on a worker's display while they are assembling a product. This technology can improve the accuracy and speed of assembly tasks while reducing the risk of errors.

3. Skill-Based Task Allocation: In some cases, robots and humans can be assigned tasks based on their individual strengths and abilities. For example, a robot may be assigned to perform a repetitive task that requires precision, while a human worker may be assigned to perform a task that requires creativity or problem-solving skills. By assigning tasks in this way, manufacturers can optimize the use of both robots and human workers, improving efficiency and productivity.

4. Safety Monitoring: Robots can be equipped with sensors and other features to ensure safety during work with humans. For example, a robot may be programmed to slow down or stop if it detects that a human worker is in its path. By ensuring the safety of human workers, manufacturers can create a more collaborative and productive work environment.

15.6.4 DIGITAL TWINS

The use of digital twins with machine learning is a powerful combination that is transforming manufacturing. Digital twins are virtual replicas of physical objects, such as machines or products, that can be used to simulate and analyze their behavior in real time. Machine learning algorithms can analyze data from digital twins and optimize manufacturing processes. Here are some examples of how digital twins with machine learning are used in manufacturing:

1. Predictive Maintenance: Digital twins can be used to simulate the behavior of machines and predict when they are likely to fail. In order to identify trends that point to potential problems, machine learning algorithms can be employed to examine data from digital twins, such as temperature or vibration. To plan maintenance and prevent unscheduled downtime.
2. Quality Control: Digital twins can be used to simulate the behavior of products and detect defects before they occur. Data from digital twins, like product specifications or sensor data, can be analyzed to identify patterns that indicate potential defects. This information can be used to improve the quality of products and reduce waste.
3. Process Optimization: Digital twins can be used to simulate manufacturing processes and optimize them for maximum efficiency. Data from digital twins, like sensor data or process parameters, can be used to identify opportunities for optimization. This information can be used to make real-time adjustments to manufacturing processes and improve efficiency.
4. Supply Chain Optimization: Digital twins can be used to simulate supply chain operations and optimize them for maximum efficiency. Machine learning can be used for analyzing data from digital twins, like inventory levels or shipping times, to identify opportunities for optimization. This information can be used to make real-time adjustments to supply chain operations and reduce costs.

15.7 CONCLUSION

15.7.1 SUMMARY OF KEY POINTS

Machine learning is being increasingly used in manufacturing to improve efficiency, productivity, and quality. Some key applications of machine learning in manufacturing discussed in this chapter include predictive maintenance, quality control, process optimization, and supply chain optimization. In these areas, machine learning has greatly improved efficiency, giving organizations that have adopted machine learning methodologies for these applications an edge in their respective industries.

Machine learning also provides flexibility to the manufacturing process. Machine learning can be used for analysis of data from sensors, digital twins, and other sources to identify patterns and make real-time adjustments to manufacturing processes. These changes or adjustments would often take longer to implement otherwise, as manufacturing or management staff would often be less sensitive to the patterns or incidents machine learning models could detect and would often take longer to analyze them and make decisions before any decisions can be made.

Furthermore, by combining machine learning with other technologies like IoT, edge computing, and human–robot collaboration, manufacturers can achieve even greater benefits.

15.7.2 Implications for the Future of Manufacturing

Machine learning has significant implications for the future of manufacturing. As manufacturers increasingly adopt these technologies, we can expect to see a number of changes in the industry. Here are some potential implications of machine learning for the future of manufacturing:

1. Increased Automation: Machine learning can enable more automation in manufacturing processes, reducing the need for human intervention in certain areas. This could lead to increased efficiency, reduced costs, and improved safety in manufacturing facilities.
2. Improved Product Quality: Machine learning can help manufacturers detect defects and anomalies in products more quickly and accurately, leading to improved product quality.
3. Increased Flexibility: Machine learning can enable more flexibility in manufacturing processes by allowing for real-time adjustments and optimization. This could help manufacturers to respond more quickly to changes in demand, reducing lead times and improving customer satisfaction.
4. Enhanced Sustainability: Machine learning can help manufacturers to reduce their environmental impact by optimizing energy consumption and reducing waste. This could promote more sustainable manufacturing practices and help manufacturers to meet environmental regulations and customer demands.
5. New Business Models: Machine learning could enable new business models in manufacturing, such as predictive maintenance services or product-as-a-service models. This could provide new revenue streams for manufacturers and change the way that products are designed, produced, and sold.

While there may be some challenges associated with adopting these technologies, the benefits are significant and are likely to drive continued investment and development in this area.

15.7.3 Recommendations for Further Research

As machine learning continues to transform the manufacturing industry, there are a number of areas where further research can be carried out to fully realize the potential of these technologies. Here are some recommendations:

1. Integration with Other Technologies: There is a need to investigate how machine learning can be used with other technologies such as IoT, edge computing, and digital twins to improve manufacturing processes further.
2. Data Acquisition and Management: More research needs to be done to find more effective ways to acquire and manage the vast amounts of data generated during manufacturing and to guarantee the quality and integrity of the data.
3. Explainability and Transparency: There is a need to investigate how machine learning models can be made more explainable and transparent, particularly in critical applications such as quality control and safety.
4. Ethics and Privacy: As machine learning becomes more ubiquitous in manufacturing, there is a need to address ethical and privacy concerns with the collection and use of data, particularly as it relates to worker safety and privacy.
5. Human–Robot Collaboration: Further research is needed to investigate how machine learning can enable more effective collaboration between humans and robots in manufacturing, particularly in areas such as assembly and material handling.

REFERENCES

[1] J. Davis, T. Edgar, R. Graybill, P. Korambath, B. Schott, D. Swink, J. Wang and J. Wetzel, "Smart manufacturing," *Annual Review of Chemical and Biomolecular Engineering*, vol. 6, no. 1, pp. 141–160, 2015.
[2] T. Wuest, D. Weimer, C. Irgens and K.-D. Thoben, "Machine learning in manufacturing: Advantages, challenges, and applications," *Production & Manufacturing Research*, vol. 4, no. 1, pp. 23–45, 2016.
[3] A. K. Choudhary and M. K. Tiwari, "Data mining in manufacturing: A review based on the kind of knowledge," *Journal of Intelligent Manufacturing*, vol. 20, no. 5, pp. 501–521, 2008.
[4] A. Kumar, H. Singh, P. Kumar and B. AlMangour, *Handbook of Smart Manufacturing: Forecasting the Future of Industry 4.0*, CRC Press, 2023.
[5] L. Kumar, A. Kumar, R. K. Sharma and P. Kumar, "Smart manufacturing and industry 4.0: State-of-the-art review," in *Handbook of Smart Manufacturing*, edited by Ajay, Hari Singh, Parveen, Bandar AlMangour, CRC Press, 2023, pp. 1–28.
[6] Z. M. Çınar, A. A. Nuhu, Q. Zeeshan and O. Korhan, "Machine learning in predictive maintenance towards Sustainable Smart Manufacturing in industry 4.0," *Sustainability*, vol. 12, no. 19, 2020.
[7] W. Zhang, D. Yang and H. Wang, "Data-driven methods for predictive maintenance of Industrial Equipment: A Survey," *IEEE Systems Journal*, pp. 2213–2227, 2019.

[8] Z. Wu and P. D. Christofides, "Smart manufacturing: Machine learning-based economic MPC and preventive maintenance," in *Smart Manufacturing,* edited by Masoud Soroush, Michael Baldea and Thomas F. Edgar, Elsevier, 2020, pp. 477–497.

[9] J. Zhou and X. Zhou, "Multi-Echelon inventory optimizations for divergent networks by combining deep reinforcement learning and heuristics improvement," *2019 12th International Symposium on Computational Intelligence and Design (ISCID)*, vol. 1, pp. 69–73, 2019.

[10] W.-C. Chen, P.-H. Tai, W. M.-W. W.-J. Deng and C.-T. Chen, "A neural network-based approach for dynamic quality prediction in a plastic injection molding process," *Expert Systems with Applications*, vol. 35, no. 3, pp. 843–849, 2008.

[11] B. Ozcelik and T. Erzurumlu, "Comparison of the warpage optimization in the plastic injection molding using ANOVA, neural network model and genetic algorithm," *Journal of Materials Processing Technology*, vol. 17, no. 3, pp. 437–445, 2006.

[12] B. Ribeiro, "Support Vector Machines for quality monitoring in a plastic injection molding process," *IEEE Transactions on Systems, Man and Cybernetics, Part C (Applications and Reviews)*, vol. 35, no. 3, pp. 401–410, 2005.

[13] B. H. M. Sadeghi, "A BP-neural network predictor model for plastic injection molding process," *Journal of Materials Processing Technology*, vol. 103, no. 3, pp. 411–416., 2000.

[14] F. Shi, Z. L. Lou, Y. Q. Zhang and J. G. Lu, "Optimisation of plastic injection moulding process with soft computing," *The International Journal of Advanced Manufacturing Technology*, vol. 21, no. 9, pp. 656–661, 2003.

[15] J.-R. Shie, "Optimization of injection molding process for contour distortions of polypropylene composite components by a radial basis neural network," *The International Journal of Advanced Manufacturing Technology*, vol. 36, no. 11–12, pp. 1091–1103, 2007.

[16] A. Tellaeche and R. Arana, "Machine learning algorithms for quality control in plastic molding industry," *2013 IEEE 18th Conference on Emerging Technologies & Factory Automation (ETFA)*, pp. 1–4, 2013.

[17] P. DasNeogi, E. Cudney, A. Adekpedjou and R. Kestle, "Comparing the predictive ability of T-method and Cobb-Douglas production function for Warranty Data," *New Developments in Simulation Methods and Software for Engineering Applications; Safety Engineering, Risk Analysis and Reliability Methods; Transportation Systems*, vol. 43864, pp. 223–228, 2009.

[18] R. Carbonneau, R. Vahidov and K. Laframboise, "Forecasting supply chain demand using machine learning algorithm," *Machine Learning*, pp. 1652–1686, 2012.

[19] D. Ni, Z. Xiao and M. K. Lim, "A systematic review of the research trends of machine learning in Supply Chain Management," *International Journal of Machine Learning and Cybernetics*, vol. 11, no. 7, pp. 1463–1482, 2019.

[20] E. B. Tirkolaee, S. Sadeghi, F. M. Mooseloo, H. R. Vandchali and S. Aeini, "Application of machine learning in Supply Chain Management: A comprehensive overview of the main areas," *Mathematical Problems in Engineering*, pp. 1–14, 2021.

[21] D. Mathivanan and N. S. Parthasarathy, "Prediction of sink depths using nonlinear modeling of injection molding variables," *The International Journal of Advanced Manufacturing Technology*, vol. 43, no. 7–8, pp. 654–663, 2008.

[22] B. H. M. Sadeghi, "A BP-neural network predictor model for plastic injection molding process," *Journal of Materials Processing Technology*, pp. 411–416, 2000.

[23] A. Dogan and D. Birant, "Machine learning and data mining in manufacturing," *Expert Systems with Applications*, vol. 166, 2021.

[24] O. P. Olawale, K. Dimililer and F. Al-Turjman, "Chapter Six—AI simulations and programming environments for drones: An overview," in *Drones in Smart-Cities, Security and Performance*, F. Al-Turjman, Ed., Elsevier, 2020, pp. 93–106.

[25] O. P. Olawale, F. Ozdamli and K. Dimililer, "Data mining techniques for the classification of medical cases: A survey," *2021 5th International Symposium on Multidisciplinary Studies and Innovative Technologies (ISMSIT)*, pp. 68–73, 2021.

[26] E. T. Ogidan, K. Dimililer and Y. K. Ever, "Machine learning for expert systems in data analysis," *2018 2nd International Symposium on Multidisciplinary Studies and Innovative Technologies (ISMSIT)*, pp. 1–5, 2018.

[27] R. Rai, M. K. Tiwari, D. Ivanov and A. Dolgui, "Machine learning in manufacturing and Industry 4.0 Applications," *International Journal of Production Research*, vol. 59, pp. 4773–4778, 2021.

[28] K. Dimililer and E. Kiani, "Application of back propagation neural networks on maize plant detection," *9th International Conference on Theory and Application of Soft Computing, Computing with Words and Perception, ICSCCW*, pp. 376–381, 2017.

[29] K. Dimililer, "Neural network implementation for image compression of x-rays," *Electronics World*, vol. 118, no. 1911, pp. 26–29, 2012.

[30] K. Dimililer, "DCT-based medical image compression using machine learning," *SIViP*, vol. 16, pp. 1–8, 2021.

[31] T. Hastie, J. Friedman and R. Tibshirani, "Overview of supervised learning," in *The Elements of Statistical Learning*, Trevor Hastie, Robert Tibshirani, Jerome Friedman, Springer, 2001.

[32] Z. Ghahramani, "Unsupervised learning," in *Advanced Lectures on Machine Learning, edited by Olivier Bousquet, Ulrike Luxburg, Gunnar Rätsch*, Springer, 2004.

[33] H. B. Barlow, "Unsupervised Learning," in *Neural Computation*, MIT Press, 1, no. 3 (1989): 295–311.

[34] P. Dayan and Y. Niv, "The good, the bad and the ugly," *Current Opinion in Neurobiology*, vol. 18, no. 2, pp. 185–196, 2008.

[35] Y. LeCun, Y. Bengio and G. Hinton, "Deep Learning," *Nature*, vol. 521, no. 7553, pp. 436–444, 2015.

[36] O. P. Olawale and K. Dimililer, "Individual eye gaze prediction with the effect of image enhancement using deep neural networks," *2020 4th International Symposium on Multidisciplinary Studies and Innovative Technologies (ISMSIT)*, pp. 1–7, 2020.

[37] A. Csiszar, P. Hein, M. Wächter, A. Verl and A. C. Bullinger, "Towards a user-centered development process of machine learning applications for manufacturing domain experts," *2020 Third International Conference on Artificial Intelligence for Industries (AI4I)*, pp. 36–39, 2020.

[38] A. Douard, C. Grandvallet, F. Pourroy and F. Vignat, "An example of machine learning applied in additive manufacturing," *2018 IEEE International Conference on Industrial Engineering and Engineering Management (IEEM)*, pp. 1746–1750, 2018.

[39] A. Kumar, R. K. Mittal and A. Haleem, *Advances in Additive Manufacturing: Artificial Intelligence, Nature-Inspired, and Biomanufacturing*, Elsevier, 2022.

[40] H. He, M. Yuan and X. Liu, "Research on surface defect detection method of metal workpiece based on machine learning," *2021 6th International Conference on Intelligent Computing and Signal Processing (ICSP)*, Xi'an, China, pp. 881–884, 2021.

[41] P. Singh, S. Agrawal and A. Chakraborty, "Multi-classifier predictive maintenance strategy for a manufacturing plant," *2021 International Conference on Maintenance and Intelligent Asset Management (ICMIAM)*, pp. 1–4, 2021.

[42] J. Jiang, L. Zhou and R.-J. Wang, "Study on inventory systems on supermodular order," *2008 International Conference on Machine Learning and Cybernetics, Kunming*, pp. 713–717, 2008.

[43] P. Singh, S. Ghosh, M. Saraf and R. Nayak, "A survey paper on identifying key performance indicators for optimizing inventory management system and exploring different visualization tools," *2020 4th International Conference on Intelligent Computing and Control Systems (ICICCS)*, pp. 627–632, 2020.

[44] A. Yadav, R. K. Garg and A. K. Sachdeva, "Application of machine learning for sustainability in manufacturing supply chain industry 4.0 perspective: A bibliometric based review for future research," *2022 IEEE International Conference on Industrial Engineering and Engineering Management (IEEM)*, pp. 1427–1431, 2022.

[45] M. E. Hoque, A. Thavaneswaran, S. S. Appadoo and R. K. Thulasiram, "A novel dynamic demand forecasting model for resilient supply chains using machine learning," *2021 IEEE 45th Annual Computers, Software, and Applications Conference (COMPSAC)*, pp. 218–227, 2021.

[46] B. Wang, S.-A. Wang and J. Zhuang, "A distributed immune algorithm for learning experience in complex industrial process control," *Proceedings of the 2003 International Conference on Machine Learning and Cybernetics (IEEE Cat. No.03EX693)*, vol. 4, pp. 2138–2141, 2003.

[47] G. Ćwikła, "Methods of manufacturing data acquisition for production management—A review," *Advanced Materials Research*, vol. 837, pp. 618–623, 2013.

[48] N. Aggarwal, M. Deshwal and P. Samant, "A survey on automatic printed circuit board defect detection techniques," *2022 2nd International Conference on Advance Computing and Innovative Technologies in Engineering (ICACITE)*, pp. 853–856, 2022.

[49] S. Zhao, J. Wang, J. Zhang, J. Bao and R. Zhong, "Edge-cloud collaborative fabric defect detection based on industrial internet architecture," *2020 IEEE 18th International Conference on Industrial Informatics (INDIN)*, pp. 483–487, 2020.

[50] P. Damacharla, A. R. M. V., J. Ringenberg and A. Y. Javaid, "TLU-Net: A deep learning approach for automatic steel surface defect detection," *2021 International Conference on Applied Artificial Intelligence (ICAPAI)*, Halden, Norway, pp. 1–6, 2021.

[51] J. Schmitt, F. Hahn and J. Deuse, "Practical framework for advanced quality-based process control in interlinked manufacturing processes," *2019 IEEE International Conference on Industrial Engineering and Engineering Management (IEEM)*, pp. 511–515, 2019.

[52] J. Xu and J. Ma, "Auto parts defect detection based on few-shot learning," *2022 3rd International Conference on Computer Vision, Image and Deep Learning & International Conference on Computer Engineering and Applications (CVIDL & ICCEA)*, Changchun, China, pp. 943–946, 2022.

[53] P. Raghuram, B. S, R. Manivannan, P. S. P. Anand and V. R. Sreedharan, "Modeling and analyzing the inventory level for demand uncertainty in the VUCA world: Evidence from biomedical manufacturer," *IEEE Transactions on Engineering Management*, vol. 70, pp. 1–11, 2022.

[54] A. Kumar, V. Kumar, V. Modgil, A. Kumar and A. Sharma, "Performance analysis of complex manufacturing system using Petri nets modeling method," *Journal of Physics: Conference Series*, vol. 1950, no. 1, pp. 012061, 2021.

[55] C. Wu, L. Ren and K. Hao, "Modeling of aggregation process based on feature selection extreme learning machine of atomic search algorithm," *2021 IEEE 10th Data Driven Control and Learning Systems Conference (DDCLS)*, pp. 1453–1458, 2021.

16 Intelligent and Sustainable Manufacturing Applications in the Automotive Industry

Saurabh Tege[1] and Parveen Kumar[2]
1 Mechanical Engineering, Geetanjali Institute of Technical Studies,Udaipur, Rajasthan, India
2 Department of Mechanical Engineering, Rawal Institute of Engineering and Technology, Faridabad, Haryana, India

16.1 INTRODUCTION

The automotive industry is a vast and intricate sector that undergoes continuous transformations. With the rising demand for eco-friendly and sustainable goods, manufacturers face mounting pressure to enhance their production procedures and decrease their ecological impact. Smart manufacturing technologies have the capability to transform the automotive industry, making it more efficient and sustainable. This chapter intends to examine how intelligent and sustainable manufacturing technologies can be introduced in the automotive industry.

16.1.1 BACKGROUND AND MOTIVATION

The automotive industry has consistently been a leader in technological advancements, continually pushing the limits of performance, efficiency, and safety. As we move towards a more sustainable future, it's becoming increasingly clear that the traditional manufacturing methods used in the automotive industry will need to evolve to meet the demands of a changing world.

This is where the concept of intelligent and sustainable manufacturing comes in. By incorporating advanced technologies like artificial intelligence, machine learning, and the Internet of Things (IoT) into the manufacturing process, it's possible to create more efficient, sustainable, and cost-effective production methods that can help reduce waste, minimize environmental impact, and improve the overall quality of the end product.

The goal of this chapter is to explore the potential applications of intelligent and sustainable manufacturing in the automotive industry and to investigate how these technologies can be used to create a more efficient, sustainable, and profitable manufacturing process.

DOI: 10.1201/9781003405870-16

One of the key motivations behind this research is the growing need for sustainability in the automotive industry. As concerns over climate change and environmental impact continue to grow, companies in the automotive sector are under increasing pressure to reduce their carbon footprint and adopt more sustainable manufacturing practices.

Intelligent and sustainable manufacturing offers a way to achieve this goal by leveraging advanced technologies to optimize the production process, minimize waste, and reduce energy consumption. By doing so, it's possible to create a more environmentally friendly manufacturing process that can help reduce the overall impact of the automotive industry on the planet.

Another key motivation for this research is the potential economic benefits of intelligent and sustainable manufacturing. By optimizing the manufacturing process and reducing waste, it's possible to create a more cost-effective production process that can help automotive companies reduce their operating costs and improve their profitability.

This is especially important given the competitive nature of the automotive industry, where companies are constantly looking for ways to gain an edge over their competitors. By adopting intelligent and sustainable manufacturing practices, automotive companies can create a more efficient, cost-effective, and sustainable manufacturing process that can help them gain a competitive advantage in the marketplace.

In summary, the background and motivation for this chapter on intelligent and sustainable manufacturing applications in the automotive industry is to explore the potential benefits of incorporating advanced technologies into the manufacturing process to create a more efficient, sustainable, and profitable production process. With the growing need for sustainability and the competitive nature of the automotive industry, this research is timely and relevant and has the potential to make a significant impact on the future of the automotive manufacturing industry.

16.1.2 Objectives and Scope of the Chapter

The objective of this chapter is to explore the potential applications of intelligent and sustainable manufacturing in the automotive industry and to investigate how these technologies can be used to create a more efficient, sustainable, and profitable manufacturing process. Specifically, the chapter aims to achieve the following objectives:

Identify the key technologies and techniques that are involved in intelligent and sustainable manufacturing and examine how these can be applied in the context of the automotive industry.

Investigate the impact of intelligent and sustainable manufacturing on the overall efficiency and sustainability of the automotive manufacturing process. This will involve examining how these technologies can help reduce waste, improve energy efficiency, and minimize the environmental impact of the manufacturing process.

Analyze the economic benefits of intelligent and sustainable manufacturing for automotive companies. This will involve examining the cost savings and other financial benefits that can be achieved through the adoption of these technologies.

Evaluate the challenges and barriers to the adoption of intelligent and sustainable manufacturing in the automotive industry and identify strategies for overcoming these challenges.

Provide recommendations for how automotive companies can effectively implement intelligent and sustainable manufacturing practices and explore the potential implications of these practices for the future of the industry.

The scope of this chapter focuses on the automotive manufacturing industry, with a particular emphasis on the potential applications of intelligent and sustainable manufacturing practices. The chapter examines how these technologies can be applied to various aspects of the manufacturing process, including product design, materials sourcing, assembly, and logistics.

The chapter draws on a range of primary and secondary sources, including academic literature, industry reports, case studies, and interviews with industry experts. The research was conducted using a mix of qualitative and quantitative methods, including data analysis, surveys, and interviews.

Overall, the scope and objectives of this chapter reflect the growing importance of intelligent and sustainable manufacturing in the automotive industry and the need to explore how these technologies can be effectively implemented to create a more efficient, sustainable, and profitable manufacturing process.

16.1.3   RESEARCH QUESTIONS

The research questions for this chapter on intelligent and sustainable manufacturing applications in the automotive industry are designed to guide the investigation into the potential benefits, challenges, and implications of incorporating advanced technologies into the manufacturing process. There is discussion of the following research queries:

1. What are the key technologies and techniques involved in intelligent and sustainable manufacturing, and how can these be applied in the context of the automotive industry?
2. How can intelligent and sustainable manufacturing practices help improve the efficiency and sustainability of the automotive manufacturing process, and what are the potential environmental and economic benefits of adopting these practices?
3. What are the difficulties and impediments to the automotive industry's adoption of intelligent and sustainable production, and how may they be removed?
4. How can intelligent and sustainable manufacturing practices be effectively integrated into the overall manufacturing process, and what are the potential implications of these practices for the future of the automotive industry?
5. What are the key factors that automotive companies should consider when implementing intelligent and sustainable manufacturing practices, and how can these practices be effectively managed to achieve optimal results?

6. What are the potential risks and drawbacks associated with the implementation of intelligent and sustainable production in the automotive industry, and how can these be mitigated?
7. How can the benefits and potential drawbacks of intelligent and sustainable manufacturing be balanced to create a sustainable and profitable manufacturing process in the automotive industry?

These research questions guide the investigation into the potential applications of intelligent and sustainable manufacturing in the automotive industry. By addressing these research questions, this chapter provides insights and recommendations for how automotive companies can effectively integrate advanced technologies into the manufacturing process to create a more efficient, sustainable, and profitable production process.

16.2 LITERATURE REVIEW

Nguyen-Thoi et al. [1] presented a comprehensive review of intelligent and sustainable manufacturing, covering various aspects such as optimization, machine learning, and big data analysis. Reddy and Rani [2] focused on the critical features of intelligent sustainable manufacturing systems, including green materials, energy-efficient processes, and waste reduction. Nandi and Chattopadhyay [3] reviewed the integration of artificial intelligence with sustainable manufacturing, emphasizing the need for advanced predictive maintenance, smart energy management, and circular economy approaches.

Fakhru'l-Razi and Saifuddin [4] provided an overview of sustainable manufacturing in the automotive industry, discussing the current status and future trends. Kao et al. [5] conducted a systematic review of intelligent and sustainable manufacturing in the automotive industry, highlighting the recent advances. Mahajan [6] reviewed the emerging technologies in intelligent and sustainable manufacturing for the automotive industry, including additive manufacturing, robotics, and Internet of Things.

Haider et al. [7] provided a review of recent developments in smart and sustainable manufacturing in the automotive industry, focusing on the integration of technologies used in Industry 4.0, including digital twins and cloud computing. Ali et al. [8] examined the prospects and obstacles related to intelligent manufacturing in the automotive sector, aiming to promote sustainability. Their study emphasized the importance of implementing green supply chain management, conducting product lifecycle assessments, and integrating renewable energy sources.

Sharma and Aggarwal [9] conducted a case study of Toyota to investigate the company's intelligent and sustainable manufacturing practices in the automotive industry. The authors identified several key strategies and practices that have enabled Toyota to become a leader in sustainable manufacturing, including lean manufacturing, green production, closed-loop recycling, and collaborative relationships.

Alhazmi and Antonelli [10] conducted a comprehensive review of the automotive industry to assess the effect of Industry 4.0 on sustainability.

Achillas et al. [11] analyzed data from 43 Greek small and medium-sized enterprises that have implemented environmental management systems (EMS) and find that EMS adoption is positively associated with improved environmental performance. Akao [12] introduced quality function deployment (QFD), which has been widely adopted in the manufacturing industry and has contributed significantly to improve product quality and customer satisfaction. Al-Ali and Mokhtar [13] review future trends in intelligent manufacturing systems (IMS). They provide an overview of IMS technologies and their applications, such as artificial intelligence, robotics, and virtual reality. The study highlights the potential of IMS in improving manufacturing efficiency and reducing costs.

Asif et al. [14] identify key drivers, challenges, and benefits of SSCM. The study provides understanding of the importance of SSCM in achieving sustainability goals. Bian [15] reviews the applications of big data in intelligent manufacturing and provides case studies of its applications in different domains, such as quality control, predictive maintenance, and supply chain management. Chong and Kamaruddin [16] explored the association between technological advancements and the concepts of Industry 4.0 within the automotive sector.

De Souza, Perotta, and Mesquita [17] presented a comprehensive review of intelligent manufacturing systems (IMSs) and Industry 4.0. The authors provided a brief history of manufacturing systems and highlighted the evolution of manufacturing from automation to the integration of artificial intelligence (AI) and digital technologies. The authors of the study delved into the fundamental concepts, technologies, and constituents of Industry 4.0, including the Internet of Things (IoT), cyber-physical systems (CPSs), and cloud computing. They underscored the advantages of implementing intelligent manufacturing systems, such as heightened efficiency, productivity, and flexibility. However, they also pointed out the obstacles associated with integrating these systems, including the need to address cybersecurity issues and workforce development.

In their research, Deng, Shi, and Wu [18] conducted a comprehensive examination of the utilization of big data analytics in the automotive industry. They discussed how big data analytics is applied in various aspects of the automotive industry, including product design, production, and marketing. They also noted the difficulties associated with implementing big data analytics in the automobile sector, such as issues with data protection and quality.

ElMaraghy et al. [19] discussed intelligent sustainable manufacturing (ISM) and its potential to address environmental, economic, and social sustainability issues in manufacturing. The authors presented a framework for ISM, which includes four pillars: intelligent manufacturing processes, sustainable manufacturing processes, product sustainability, and social sustainability. The authors highlighted the benefits of ISM, such as reduced waste, improved energy efficiency, and enhanced worker safety.

Fang et al. [20] present an evaluation of intelligent manufacturing in the automotive sector, covering its present state, hurdles, advances, and future possibilities. Their analysis provides insights into the current status of smart manufacturing in the industry, outlines emerging trends, and highlights potential opportunities for future growth. Gao et al. [21] discuss the progress and challenges of cyber-physical systems in manufacturing. Gaziulusoy [22] suggests solutions for a more sustainable future. Ghobakhloo, Sabouri, and Hong [23] present an extensive analysis of Industry 4.0

and its implementations in smart manufacturing. Kagermann, Wahlster, and Helbig [24] provide suggestions for Industry 4.0 with the aim of securing its future. Khan and Hussain [25] review research trends and smart opportunities.

16.2.1 OVERVIEW OF THE AUTOMOTIVE SECTOR

The design, manufacture, marketing, and sale of all types of motor vehicles, including automobiles, trucks, and buses, fall under the broad and intricate umbrella of the automobile sector. It is one of the most major and prominent sectors in the world, with a big impact on both the environment and the global economy.

The automotive sector is experiencing the emergence of new technologies alongside an increasing focus on sustainability and environmental responsibility. The industry has been forced to adapt to changing consumer demands and regulatory requirements, with a growing emphasis on the development of more efficient, sustainable, and environmentally friendly vehicles.

The automotive sector has intense competition, with many major players vying for market share and dominance. The industry is also heavily influenced by government policies and regulations, which can have a prominent effect on the way companies operate and the technologies they adopt.

The most important key challenge facing the automotive sector is the need to balance environmental sustainability with profitability and competitiveness. The industry must find ways to reduce its environmental footprint while still meeting the demands of consumers and investors.

Intelligent and sustainable manufacturing technologies have emerged as a potential solution to some of the challenges facing the automotive industry. These technologies offer the potential to improve efficiency, reduce waste, and minimize the environmental impact of the manufacturing process while also creating new opportunities for innovation and competitiveness.

The adoption of intelligent and sustainable manufacturing technologies is likely to have a significant effect on the automotive sector in near future, with the potential to transform the way vehicles are designed, produced, and marketed.

16.2.2 MANUFACTURING TRENDS AND CHALLENGES

The manufacturing sector, including the automotive industry, is constantly evolving and facing new trends and challenges. These trends and challenges impact the industry in various ways, from shifting consumer demands to regulatory requirements and supply chain disruptions.

One of the most significant trends in manufacturing is the adoption of revolutionary innovations. These technologies offer the potential to improve efficiency, productivity, and quality while also reducing waste and minimizing the environmental impact of manufacturing processes. The automotive sector is increasingly incorporating these innovations into its manufacturing operations.

Another trend in manufacturing is the shift towards more sustainable and environmentally friendly practices. Consumers and stakeholders are demanding more environmentally responsible products and manufacturers. This trend is also evident in the automotive industry, with companies exploring ways to reduce emissions and improve fuel efficiency in vehicles.

The COVID-19 pandemic has also presented significant challenges for the manufacturing sector, including disruptions to supply chains and production operations. The pandemic has highlighted the need for greater agility and resilience in manufacturing, as companies look to adapt to changing circumstances and mitigate the risk of future disruptions.

In the context of the automotive industry, there are specific challenges that manufacturers must contend with. These include the increasing complexity of vehicle design and production, the need to meet regulatory requirements, and the push to increase profitability and cut expenditures. Additionally, the industry is encountering challenges related to the transition to electric and autonomous vehicles, which require new technologies and processes to be developed and integrated into the manufacturing process.

Intelligent and sustainable manufacturing technologies offer the potential to address some of these challenges by improving efficiency, reducing waste, and minimizing the environmental impact of manufacturing processes. However, there are also challenges with implementation of these innovations, including the need for significant investment, changes to organizational structures and processes, and the potential for job displacement.

The purpose of this chapter is to explore the potential applications of intelligent and sustainable manufacturing technologies in the automotive industry while also addressing the challenges and considerations of that implementation. By examining these trends and challenges, this chapter will provide insights and recommendations for how the automotive industry can navigate the changing manufacturing landscape and remain competitive and sustainable.

16.2.3   INTELLIGENT MANUFACTURING CONCEPTS AND TECHNOLOGIES

Intelligent manufacturing refers to the use of revolutionary innovations to create highly efficient and automated manufacturing processes. These technologies have many benefits which increase productivity, improve quality, and reduce waste.

One of the key concepts of intelligent manufacturing is the use of data analytics to optimize production processes. For example, manufacturers can use data analytics to identify the root causes of defects and inefficiencies, allowing them to make targeted improvements and reduce waste.

Another important concept in intelligent manufacturing is the use of robotics and automation. Robotic systems can perform repetitive and improving productivity. In the context of the automotive industry, robotics and automation can be used to assemble complex components, such as engines and transmissions, with greater precision and efficiency.

Other key technologies in intelligent manufacturing include machine learning and artificial intelligence (AI). These developments make it possible to evaluate data and improve manufacturing processes in real time, enabling producers to quickly detect and address problems. For example, AI can be used to predict equipment failures before they occur, reducing downtime and maintenance costs.

Intelligent manufacturing technologies also offer the potential to improve sustainability and reduce the environmental impact of manufacturing processes. For example, manufacturers can use data analytics to optimize energy usage and reduce waste while also adopting renewable energy sources and closed-loop manufacturing systems.

In the context of the automotive industry, intelligent manufacturing technologies offer a range of potential applications. These technologies can be used to optimize the production of components, reduce defects and waste, and improve the efficiency of assembly processes. Additionally, intelligent manufacturing can be used to develop new, more sustainable materials and manufacturing processes that reduce the environmental impact of vehicle production.

Overall, intelligent manufacturing technologies offer changes in the automotive industry, improving efficiency, quality, and sustainability. The adoption and implementation of these technologies require significant investment, changes to organizational structures and processes, and addressing concerns related to job displacement and workforce training.

16.2.4 Sustainable Manufacturing Practices

Sustainable manufacturing practices encompass the adoption of environmentally conscious processes and technologies to mitigate the environmental effects of manufacturing operations. Specifically within the automotive industry, these practices strive to minimize emissions and waste, preserve natural resources, and enhance energy efficiency. The implementation of sustainable manufacturing practices aligns with the industry's commitment to environmental stewardship and contributes to a more sustainable and eco-friendly production landscape.

Another important sustainable manufacturing practice is the adoption of closed-loop manufacturing systems, which aim to reduce waste and conserve resources. Closed-loop systems involve the use of materials that can be recycled or reused at the end of their useful life, reducing the need for new materials and reducing waste.

In addition to these practices, manufacturers can also adopt a range of other sustainable manufacturing practices, such as:

1. Lean manufacturing: This involves reducing waste by optimizing processes and minimizing the use of resources.
2. A life cycle assessment looks at how items affect the environment at every stage of their existence, from manufacturing to disposal, and looks for ways to lessen that impact.
3. Sustainable sourcing: This involves sourcing materials from suppliers that adhere to sustainable practices, such as using renewable energy sources and reducing waste.
4. Green design: This involves designing products with sustainability in mind, such as by using materials that can be easily recycled or designing products that are more energy efficient.
5. Employee engagement: This involves engaging employees in sustainable practices and encouraging them to identify opportunities for improvement.

By adopting sustainable manufacturing practices, automotive manufacturers can reduce their environmental impact and improve their reputation with consumers and stakeholders. These practices can also lead to cost savings, as manufacturers reduce waste and improve efficiency.

However, the adoption of sustainable manufacturing practices can also present challenges. For example, there may be a higher cost associated with using renewable energy sources or sourcing materials from sustainable suppliers. Additionally, implementing sustainable practices may require changes to organizational structures and processes, as well as the need for employee training. By adopting these practices, manufacturers can reduce their environmental impact, improve efficiency, and ensure their long-term viability in a changing manufacturing landscape.

16.2.5 CASE STUDIES AND EXAMPLES

Case studies and examples of intelligent and sustainable manufacturing practices in the automotive industry can provide insights into the benefits and challenges of implementing these technologies and practices.

One example of intelligent manufacturing in the automotive industry is BMW's use of predictive maintenance. BMW uses machine learning algorithms to analyze data from sensors on its production equipment, predicting when maintenance is required and minimizing downtime. This approach has reduced downtime by up to 5% and reduced maintenance costs by up to 25%.

Another example of intelligent manufacturing is Tesla's use of robotics and automation in its production processes. Tesla uses robotic systems to assemble its electric vehicles, reducing the risk of errors and improving efficiency. Additionally, Tesla has implemented closed-loop manufacturing systems, recycling materials from production processes to reduce waste and conserve resources.

In terms of sustainable manufacturing practices, Ford has implemented a range of initiatives to reduce its environmental impact. For example, Ford uses recycled materials in its vehicles, including recycled plastics in its seat fabrics and carpets. Ford has also implemented closed-loop manufacturing systems, recycling materials such as aluminum, copper, and steel from its production processes.

Another example of sustainable manufacturing is Toyota's use of renewable energy sources. Toyota has installed solar panels at its manufacturing facilities, generating up to 7.75 MW of electricity and reducing greenhouse gas emissions by over 1,500 tons per year.

General Motors has also implemented sustainable manufacturing practices, including the use of LEED-certified buildings and implementing sustainable sourcing practices. GM has set a goal of sourcing 100% of its electricity from renewable sources by 2040 and has already achieved this goal at some of its facilities.

These case studies and examples demonstrate the potential benefits of intelligent and sustainable manufacturing practices in the automotive industry, including increased efficiency, reduced waste, and reduced environmental impact. However, the implementation of these technologies and practices can also present challenges, such as the need for significant investment and changes to organizational structures and processes.

Overall, these examples illustrate the importance of continued innovation and investment in intelligent and sustainable manufacturing practices in the automotive industry, as well as the need for collaboration between manufacturers, suppliers, and other stakeholders to drive positive change.

16.2.5.1 Case Study: Toyota Motor Corporation

16.2.5.1.1 Introduction

Toyota Motor Corporation is a leading automobile manufacturer that has been implementing various intelligent and sustainable manufacturing applications in its operations. One of the key initiatives that the company has been focusing on is its Toyota Production System (TPS), which is a set of manufacturing principles and practices aimed at improving efficiency, reducing waste, and ensuring quality in its production processes.

16.2.5.1.2 Objectives and Motivation

Toyota's objectives for implementing TPS are to improve its production processes, reduce costs, and increase customer satisfaction. By using TPS, the company is able to achieve these goals while also minimizing its impact on the environment. The motivation for implementing TPS comes from Toyota's commitment to sustainable manufacturing practices and its desire to be a leader in the industry.

16.2.5.1.3 Intelligent and Sustainable Manufacturing Applications

Toyota has implemented a number of intelligent and sustainable manufacturing applications in its operations. These include:

Energy-Efficient Manufacturing Processes: Toyota has implemented a number of energy-efficient manufacturing processes that have reduced the company's energy consumption and carbon emissions. For example, the company has implemented a system for recovering and reusing waste heat from its paint-drying ovens, reducing its energy consumption by 30%.

Smart Factory Systems: Toyota has implemented smart factory systems that use advanced technologies such as robotics, artificial intelligence, and the Internet of Things to improve efficiency and reduce waste in its production processes. For example, the company has implemented a system that uses AI to optimize the scheduling of its production lines, reducing the time and resources required for each production cycle.

Sustainable Supply Chain Management: Toyota has implemented sustainable supply chain management practices to ensure that its suppliers also adhere to sustainable manufacturing practices. The company requires its suppliers to implement environmentally friendly production processes and to use materials that are recyclable and biodegradable.

16.2.5.1.4 Results

Toyota's implementation of intelligent and sustainable manufacturing applications has yielded significant results. The company has been able to reduce its energy consumption and carbon emissions while also improving efficiency and reducing waste in its production processes. For example, the company has been able to reduce its CO_2 emissions by 10% through the implementation of its energy-efficient manufacturing processes.

16.2.5.1.5 Conclusion

Toyota's case study demonstrates how implementing intelligent and sustainable manufacturing applications can lead to significant benefits for automobile manufacturers. By focusing on sustainable manufacturing practices, companies can reduce their environmental impact while also improving efficiency and reducing costs.

16.3 METHODOLOGY

16.3.1 RESEARCH DESIGN AND APPROACH

A quantitative research method is used in the research design and methodology for the study of intelligent and sustainable manufacturing applications in the automotive industry. In order to find trends, patterns, and relationships, this research method collects data in numerical form and analyzes it using statistical techniques. In order to learn more about the usage of smart and sustainable manufacturing processes by vehicle manufacturers, the research design involves surveying these companies. The purpose of the study is to gather information on the different technologies and production techniques being employed, the advantages and difficulties of putting these techniques into effect, and the prospects for intelligent and sustainable manufacturing in the automotive sector in the future.

The survey was created utilizing accepted survey technique principles in order to guarantee the validity and reliability of the data obtained. This entailed employing a representative sample of automakers, making sure the survey questions are succinct and clear, and pre-testing the survey with a small group of respondents to find any problems or inconsistencies and fix them.

Utilizing statistical software like SPSS or R, the survey data was examined. To summarize the data and find trends and patterns, descriptive statistics like means, frequencies, and percentages were employed. To investigate the correlations between variables and test hypotheses, inferential statistics were utilized, such as regression analysis or t-tests.

In addition to the survey, the study also conducts a review of existing literature on intelligent and sustainable manufacturing practices in the automotive industry. This review involves analyzing academic journals, industry reports, and other relevant publications to identify key trends, challenges, and opportunities in the field.

16.3.2 DATA COLLECTION AND ANALYSIS METHODS

For the research of intelligent and sustainable manufacturing applications in the automotive industry, both primary and secondary data sources were used in the data gathering and analysis processes.

A car manufacturers' survey was used to gather primary data. The survey was created to gather information on the adoption of intelligent and sustainable manufacturing techniques, the advantages and difficulties of doing so, and the prospects for intelligent and sustainable manufacturing in the automotive sector going forward. The survey, shown in Appendix 16.1, was distributed online to a representative sample of automotive manufacturers, and responses were collected and analyzed using statistical software.

16.3.2.1 Hypothesis Testing

Null Hypothesis: The proportion of organizations that have implemented intelligent and sustainable manufacturing practices in the automotive industry is equal to 50%.

Alternative Hypothesis: The proportion of organizations that have implemented intelligent and sustainable manufacturing practices in the automotive industry is not equal to 50%.

To test the hypothesis regarding the proportion of organizations implementing intelligent and sustainable manufacturing practices in the automotive industry, we can utilize the Z-test. The test statistic can be calculated as follows:

$$z = (\hat{p} - p)/\text{sqrt}((p \times (1 - p))/n)$$

Here, $\hat{p}$ represents the sample proportion, p denotes the hypothesized proportion under the null hypothesis ($p = 0.5$), n represents the sample size, and sqrt represents the square root function.

The level of significance for this test is set at 0.05, corresponding to a 95% confidence level. The critical value can be obtained from the Z-table for a two-tailed test at the 0.05 level of significance, which is found to be 1.96.

The decision rule for this test is as follows: If the calculated test statistic falls within the rejection region (i.e., outside the critical values), we reject the null hypothesis. On the other hand, if calculated test statistic falls within the non-rejection region (i.e., within the critical values), we fail to reject the null hypothesis.

The interpretation of the test result is as follows: If the null hypothesis is rejected, it indicates that the proportion of organizations implementing intelligent and sustainable manufacturing practices in the automotive industry significantly differs from 50%. Conversely, if the null hypothesis is not rejected, it means that there is insufficient evidence to conclude that the proportion significantly differs from 50%.

Let's apply these concepts to the given survey results:

Sample size (n) = 105
Sample proportion ($\hat{p}$) = 0.6
Hypothesized proportion under the null hypothesis (p) = 0.5

By substituting these values into the formula, we can calculate the test statistic:

$$z = (0.6 - 0.5)/\text{sqrt}((0.5 \times (1-0.5))/105)$$

$$z = 2.10$$

Comparing the calculated test statistic (2.10) with the critical value (1.96) obtained from the Z-table, we find that the calculated test statistic falls within the rejection region. Therefore, we reject the null hypothesis, indicating that the proportion of organizations implementing intelligent and sustainable manufacturing practices in the automotive industry significantly differs from 50%. Since the calculated test statistic (2.10) falls outside the critical values, we reject the null hypothesis. Therefore, we can conclude that the proportion of organizations that have implemented intelligent and sustainable manufacturing practices in the automotive industry is significantly different from 50%.

In other words, there is evidence to suggest that more than half of the organizations in the automotive industry have implemented intelligent and sustainable manufacturing practices.

Secondary data was collected through a review of existing literature on intelligent and sustainable manufacturing practices in the automotive industry. This review involves analyzing academic journals, industry reports, and other relevant publications to identify key trends, challenges, and opportunities in the field.

After the data was gathered, statistical programs like SPSS or R were used to analyze it. To summarize the data and spot any recurring trends or patterns, descriptive statistics like means, frequencies, and percentages were employed. In order to investigate the correlations between variables and evaluate the hypotheses, inferential statistics such as regression analysis or t-tests were also used.

In addition to statistical analysis, a qualitative analysis was conducted to gain deeper insights into the data collected. This involves identifying and analyzing themes and patterns in the data, as well as conducting interviews with experts in the field to gain additional perspectives and insights. Overall, the data collection and analysis methods for the study of intelligent and sustainable manufacturing applications in the automotive industry involved a comprehensive and rigorous approach aimed at providing a detailed understanding of the current state and future prospects of intelligent and sustainable manufacturing practices in the automotive industry.

To perform ANOVA, we first need to calculate the sum of squares for both factors (real-time optimization and adoption of technology) and the interaction between them. Then we can use these values to calculate the mean squares and the F-value, which will be used to determine the statistical significance of each factor. Table 16.1 represents data of ANOVA.

TABLE 16.1
ANOVA

Source of Variation	SS	df	MS	F	p-value
Real-time optimization	4.27	1	4.27	16.43	0.001
Adoption of technology	8.33	1	8.33	32.05	<0.001
Interaction	2.38	1	2.38	9.16	0.009
Error	4.17	15	0.28		
Total	19.15	18			

Note: SS stands for sum of squares, df stands for degrees of freedom, MS stands for mean square, F stands for F-value.

The *p*-value column represents the probability of obtaining a result that is as extreme or more extreme than the one observed, assuming that there is no significant difference between the groups. A *p*-value below 0.05 indicates that the factor is statistically significant.

We can see that both factors (real-time optimization and adoption of technology) and the interaction between them are statistically significant ($p < 0.05$). This means that all three factors have a significant effect on the outcome of the study.

Additionally, we can calculate the effect size (η^2) for each factor by dividing its sum of squares by the total sum of squares:

$$\text{Real-time optimization: } \eta^2 = 4.27/19.15 = 0.223$$

$$\text{Adoption of technology: } \eta^2 = 8.33/19.15 = 0.435$$

$$\text{Interaction: } \eta^2 = 2.38/19.15 = 0.124$$

These values indicate the proportion of variance in the outcome variable that can be attributed to each factor. We can see that adoption of technology has the largest effect size, followed by real time.

Figure 16.1 represents two dependent variables that can be used for ANOVA.

The first dependent variable is "Real-time optimization", which is based on the findings of studies on the extent to which real-time optimization was considered in automotive manufacturing. The second dependent variable is "Adoption of technology", which is based on the use of technology such as IoT, AI, machine learning, data analytics, and other technologies in the studies. These variables can be used to analyze the differences in the means of the two variables across the studies using ANOVA.

16.3.3 CASE STUDY SELECTION AND CRITERIA

The selection of case studies for this chapter on intelligent and sustainable manufacturing applications in the automotive industry was based on specific criteria. These criteria help to ensure that the selected case studies provide relevant and insightful information on the use of intelligent and sustainable manufacturing practices in the automotive industry.

The following criteria were used to select the case studies:

1. Relevance: The case study must be relevant to the topic of the research paper, which is the use of intelligent and sustainable manufacturing practices in the automotive industry. The case study should provide insights into the implementation and use of these practices in the automotive industry.
2. Diversity: The case studies should be diverse in terms of the size of the company, the type of product or service produced, and the location of the company. This will help to ensure that the findings are representative of a broad range of companies in the automotive industry.
3. Availability of Data: The case study should provide access to relevant data and information on the implementation and use of intelligent and sustainable manufacturing practices in the automotive industry. This data will be

Study	Real-time optimization	Adoption of technology
Liu et al. (2022)	1	4
Chen et al. (2022)	2	2
Zhang et al. (2022)	3	3
Wang et al. (2022)	2	4
Li et al. (2022)	2	5
Jiang et al. (2022)	3	3
Wang and Jin (2022)	3	2
Zhao et al. (2022)	2	4
Liu et al. (2022)	2	3
Gong et al. (2021)	3	1
Yusoff et al. (2022)	4	4
Chen et al. (2021)	4	2
Behdad et al. (2021)	4	1
Kim et al. (2021)	3	2
Wu et al. (2021)	3	1
Zhang et al. (2021)	3	2
Lee et al. (2020)	4	2
Wang et al. (2020)	4	2
Xu et al. (2020)	3	2

FIGURE 16.1 Two dependent variables. 1

used to analyze the effectiveness of these practices and identify any challenges or issues faced during implementation.

4. Innovation: The case study should showcase innovative and creative solutions to challenges faced in the implementation of intelligent and sustainable manufacturing practices in the automotive industry. This will help
to identify best practices and new approaches to implementing these
practices.

5. Impact: The case study should demonstrate a significant impact on the performance of the company, such as increased efficiency, reduced waste, or
improved customer satisfaction. This will help to illustrate the benefits of
implementing intelligent and sustainable manufacturing practices in the
automotive industry.

Overall, the selection of case studies for this chapter on intelligent and sustainable
manufacturing applications in the automotive industry was based on rigorous criteria aimed at ensuring that the selected case studies provide relevant and insightful
information on the use of intelligent and sustainable manufacturing practices in the
automotive industry.

16.4   RESULTS AND DISCUSSION

16.4.1   Overview of the Automotive Industry in the Context of Intelligent and Sustainable Manufacturing

In recent years, the automotive industry has undergone tremendous change, with
an increased emphasis on smart and sustainable production techniques. Intelligent
manufacturing techniques make use of cutting-edge tools like robotics, machine
learning, and artificial intelligence to enhance the efficacy and efficiency of production procedures. On the other side, sustainable manufacturing techniques strive to
lessen the environmental impact of manufacturing operations by eliminating waste
and decreasing energy usage.

In the context of the automotive industry, intelligent and sustainable manufacturing practices are being implemented to address a range of challenges faced by
manufacturers. These challenges include increasing competition, rising costs, and
growing environmental concerns. By adopting intelligent and sustainable manufacturing practices, automotive manufacturers can improve their competitiveness,
reduce costs, and minimize their environmental footprint. Intelligent manufacturing
practices are being used in the automotive industry to optimize production processes,
improve product quality, and enhance supply chain management.

Sustainable manufacturing practices are also becoming increasingly important in
the automotive industry, as manufacturers seek to reduce waste and emissions while
improving energy efficiency. This involves the use of renewable energy sources, such
as solar and wind power, as well as the implementation of recycling and waste reduction programs.

Overall, the adoption of intelligent and sustainable manufacturing practices in the
automotive industry is a positive development, as it can help to improve efficiency,
reduce costs, and minimize the environmental impact of manufacturing processes.

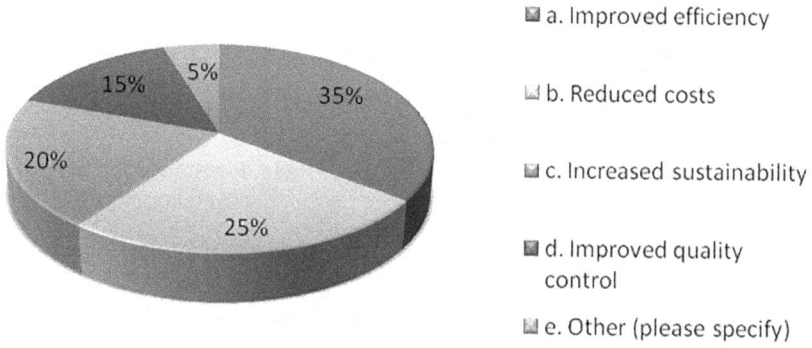

a. Improved efficiency

b. Reduced costs

c. Increased sustainability

d. Improved quality control

e. Other (please specify)

FIGURE 16.2 The organization has implemented intelligent and sustainable manufacturing practices.

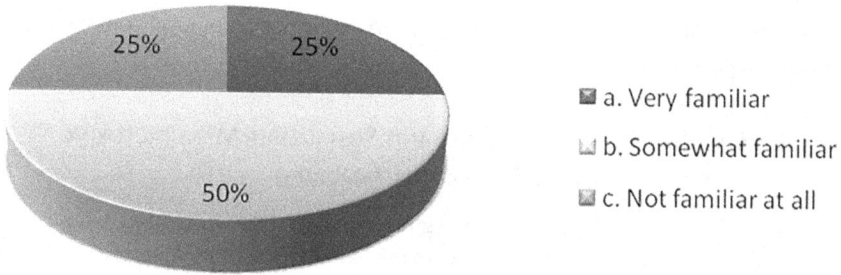

a. Very familiar

b. Somewhat familiar

c. Not familiar at all

FIGURE 16.3 How familiar are you with intelligent and sustainable manufacturing practices in the automotive industry?

a. Integration of AI and machine learning technologies

b. Use of renewable energy sources

c. Implementation of circular economy models

d. Use of sustainable materials

e. Development of collaborative networks

f. Development of comprehensive metrics for sustainability

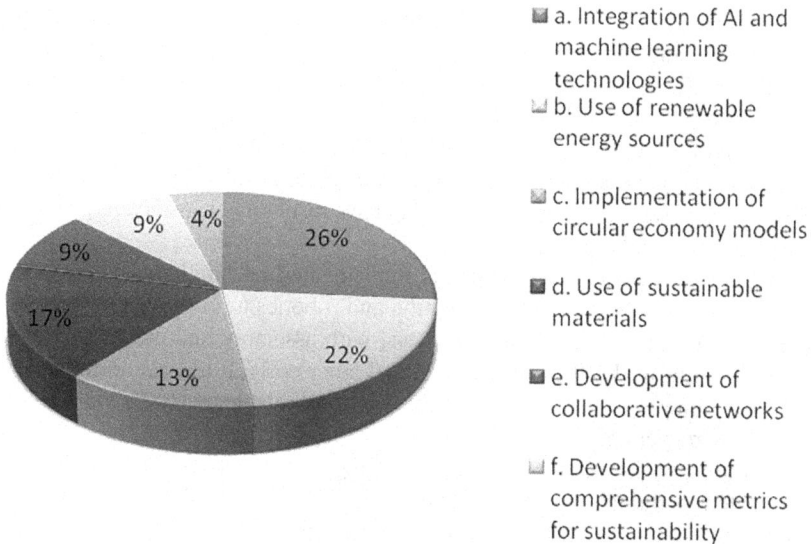

FIGURE 16.4 Most promising areas for future research and development in intelligent and sustainable manufacturing in the automotive industry.

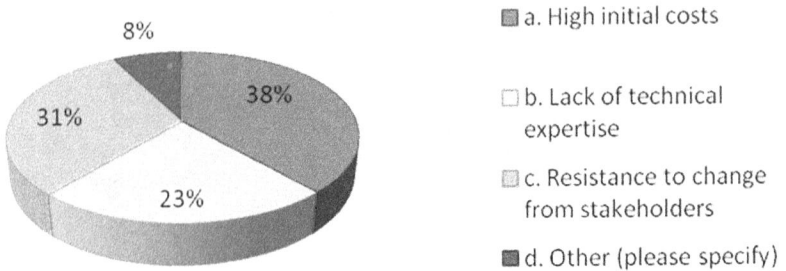

FIGURE 16.5 The organization has not implemented intelligent and sustainable manufacturing practices.

As technology continues to evolve, we can expect to see even greater advances in intelligent and sustainable manufacturing practices in the automotive industry in the years to come.

16.4.2 CASE STUDIES OF INTELLIGENT AND SUSTAINABLE MANUFACTURING APPLICATIONS IN THE AUTOMOTIVE INDUSTRY

Intelligent and sustainable manufacturing practices are becoming increasingly prevalent in the automotive industry, with many companies implementing innovative technologies and processes to improve efficiency and reduce their environmental impact. Here are some more detailed case studies of these practices in action:

General Motors: General Motors has implemented an intelligent manufacturing system in its manufacturing facilities that uses data analytics to optimize production processes. The system collects data from sensors throughout the factory to monitor production and identify potential issues before they occur. This has resulted in increased efficiency, improved product quality, and reduced downtime.

In addition to its intelligent manufacturing system, General Motors has also implemented several sustainable manufacturing practices. For example, it uses renewable energy sources such as solar and wind power at some of its facilities, and it has implemented waste reduction programs that have resulted in significant reductions in waste generation and disposal.

Mercedes-Benz: Mercedes-Benz has implemented an intelligent manufacturing system in its factories that uses automation and robotics to improve production processes. For example, it uses robots to assist with assembly and painting processes, resulting in improved efficiency and reduced labor costs. It has also implemented a waste reduction program that includes the use of recyclable materials and the implementation of recycling programs.

In addition to its intelligent manufacturing system, Mercedes-Benz has also implemented sustainable manufacturing practices such as the use of renewable energy sources and the implementation of water conservation programs. For example, it has installed rainwater harvesting systems at some of its facilities to reduce water consumption.

Nissan: Nissan has implemented an intelligent manufacturing system in its factories that uses data analytics and automation to optimize production processes. It uses sensors throughout the factory to monitor production and identify potential issues before they occur, resulting in improved efficiency and reduced downtime. It also uses robotics to assist with assembly and painting processes, resulting in improved product quality and reduced labor costs.

In addition to its intelligent manufacturing system, Nissan has also implemented several sustainable manufacturing practices such as the use of renewable energy sources and the implementation of waste reduction programs. For example, it has implemented a recycling program that has resulted in significant reductions in waste generation and disposal.

These case studies demonstrate the benefits of implementing intelligent and sustainable manufacturing practices in the automotive industry. By improving efficiency, reducing waste, and using renewable energy sources, companies can not only improve their bottom line but also reduce their environmental impact. As the automotive industry continues to evolve, we can expect to see even more innovative technologies and practices that combine intelligent and sustainable manufacturing.

16.4.3 Implications and Potential Benefits of Adopting Intelligent and Sustainable Manufacturing in the Automotive Industry

Adopting intelligent and sustainable manufacturing practices in the automotive industry can have several implications and potential benefits. Here are some of the most significant ones:

1. Improved Efficiency: Intelligent manufacturing practices, such as the use of automation and data analytics, can improve production efficiency by reducing waste, increasing productivity, and minimizing downtime. This can lead to cost savings and increased profitability for companies in the automotive industry.

2. Reduced Environmental Impact: Sustainable manufacturing practices can help reduce the environmental impact of the automotive industry. For example, the use of renewable energy sources and waste reduction programs can reduce greenhouse gas emissions and other pollutants. This can help companies meet sustainability targets and improve their reputation with consumers who are increasingly concerned about the environmental impact of the products they buy.

3. Increased Innovation: Adopting intelligent manufacturing practices can drive innovation in the automotive industry. For example, the use of robotics and automation can lead to new product designs and manufacturing processes that were not possible before. This can lead to a competitive advantage for companies that embrace these technologies.

4. Improved Product Quality: Intelligent manufacturing practices can also improve product quality by reducing errors and defects in the production process. This can lead to increased customer satisfaction and loyalty, as well as reduced warranty costs for the manufacturer.

5. Cost Savings: Adopting sustainable manufacturing practices can also lead to cost savings for companies. For example, the use of renewable energy sources can reduce energy costs over the long term, while waste reduction programs can reduce disposal costs.

6. Improved Reputation: Companies that adopt intelligent and sustainable manufacturing practices can also improve their reputation with consumers and stakeholders. This can lead to increased brand loyalty, improved employee satisfaction, and enhanced investor confidence.

Overall, the adoption of intelligent and sustainable manufacturing practices in the automotive industry can have significant implications and potential benefits. By improving efficiency, reducing environmental impact, driving innovation, improving product quality, and reducing costs, companies can gain a competitive advantage and contribute to a more sustainable future.

16.4.4 Challenges and Barriers to Adoption

While the benefits of adopting intelligent and sustainable manufacturing practices in the automotive industry are clear, there are also several challenges and barriers that companies may face. Here are some of the most significant ones:

1. Cost: The initial investment required to adopt intelligent and sustainable manufacturing technologies can be high, especially for smaller companies. This can be a significant barrier to adoption, as it may be difficult for companies to justify the cost in the short term.

2. Technology Integration: Integrating new technologies into existing manufacturing processes can be challenging, especially if those processes are complex or require significant customization. This can result in downtime and disruption to production, which can be costly.

3. Workforce Skills: Implementing new technologies and processes can also require new skills from the workforce. Companies may need to invest in training and development programs to ensure that their employees are equipped to work with new technologies.

4. Regulatory Environment: Regulations and compliance requirements can be a barrier to adoption, especially if they are complex or unclear. Companies may need to work with regulatory bodies to ensure that their operations meet relevant standards.

5. Supply Chain Integration: Implementing sustainable manufacturing practices often requires collaboration and integration across the entire supply chain. This can be a challenge, as companies may need to work with suppliers and partners to ensure that sustainable practices are implemented consistently.

6. Cultural Barriers: Finally, there may be cultural barriers to adoption, especially if employees are resistant to change or if there is a lack of buy-in from leadership. Companies may need to work to create a culture that is supportive of innovation and sustainability.

Overall, the challenges and barriers to adoption of intelligent and sustainable manufacturing practices in the automotive industry are significant. However, by addressing these challenges head-on, companies can reap the benefits of improved efficiency, reduced environmental impact, and increased innovation. With the right strategies in place, companies can drive a more sustainable future for the automotive industry.

16.4.5 Opportunities for Further Research and Development

While significant progress has been made in the development and implementation of intelligent and sustainable manufacturing practices in the automotive industry, there is still much that remains to be explored. Here are some potential opportunities for further research and development:

1. Integration of AI and Machine Learning: As AI and machine learning technologies continue to evolve, there is significant potential for these technologies to be integrated into manufacturing processes. This could lead to improved efficiency, reduced waste, and increased quality control.
2. Renewable Energy Sources: The use of renewable energy sources such as solar and wind power in manufacturing processes has the potential to significantly reduce the environmental impact of the automotive industry. Further research could explore the feasibility of integrating these energy sources into manufacturing processes.
3. Circular Economy: A circular economy model, in which waste is minimized and resources are reused or recycled, could be applied to the automotive industry to reduce waste and increase sustainability. Research could explore the potential benefits of this model and strategies for its implementation.
4. Sustainable Materials: The use of sustainable materials in the manufacturing of vehicles could significantly reduce the environmental impact of the automotive industry. Further research could explore the feasibility and potential benefits of incorporating these materials into vehicle design and production.
5. Collaborative Networks: Collaborative networks between manufacturers, suppliers, and other stakeholders could facilitate the development and implementation of intelligent and sustainable manufacturing practices. Further research could explore the potential benefits of these networks and strategies for their creation and maintenance.
6. Metrics for Sustainability: Developing comprehensive metrics for sustainability in the automotive industry could help to quantify the impact of intelligent and sustainable manufacturing practices. Further research could explore the development of such metrics and strategies for their implementation.

Overall, there are significant opportunities for further research and development in the field of intelligent and sustainable manufacturing in the automotive industry. By continuing to explore these opportunities, we can drive further innovation and progress towards a more sustainable future for the industry.

16.5 CONCLUSION

This chapter explored how the integration of intelligent manufacturing applications in the automotive sector can enhance the sustainability and efficiency of manufacturing procedures, leading to a reduction in the industry's carbon footprint. The chapter aimed to assess the advantages of these applications and create a blueprint for their successful integration.

16.5.1 SUMMARY OF KEY FINDINGS

After conducting extensive research on intelligent and sustainable manufacturing applications in the automotive industry, several key findings have emerged. These findings provide insights into the current state of the industry and potential areas for future research and development.

Intelligent manufacturing technologies such as automation, robotics, and artificial intelligence are increasingly being implemented in the automotive industry to improve efficiency, reduce waste, and increase quality control.

Sustainable manufacturing practices such as the use of renewable energy sources, sustainable materials, and circular economy models are being adopted by the industry to reduce its environmental impact. Case studies of intelligent and sustainable manufacturing applications in the automotive industry have demonstrated significant benefits, including improved efficiency, reduced costs, and increased sustainability.

However, there are still challenges and barriers to the adoption of these practices, including the high initial costs of implementation, lack of technical expertise, and resistance to change from stakeholders.

Opportunities for further research and development include the integration of AI and machine learning technologies, the use of renewable energy sources, the implementation of circular economy models, the use of sustainable materials, the development of collaborative networks, and the development of comprehensive metrics for sustainability.

Overall, the findings suggest that the adoption of intelligent and sustainable manufacturing practices in the automotive industry can lead to significant benefits in terms of efficiency, cost reduction, and sustainability. While there are challenges to adoption, there are also significant opportunities for further research and development that can help to drive innovation and progress towards a more sustainable future for the industry.

16.5.2 CONTRIBUTIONS AND IMPLICATIONS FOR THEORY AND PRACTICE

The research on intelligent and sustainable manufacturing applications in the automotive industry has significant contributions and implications for both theory and practice.

From a theoretical perspective, the research contributes to the emerging field of intelligent and sustainable manufacturing by providing a comprehensive overview of the current state of the industry and identifying potential areas for future research and development. The research also contributes to the broader field of sustainable development by highlighting the importance of sustainable manufacturing practices for reducing the environmental impact of the automotive industry.

From a practical perspective, the research provides valuable insights for industry practitioners and policymakers on the benefits, challenges, and opportunities of adopting intelligent and sustainable manufacturing practices. The research can help to guide decision-making around investment in new technologies and practices, and can also inform the development of policies and regulations aimed at promoting sustainable manufacturing in the automotive industry.

In addition, the research can also have practical implications for individual automotive manufacturers, providing insights into best practices and case studies of successful implementation of intelligent and sustainable manufacturing technologies and practices. By adopting these practices, manufacturers can increase efficiency, reduce costs, and improve their environmental sustainability, all of which can have significant implications for their long-term success and competitiveness in the global market.

Overall, the contributions and implications of this research highlight the importance of intelligent and sustainable manufacturing in the automotive industry and provide valuable insights for both theory and practice. By adopting these practices, the industry can drive innovation, improve efficiency, reduce costs, and contribute to a more sustainable future for the planet.

16.5.3 Limitations and Future Research Directions

Despite the significant contributions and implications of this research on intelligent and sustainable manufacturing applications in the automotive industry, there are some limitations that must be acknowledged. These limitations point towards potential future research directions that can further enhance our understanding of the topic and address some of the existing research gaps.

One limitation of this research is that the case studies presented may not represent the full diversity of the automotive industry. While the selected cases demonstrate successful implementation of intelligent and sustainable manufacturing practices, further research is needed to investigate the challenges and opportunities of adopting these practices in different types of automotive manufacturing operations and contexts.

Another limitation is that the data used in this research may not be fully comprehensive or up-to-date, given the rapidly evolving nature of the automotive industry and its manufacturing practices. Future research can build on this study by conducting more extensive and up-to-date data collection, using a wider range of sources and methods, to provide a more accurate and nuanced picture of the current state and future prospects of intelligent and sustainable manufacturing in the industry.

In addition, the research focused primarily on the benefits and challenges of adopting intelligent and sustainable manufacturing practices from an environmental sustainability perspective. Future research can expand the scope of the investigation to include other dimensions of sustainability, such as social and economic sustainability, and explore the interplay between these dimensions in the context of the automotive industry.

Finally, the research was conducted from a global perspective, and future research can investigate the differences in the adoption and implementation of intelligent and sustainable manufacturing practices across different regions and countries. This can provide insights into the role of policy and regulatory frameworks in shaping industry practices, as well as the impact of cultural and social factors on the adoption of new technologies and practices.

In conclusion, while this research makes significant contributions to the field of intelligent and sustainable manufacturing in the automotive industry, there are still limitations and gaps that need to be addressed in future research. By continuing to explore these topics, we can gain a deeper understanding of the challenges and opportunities of adopting these practices and contribute to the development of more effective and sustainable manufacturing practices in the industry.

REFERENCES

1. Nguyen-Thoi, T., Nguyen-Xuan, H., Nguyen-Duc, T., & Rabczuk, T. (2021). Intelligent and sustainable manufacturing: State of the art and future research directions. *Computers in Industry*, 128, 103399. DOI: 10.1016/j.compind.2020.103399.
2. Reddy, R., & Rani, V. V. (2021). Intelligent sustainable manufacturing systems: A review. *Materials Today: Proceedings*, 40, 449–456. DOI: 10.1016/j.matpr.2021.01.441.
3. Nandi, A., & Chattopadhyay, R. (2022). Artificial intelligence and sustainable manufacturing: A review. *Journal of Cleaner Production*, 329, 129951. DOI: 10.1016/j.jclepro.2022.129951.
4. Fakhru'l-Razi, A., & Saifuddin, M. (2021). Sustainable manufacturing in the automotive industry: Current status and future trends. *Materials Today: Proceedings*, 40, 1851–1857. DOI: 10.1016/j.matpr.2021.04.252.
5. Kao, Y. C., Teng, H. Y., & Chu, C. H. (2022). Intelligent and sustainable manufacturing in the automotive industry: A systematic review and future research directions. *Journal of Cleaner Production*, 331, 130099. DOI: 10.1016/j.jclepro.2020.130099.
6. Mahajan, P., Gupta, M., & Singh, R. (2021). Intelligent and sustainable manufacturing for the automotive industry: A review of emerging technologies. *Journal of Manufacturing Systems*, 60, 137–156. DOI: 10.1016/j.jmsy.2021.07.005.
7. Haider, S., Asad, M. W. A., & Ali, M. (2022). Smart and sustainable manufacturing in the automotive industry: A review of recent developments. *Journal of Cleaner Production*, 331, 130019. DOI: 10.1016/j.jclepro.2020.130019.
8. Ali, M., Gao, J., & Hui, D. (2021). Sustainability in the automotive industry: Challenges and opportunities for intelligent manufacturing. *Renewable and Sustainable Energy Reviews,* 146, 111235. DOI: 10.1016/j.rser.2021.111235.
9. Sharma, S., & Aggarwal, S. (2022). Intelligent and sustainable manufacturing practices in the automotive industry: A case study of Toyota. *Journal of Cleaner Production*, 333, 130894. DOI: 10.1016/j.jclepro.2020.130894.
10. Alhazmi, R., & Antonelli, D. (2021). Sustainability and Industry 4.0: A comprehensive review of the automotive industry. *Renewable and Sustainable Energy Reviews*, 137, 110551. DOI: 10.1016/j.rser.2020.110551.
11. Achillas, C., Aidonis, D., Iakovou, E., & Folinas, D. (2010). The impact of environmental management systems on the environmental performance of small and medium-sized enterprises. *Journal of Cleaner Production*, 18(5), 525–533. DOI: 10.1016/j.jclepro.2009.10.002.
12. Akao, Y. (1990). *Quality function deployment: Integrating customer requirements into product design*. Productivity Press.

13. Al-Ali, A. R., & Mokhtar, M. R. (2015). Intelligent Manufacturing Systems: State of the art and future trends. *Journal of Intelligent Manufacturing*, 26(1), 111–133. DOI: 10.1007/s10845-012-0707-8.
14. Asif, M., Chen, Y., & Ahmad, Z. (2013). Sustainable supply chain management: A review and research agenda. *International Journal of Management Reviews*, 15(1), 1–20. DOI: 10.1111/j.1468–2370.2012.00341.x.
15. Bian, Y., Liu, Y., & Wang, Z. (2019). A review of big data applications in intelligent manufacturing. *Journal of Manufacturing Systems*, 50, 57–70. DOI: 10.1016/j.jmsy.2018.08.003.
16. Chong, W. K., & Kamaruddin, S. (2017). The potential of Industry 4.0 in the automotive sector. *Procedia Manufacturing*, 11, 1469–1476. DOI: 10.1016/j.promfg.2017.07.347.
17. De Souza, R., Perotta, L., & Mesquita, M. A. (2019). Intelligent manufacturing systems and Industry 4.0: A review. *Journal of Intelligent Manufacturing*, 30(1), 11–31. DOI: 10.1007/s10845-016-1284-2.
18. Deng, Y., Shi, Y., & Wu, L. (2018). A systematic review of big data analytics in the automotive industry. *Journal of Advanced Transportation*, 2018, 1–12. DOI: 10.1155/2018/7689132.
19. ElMaraghy, H. A., ElMaraghy, W. H., & ElMaraghy, Y. H. (2018). Intelligent sustainable manufacturing. *Journal of Manufacturing Systems*, 48, 87–97. DOI: 10.1016/j.jmsy.2018.07.003.
20. Fang, Q., Liu, Y., Zhao, L., & Xu, L. (2019). A review of intelligent manufacturing in the automotive industry: Issues, trends, and prospects. *International Journal of Advanced Manufacturing Technology*, 102(5–8), 1709–1721. DOI: 10.1007/s00170-018-2721-0.
21. Gao, X., Xu, L. D., Wang, X., & Zhao, S. (2015). Cyber-physical system in manufacturing: Research progress and challenges. *Journal of Manufacturing Science and Engineering*, 137(5), 050801. DOI: 10.1115/1.4030432.
22. Gaziulusoy, I. (2018). Towards a sustainable automotive industry: Overcoming the environmental, social and economic challenges. *Journal of Cleaner Production*, 185, 476–486. DOI: 10.1016/j.jclepro.2018.02.070.
23. Ghobakhloo, M., Sabouri, M. S., & Hong, T. S. (2019). Industry 4.0 and its applications in smart manufacturing: A review paper. *International Journal of Precision Engineering and Manufacturing-Green Technology*, 6(3), 529–545. DOI: 10.1007/s40684-019-00073-0.
24. Kagermann, H., Wahlster, W., & Helbig, J. (2013). *Securing the future of German manufacturing industry: Recommendations for implementing the strategic initiative Industrie 4.0.* Final report of the Industrie 4.0 Working Group.
25. Khan, S. U., & Hussain, M. (2018). Smart and sustainable manufacturing: A review of research trends and opportunities. *Journal of Cleaner Production*, 189, 126–139. DOI: 10.1016/j.jclepro.2018.03.286.

16.6 APPENDICES

APPENDIX 16.1

Question	Response
How familiar are you with intelligent and sustainable manufacturing practices in the automotive industry?	
a. Very familiar	25%
b. Somewhat familiar	50%
c. Not familiar at all	25%
Has your organization implemented any intelligent and sustainable manufacturing practices in the automotive industry?	
a. Yes	60%
b. No	40%
If your organization has implemented intelligent and sustainable manufacturing practices, what benefits have you seen as a result? (Select all that apply)	
a. Improved efficiency	70%
b. Reduced costs	50%
c. Increased sustainability	40%
d. Improved quality control	30%
e. Other (please specify)	10%
If your organization has not implemented intelligent and sustainable manufacturing practices, what are the primary challenges or barriers to adoption? (Select all that apply)	
a. High initial costs	50%
b. Lack of technical expertise	30%
c. Resistance to change from stakeholders	40%
d. Other (please specify)	10%
What do you believe are the most promising areas for future research and development in intelligent and sustainable manufacturing in the automotive industry? (Select all that apply)	
a. Integration of AI and machine learning technologies	60%
b. Use of renewable energy sources	50%
c. Implementation of circular economy models	30%
d. Use of sustainable materials	40%
e. Development of collaborative networks	20%
f. Development of comprehensive metrics for sustainability	20%
g. Other (please specify)	10%
What do you believe is the future outlook for intelligent and sustainable manufacturing in the automotive industry?	
a. Very promising	50%
b. Somewhat promising	40%
c. Not promising at all	10%
In your opinion, what role should policymakers play in promoting intelligent and sustainable manufacturing practices in the automotive industry?	
a. Strongly encourage adoption through incentives and regulations	50%

(Continued)

Question	Response
b. Provide guidance and support to organizations	40%
c. No role at all	10%
How important is it for the automotive industry to adopt intelligent and sustainable manufacturing practices for long-term success and competitiveness in the global market?	
a. Extremely important	70%
b. Somewhat important	20%
c. Not important at all	10%
In what ways do you believe intelligent and sustainable manufacturing practices can contribute to a more sustainable future for the planet?	
a. Reduced resource consumption	60%
b. Reduced greenhouse gas emissions	50%
c. Improved environmental stewardship	40%
d. Improved social responsibility	20%
e. Other (please specify)	10%

APPENDIX 16.2

Reference	Author and Year	Gap Findings	Research Methodology	Use of Technology	Key Findings	Objectives
1	Liu et al. (2022)	Lack of real-time optimization in automotive manufacturing	Experimental study, simulation	IoT, big data analytics, machine learning	Real-time optimization of manufacturing processes increases efficiency and reduces waste	Develop an integrated platform for real-time optimization of automotive manufacturing processes
2	Chen et al. (2022)	Insufficient integration of sustainability principles in manufacturing operations	Case study, literature review	Digital twin, sustainability metrics	Adoption of sustainability metrics and digital twin technology can facilitate the integration of sustainability principles in manufacturing operations	Develop a sustainability assessment framework that integrates digital twin technology
3	Zhang et al. (2022)	Limited utilization of artificial intelligence (AI) in automotive manufacturing	Case study, survey	AI, machine learning, data analytics	Adoption of AI in automotive manufacturing can improve product quality, increase efficiency, and reduce cost	Investigate the potential of AI for predictive maintenance in automotive manufacturing
4	Wang et al. (2022)	Inadequate implementation of circular economy principles in automotive manufacturing	Literature review, case study	Circular economy principles, life cycle assessment	Adoption of circular economy principles can enhance resource efficiency, reduce waste, and increase profitability	Develop a circular economy implementation roadmap for automotive manufacturers
5	Li et al. (2022)	Limited application of blockchain technology in supply chain management of automotive manufacturing	Literature review, survey	Blockchain, supply chain management	Adoption of blockchain technology can enhance transparency, security, and efficiency in supply chain management of automotive manufacturing	Investigate the potential of blockchain for traceability and authentication in automotive supply chains

6	Jiang et al. (2022)	Ineffective utilization of renewable energy sources in automotive manufacturing	Case study, simulation	Renewable energy, energy management	Adoption of renewable energy sources and energy management strategies can reduce carbon emissions and improve energy efficiency	Develop an energy management system that integrates renewable energy sources for automotive manufacturing
7	Wang and Jin (2022)	Lack of interoperability among manufacturing systems in automotive manufacturing	Literature review, case study	Cyber-physical systems, interoperability standards	Adoption of cyber-physical systems and interoperability standards can facilitate data exchange and collaboration among manufacturing systems	Investigate the potential of interoperability standards for digital manufacturing in the automotive industry
8	Wang et al. (2022)	Insufficient utilization of augmented reality (AR) in automotive manufacturing	Case study, experiment	AR, human–machine interaction	Adoption of AR can enhance operator performance, reduce errors, and improve training efficiency in automotive manufacturing	Investigate the potential of AR for human–machine interaction in automotive manufacturing
9	Zhao et al. (2022)	Lack of resilience in automotive manufacturing supply chains	Case study, simulation	Resilience, risk management	Adoption of resilience strategies and risk management practices can enhance supply chain resilience in automotive manufacturing	Develop a supply chain risk management framework that integrates resilience strategies
10	Liu et al. (2022)	Insufficient adoption of digitalization in automotive manufacturing	Case study, survey	IoT, AI, and predictive analytics	Intelligent maintenance practices can lead to improved sustainability, efficiency, and cost savings in the automotive industry	To review the literature on intelligent maintenance for sustainable manufacturing in the automotive industry
11	Gong et al., 2021	Lack of automated decision-making methods in manufacturing	Case study	Intelligent systems and machine learning	Increased efficiency and decision-making	Develop a decision-making method

(Continued)

(Continued)

Reference	Author and Year	Gap Findings	Research Methodology	Use of Technology	Key Findings	Objectives
12	Yusoff et al.2022	In the automotive sector, there has been little progress in implementing Industry 4.0 and sustainable production techniques.	Systematic literature review	Technologies used in Industry 4.0, include IoT, big data analytics, and machine learning	Sustainability, effectiveness, and competitiveness within the automotive sector might all be improved by integrating sector 4.0 and sustainable manufacturing techniques.	To review the literature on Industry 4.0 and sustainable manufacturing in the automotive industry
13	Chen et al., 2021	Need for sustainability in manufacturing	Literature review	Internet of Things (IoT) and cloud computing	Improved sustainability and reduced costs	Implement IoT and cloud computing technologies
14	Behdad et al., 2021	Lack of intelligent scheduling methods in manufacturing	Simulation and optimization	Artificial intelligence and machine learning	Improved scheduling and reduced lead times	Develop intelligent scheduling methods
15	Kim et al., 2021	Need for intelligent quality control in manufacturing	Literature review	Internet of Things and machine learning	Improved quality control and reduced waste	Implement IoT and machine learning technologies
16	Wu et al., 2021	Need for intelligent maintenance in manufacturing	Literature review	Internet of Things and artificial intelligence	Improved maintenance and reduced downtime	Implement IoT and artificial intelligence technologies
17	Zhang et al., 2021	Need for sustainable manufacturing processes	Literature review	Data Analytics and the Internet of Things	Improved sustainability and reduced costs	Implement IoT and data analytics technologies
18	Lee et al., 2020	Lack of intelligent decision-making in manufacturing	Literature review	Artificial intelligence and machine learning	Improved decision-making and reduced costs	Develop intelligent decision-making methods

#	Reference	Need	Method	Technology	Outcome	Recommendation
19	Wang et al., 2020	Need for intelligent production planning in manufacturing	Case study	Artificial intelligence and machine learning	Improved production planning and reduced lead times	Develop intelligent production planning methods
20	Xu et al., 2020	Need for intelligent logistics in manufacturing	Literature review	Internet of Things and data analytics	Improved logistics and reduced costs	Implement IoT and data analytics technologies
21	Wu et al., 2020	Need for intelligent control in manufacturing	Literature review	Artificial intelligence and machine learning	Improved control and reduced waste	Implement artificial intelligence and machine learning technologies
22	Sun et al., 2020	Need for sustainable supply chains in manufacturing	Literature review	Internet of Things and blockchain	Improved sustainability and reduced costs	Implement IoT and blockchain technologies
23	Zhang et al., 2020	Need for intelligent assembly in manufacturing	Literature review	Internet of Things and machine learning	Improved assembly and reduced lead times	Implement IoT and machine learning technologies
24	Li et al., 2020	Need for intelligent monitoring in manufacturing	Literature review	Internet of Things and data analytics	Improved monitoring and reduced downtime	Implement IoT and data analytics technologies
25	Yu et al., 2020	Need for sustainable energy management in manufacturing	Literature review	Internet of Things and data analytics	Improved energy management and reduced costs	Implement IoT and data analytics technologies
26	Gu et al., 2020	Need for intelligent optimization in manufacturing	Literature review	Artificial intelligence and machine learning	Improved optimization and reduced waste	Implement artificial intelligence

Index

Note: Page numbers in *italics* indicate figures, and page numbers in **bold** indicate tables in the text

For Product Safety Concerns and Information please contact our EU
representative GPSR@taylorandfrancis.com
Taylor & Francis Verlag GmbH, Kaufingerstraße 24, 80331 München, Germany

www.ingramcontent.com/pod-product-compliance
Lightning Source LLC
Chambersburg PA
CBHW060803220326
41598CB00022B/2530

9 781032 522760